CROP BREEDING

Bioinformatics and Preparing
for Climate Change

CROP BREEDING

Bioinformatics and Preparing for Climate Change

Edited by
Santosh Kumar, PhD

Apple Academic Press Inc. | Apple Academic Press Inc.
3333 Mistwell Crescent | 9 Spinnaker Way
Oakville, ON L6L 0A2 | Waretown, NJ 08758
Canada | USA

©2017 by Apple Academic Press, Inc.

First issued in paperback 2021

Exclusive worldwide distribution by CRC Press, a member of Taylor & Francis Group
No claim to original U.S. Government works

ISBN 13: 978-1-77463-612-1 (pbk)
ISBN 13: 978-1-77188-344-3 (hbk)

Library and Archives Canada Cataloguing in Publication

Crop breeding : bioinformatics and preparing for climate change / edited by
Santosh Kumar, PhD.

Includes bibliographical references and index.
Issued in print and electronic formats.
ISBN 978-1-77188-344-3 (hardcover).--ISBN 978-1-77188-345-0 (pdf)
1. Plant breeding. 2. Bioinformatics. 3. Genomics. 4. Crop improvement. 5. Crops and climate. 6. Food supply. 7. Food security. I. Kumar, Santosh, 1974-, author, editor

SB123.C76 2016   631.5'3   C2015-906333-7   C2015-906334-5

Library of Congress Cataloging-in-Publication Data

Names: Kumar, Santosh, 1978- editor.
Title: Crop breeding : bioinformatics and preparing for climate change/editor:
Santosh Kumar, PhD.
Description: Oakville, ON ; Waretown, NJ : Apple Academic Press, [2016] |
Includes bibliographical references and index.
Identifiers: LCCN 2015035572 | ISBN 9781771883443 (alk. paper)
Subjects: LCSH: Crops--Genetic engineering. | Plant breeding. | Crop improvement. | Crops and climate.
Classification: LCC SB123.57 .C755 2016 | DDC 631.5/233--dc23
LC record available at http://lccn.loc.gov/2015035572

Apple Academic Press also publishes its books in a variety of electronic formats. Some content that appears in print may not be available in electronic format. For information about Apple Academic Press products, visit our website at **www.appleacademicpress.com** and the CRC Press website at **www.crcpress.com**

About the Editor

SANTOSH KUMAR, PhD

Dr. Santosh Kumar obtained his BSc agriculture degree in plant breeding from Punjab Agricultural University, Ludhiana, India. He obtained his MSc working on wheat phytosiderophores from the Indian Agricultural Research Institute, New Delhi, India, and PhD working on the control of flowering and germination in barley from the University of Manitoba, Canada. At present, Dr. Kumar is a research scientist for Agriculture and Agri-Food Canada stationed at the Brandon Research and Development Centre, Manitoba, Canada. Dr. Kumar has studied various disciplines in plant sciences and has received many awards throughout his academic carrier. His current interests include wheat breeding, germplasm improvement, genomics and bioinformatics. Dr. Kumar is a member of various plant science societies in Canada and the U.S. Dr. Kumar has publications that include research articles, a review paper, and a book chapter in areas of plant physiology, genetics, and bioinformatics.

Contents

Acknowledgment and How to Cite

The editor and publisher thank each of the authors who contributed to this book. The chapters in this book were previously published elsewhere. To cite the work contained in this book and to view the individual permissions, please refer to the citation at the beginning of each chapter. Each chapter was carefully selected by the editor; the result is a book that looks at crop breeding and climate change from a variety of perspectives. The chapters included are broken into three sections, which describe the following topics:

1. Crop Genomics and Climate Change
In this section, we have chosen four articles, as follows:

- In chapter 1, Henry discusses the need to access whole genome information on all crop species, including their wild relatives, as the foundation that will make rapid crop adaptation possible in response to climate change. This gives us a good introduction to the next three chapters, which include research into specific ways in which genomics can be applied to crops under climate change.
- In chapter 2, Springate and Kover offer research that indicates the phonological sensitivity may be useful as an indicator of genotypes that can be used in both restoration and conservation projects.
- In chapter 3, Chown and colleagues provide evidence that genomic applications are a useful and cost-effective tool for the surveillance and management of invasive species.
- In chapter 4, Sengupta and Majumder indicate that transcriptomic and genomic studies can create the necessary platform biologists need to modify crops for traits such as vascular integrity and water transport.

2. Genomic Toolkit for Crop Genomics
In this section, we have chosen these three articles:

- Chapter 5, by Li et al., discusses ways in which life processes can be captured and digitized by means of computer technology.

- Chapter 6, by Kumar and my colleagues Banks and Cloutier, explains the ways in which NGS technologies have revolutionized genomics-related research.
- In chapter 7, the authors (Boutet et al.) aim to develop a comprehensive SNP resource in pea using genotyping by HiSeq sequencing of whole genome DNA and then to apply it for substantial genetic mapping.

3. Crop Improvement Using Genomics Under Changing Climate

Here, the three chosen articles include the following research:

- In chapter 8, Seaver and his colleagues' techniques for relatively large, yet accurate, metabolic reconstructions.
- In chapter 9, Sulieman and his colleagues' strategies (which include conventional breeding techniques, marker-assisted breed, and genetic engineering) for generating drought-tolerant soybean cultivar.
- In chapter 10, Shiv Kumar and his colleagues' identification of practical ways in which genomics can be applied to another specific crop (in this case, lentils).

List of Contributors

Travis W. Banks
Department of Applied Genomics, Vineland Research and Innovation Centre, Vineland Station, ON, Canada L0R 2E0

Alain Baranger
INRA, UMR 1349 IGEPP; PISOM, UMT INRA/CETIOM

Michael Baum
International Center for Agricultural Research in the Dry Areas, Amman, Jordan

Olivier Bouchez
GeT-PlaGe, Genotoul; INRA, UMR1388 INRA/ENVT/ENSAT GenPhySE

Gilles Boutet
INRA, UMR 1349 IGEPP; PISOM, UMT INRA/CETIOM

Louis M. T. Bradbury
Horticultural Sciences Department, University of Florida, Gainesville, FL, USA and Department of Biology, York College, City University of New York, New York, NY, USA

Margaret Byrne
Science and Conservation Division, Department of Parks and Wildlife, Bentley Delivery Centre, Bentley, WA, Australia

Susete Alves Carvalho
INRA, UMR 1349 IGEPP; INRIA Rennes—Bretagne Atlantique/IRISA, EPI GenScale

Steven L. Chown
School of Biological Sciences, Monash University, Clayton, Vic., Australia

Sylvie Cloutier
Ottawa Research and Development Centre, Agriculture and Agri-Food Canada, 960 Carling Ave., Ottawa, Ontario, Canada K1A 0C6

Örjan Carlborg
Swedish University of Agricultural Sciences, Department of Clinical Sciences, Division of Computational Genetics, Uppsala, Sweden

Maryam Nasr Esfahani
Signaling Pathway Research Unit, RIKEN Center for Sustainable Resource Science (CSRS), 1-7-22 Suehiro-cho, Tsurumi, Yokohama 230-0045, Japan and Department of Biology, Lorestan University, Khorramabad 68151-44316, Iran

Matthieu Falque
INRA, UMR Génétique Quantitative et Evolution; Le Moulon, INRA; Univ Paris-Sud; CNRS; AgroParisTech

Océane Frelin
Horticultural Sciences Department, University of Florida, Gainesville, FL, USA

Philippa C. Griffin
Department of Genetics, Bio21 Institute, The University of Melbourne, Parkville, Vic., Australia

Chien Van Ha
Signaling Pathway Research Unit, RIKEN Center for Sustainable Resource Science (CSRS), 1-7-22 Suehiro-cho, Tsurumi, Yokohama 230-0045, Japan and National Key Laboratory of Plant Cell Biotechnology, Agricultural Genetics Institute, Vietnamese Academy of Agricultural Science, Hanoi 100000, Vietnam

Aladdin Hamwieh
International Center for Agricultural Research in the Dry Areas, Cairo, Egypt

Andrew D. Hanson
Horticultural Sciences Department, University of Florida, Gainesville, FL, USA

Christopher S. Henry
Mathematics and Computer Science Division, Argonne National Laboratory, Argonne, IL, USA and Computation Institute, The University of Chicago, Chicago, IL, USA

Robert J. Henry
Queensland Alliance for Agriculture and Food Innovation, University of Queensland, Brisbane, QLD, Australia

Kathryn A. Hodgins
School of Biological Sciences, Monash University, Clayton, Vic., Australia

Ary A. Hoffmann
Departments of Zoology and Genetics, Bio21 Institute, The University of Melbourne, Parkville, Vic., Australia

Paula X. Kover
Department of Biology and Biochemistry, University of Bath, Bath, UK

Jitendra Kumar
Division of Crop Improvement, Indian Institute of Pulses Research, Kanpur, India

Santosh Kumar
Brandon Research and Development Centre, Agriculture and Agri-Food Canada, 2701 Grand Valley, Road, Brandon, Manitoba, Canada R7A 5Y3

Shiv Kumar
Biodiversity and Integrated Gene Management Program, International Center for Agricultural Research in the Dry Areas, Rabat, Morocco

Hon-Ming Lam
Center for Soybean Research, State Key Laboratory of Agrobiotechnology and School of Life Sciences, the Chinese University of Hong Kong, Shatin, N.T., Hong Kong

Clément Lavaud
INRA, UMR 1349 IGEPP; PISOM, UMT INRA/CETIOM

Emeline Lhuillier
GeT-PlaGe, Genotoul

Man-Wah Li
Center for Soybean Research, State Key Laboratory of Agrobiotechnology and School of Life Sciences, the Chinese University of Hong Kong, Shatin, N.T., Hong Kong

Arun Lahiri Majumder
Division of Plant Biology, Acharya J C Bose Biotechnology Innovation Centre, Bose Institute, Kolkata, India

Dong Van Nguyen
National Key Laboratory of Plant Cell Biotechnology, Agricultural Genetics Institute, Vietnamese Academy of Agricultural Science, Hanoi 100000, Vietnam

Meng Ni
Center for Soybean Research, State Key Laboratory of Agrobiotechnology and School of Life Sciences, the Chinese University of Hong Kong, Shatin, N.T., Hong Kong

Rie Nishiyama
Signaling Pathway Research Unit, RIKEN Center for Sustainable Resource Science (CSRS), 1-7-22 Suehiro-cho, Tsurumi, Yokohama 230-0045, Japan

John G. Oakeshott
CSIRO Land and Water Flagship, Black Mountain Laboratories, Canberra, ACT, Australia

Pierre Peterlongo
INRIA Rennes—Bretagne Atlantique/IRISA, EPI GenScale

Chung Thi Bao Pham
Department of Mutation and Heterosis Breeding, Agricultural Genetics Institute, Vietnamese Academy of Agricultural Science, Hanoi 100000, Vietnam

Marie-Laure Pilet-Nayel
INRA, UMR 1349 IGEPP; PISOM, UMT INRA/CETIOM

Xinpeng Qi
Center for Soybean Research, State Key Laboratory of Agrobiotechnology and School of Life Sciences, the Chinese University of Hong Kong, Shatin, N.T., Hong Kong

Karthika Rajendran
Biodiversity and Integrated Gene Management Program, International Center for Agricultural Research in the Dry Areas, Rabat, Morocco

Nathalie Rivière
Biogemma, route d'Ennezat

Eytan Ruppin
Sackler Faculty of Medicine, Tel Aviv University, Tel Aviv, Israel

Samuel M. D. Seaver
Mathematics and Computer Science Division, Argonne National Laboratory, Argonne, IL, USA and Computation Institute, The University of Chicago, Chicago, IL, USA

Sonali Sengupta
Division of Plant Biology, Acharya J C Bose Biotechnology Innovation Centre, Bose Institute, Kolkata, India

David A. Springate
School of Life Sciences, University of Manchester, Manchester, UK

Saad Sulieman
Signaling Pathway Research Unit, RIKEN Center for Sustainable Resource Science (CSRS), 1-7-22 Suehiro-cho, Tsurumi, Yokohama 230-0045, Japan and Department of Agronomy, Faculty of Agriculture, University of Khartoum, 13314 Shambat, Khartoum North, Sudan

Lam-Son Phan Tran
Signaling Pathway Research Unit, RIKEN Center for Sustainable Resource Science (CSRS), 1-7-22 Suehiro-cho, Tsurumi, Yokohama 230-0045, Japan

Yasuko Watanabe
Signaling Pathway Research Unit, RIKEN Center for Sustainable Resource Science (CSRS), 1-7-22 Suehiro-cho, Tsurumi, Yokohama 230-0045, Japan

Raphy Zarecki
Sackler Faculty of Medicine, Tel Aviv University, Tel Aviv, Israel

Introduction

Climate change is a variation in global or regional climate pattern attributed largely to increased greenhouse gases by use of fossil fuel. Considering the evolutionary timeline for existence of earth's atmosphere, the sudden and severe changes in atmospheric variables has led to unpredictable changes in climatic conditions that manifest differently in various parts of the globe. In 2014, the National Oceanic and Atmospheric Administration reported record-setting figures for increased greenhouse gases, including CO_2, and global temperature increases from around the globe. The earth's indicators, such as a warming planet, rising greenhouse gases, elevated sea and ocean levels, flash floods, and severe droughts, continue to show trends of sudden but rapid global climate change.

The extreme weather patterns influence crop production by exerting immense physiological and mechanical stresses on the plants. The changing air composition (high CO_2), soil factors (nutrient, moisture, and texture), temperature, and photoperiod severely impede crop yield and quality. The change in weather patterns also brings multitudes of biotic stresses, such as rapidly mutating plant pathogens and gregarious insects that can completely destroy the crops.

With the onus of feeding a rapidly increasing population under this changing climate, it is imperative that we develop crop varieties that can be productive in harsh and variable environments. The application of genomics to crop plants promises to be a key strategy in the world's defense against the looming threat of climate change. To that end, researchers are actively developing new breeding strategies.

Genomics-based approaches are increasing our understanding of crop biology and facilitate crop improvement by exploring available germplasm and wild crop genetic resources. Genomic tools allow researchers to target appropriate traits for optimal crop performance under changing climate conditions in different parts of the globe.

With a better understanding of how crop systems work, screening allele combinations and phenotypes to create more resilient crops is the key to sustainable agriculture. Breeding for improved crop types will reduce dependency on extensive chemical management (insecticides, pesticides, herbicides, fertilizers), thus reducing fossil fuel usages which contributes to greenhouse gas emissions, potentially mitigating further climate change.

The research included in this compendium has been chosen as a representation of current knowledge in this vital field of investigation, as well as to create a foundation for ongoing research. We sincerely hope that genomic-based crop breeding approaches have the potential to help feed the world in the years to come.

Santosh Kumar, PhD

Food security requires the development and deployment of crop varieties resilient to climate variation and change. The study of variations in the genome of wild plant populations can be used to guide crop improvement. Genome variation found in wild crop relatives may be directly relevant to the breeding of environmentally adapted and climate resilient crops. In Chapter 1, Henry uses analysis of the genomes of populations growing in contrasting environments to reveal the genes subject to natural selection in adaptation to climate variations. Whole genome sequencing of these populations should define the numbers and types of genes associated with climate adaptation. This strategy is facilitated by recent advances in sequencing technologies. Wild relatives of rice and barley have been used to assess these approaches. This strategy is most easily applied to species for which a high quality reference genome sequence is available and where populations of wild relatives can be found growing in diverse environments or across environmental gradients.

Significant changes in plant phenology have been observed in response to increases in mean global temperatures. There are concerns that accelerated phenologies can negatively impact plant populations. However, the fitness consequence of changes in phenology in response to elevated temperature is not well understood, particularly under field conditions. The

authors of Chapter 2, Springate and Kover, address this issue by exposing
a set of recombinant inbred lines of *Arabidopsis thaliana* to a simulated
global warming treatment in the field. We find that plants exposed to elevated temperatures flower earlier, as predicted by photothermal models.
However, contrary to life-history trade-off expectations, they also flower
at a larger vegetative size, suggesting that warming probably causes acceleration in vegetative development. Although warming increases mean
fitness (fruit production) by ca. 25%, there is a significant genotype-by-
environment interaction. Changes in fitness rank indicate that imminent
climate change can cause populations to be maladapted in their new environment, if adaptive evolution is limited. Thus, changes in the genetic
composition of populations are likely, depending on the species' generation time and the speed of temperature change. Interestingly, genotypes
that show stronger phenological responses have higher fitness under elevated temperatures, suggesting that phenological sensitivity might be
a good indicator of success under elevated temperature at the genotypic
level as well as at the species level.

The rate of biological invasions is expected to increase as the effects of
climate change on biological communities become widespread. Climate
change enhances habitat disturbance which facilitates the establishment
of invasive species, which in turn provides opportunities for hybridization and introgression. These effects influence local biodiversity that can
be tracked through genetic and genomic approaches. Metabarcoding and
metagenomic approaches provide a way of monitoring some types of
communities under climate change for the appearance of invasives. Introgression and hybridization can be followed by the analysis of entire
genomes so that rapidly changing areas of the genome are identified and
instances of genetic pollution monitored. Genomic markers enable accurate tracking of invasive species' geographic origin well beyond what was
previously possible. New genomic tools are promoting fresh insights into
classic questions about invading organisms under climate change, such as
the role of genetic variation, local adaptation and climate pre-adaptation
in successful invasions. These tools are providing managers with often
more effective means to identify potential threats, improve surveillance
and assess impacts on communities. In Chapter 3, Chown and colleagues
provide a framework for the application of genomic techniques within a

management context and also indicate some important limitations in what can be achieved.

Some areas in plant abiotic stress research are not frequently addressed by genomic and molecular tools. One such area is the cross reaction of gravitational force with upward capillary pull of water and the mechanical-functional trade-off in plant vasculature. Although frost, drought and flooding stress greatly impact these physiological processes and consequently plant performance, the genomic and molecular basis of such trade-off is only sporadically addressed and so is its adaptive value. Embolism resistance is an important multiple stress-opposition trait and offers scopes for critical insight to unravel and modify the input of living cells in the process and their biotechnological intervention may be of great importance. Vascular plants employ different physiological strategies to cope with embolism and variation is observed across the kingdom. The genomic resources in this area have started to emerge and open up possibilities of synthesis, validation and utilization of the new knowledge-base. Chapter 4, by Sengupta and Majumder, assesses the research on this issue and discusses new possibilities for bridging physiology and genomics of a plant, and foresees its implementation in crop science.

Abiotic and biotic stresses lead to massive reprogramming of different life processes and are the major limiting factors hampering crop productivity. Omics-based research platforms allow for a holistic and comprehensive survey on crop stress responses and hence may bring forth better crop improvement strategies. Since high-throughput approaches generate considerable amounts of data, bioinformatics tools will play an essential role in storing, retrieving, sharing, processing, and analyzing them. Genomic and functional genomic studies in crops still lag far behind similar studies in humans and other animals. In Chapter 5, Li and colleagues summarize some useful genomics and bioinformatics resources available to crop scientists. In addition, the authors discuss the major challenges and advancements in the "-omics" studies, with an emphasis on their possible impacts on crop stress research and crop improvement.

The decreasing cost along with rapid progress in next-generation sequencing and related bioinformatics computing resources has facilitated large-scale discovery of SNPs in various model and nonmodel plant species. Large numbers and genome-wide availability of SNPs make them the marker of choice in partially or completely sequenced genomes. Although excellent reviews have been published on next-generation sequencing, its associated bioinformatics challenges, and the applications of SNPs in genetic studies, a comprehensive review connecting these three intertwined research areas is needed. Chapter 6, by Kumar and colleagues, touches upon various aspects of SNP discovery, highlighting key points in availability and selection of appropriate sequencing platforms, bioinformatics pipelines, SNP filtering criteria, and applications of SNPs in genetic analyses. The use of next-generation sequencing methodologies in many non-model crops leading to discovery and implementation of SNPs in various genetic studies is discussed. Development and improvement of bioinformatics software that are open source and freely available have accelerated the SNP discovery while reducing the associated cost. Key considerations for SNP filtering and associated pipelines are discussed in specific topics. A list of commonly used software and their sources is compiled for easy access and reference.

Progress in genetics and breeding in pea still suffers from the limited availability of molecular resources. SNP markers that can be identified through affordable sequencing processes, without the need for prior genome reduction or a reference genome to assemble sequencing data would allow the discovery and genetic mapping of thousands of molecular markers. Such an approach could significantly speed up genetic studies and marker assisted breeding for non-model species. The authors of chapter 7 discovered a total of 419,024 SNPs using HiSeq whole genome sequencing of four pea lines, followed by direct identification of SNP markers without assembly using the discoSnp tool. Subsequent filtering led to the identification of 131,850 highly designable SNPs, polymorphic between at least two of the four pea lines. A subset of 64,754 SNPs was called and genotyped by short read sequencing on a subpopulation of 48 RILs from the cross 'Baccara' x 'PI180693'. This data was used to construct a WGGBS-derived pea genetic map comprising 64,263 markers. This map is collinear with previous pea consensus maps and therefore with the *Medicago truncatula* genome. Sequencing of four additional pea lines showed that 33 % to 64 % of the mapped SNPs, depending on the pairs of lines considered, are polymorphic and can therefore be useful in other crosses. The subsequent genotyping of a subset of 1000 SNPs, chosen for their mapping positions using a KASP™ assay, showed that almost all generated SNPs are highly designable and that most (95 %) deliver highly qualitative genotyping results. Using rather low sequencing coverages in SNP discovery and in SNP inferring did not hinder the identification of hundreds of thousands of high quality SNPs. The development and optimization of appropriate tools in SNP discovery and genetic mapping have allowed the authors of chapter 7 to make available a massive new genomic resource in pea. It will be useful for both fine mapping within chosen QTL confidence intervals and marker assisted breeding for important traits in pea improvement.

There is a growing demand for genome-scale metabolic reconstructions for plants, fueled by the need to understand the metabolic basis of crop yield and by progress in genome and transcriptome sequencing. Methods are also required to enable the interpretation of plant transcriptome data to study how cellular metabolic activity varies under different growth conditions or even within different organs, tissues, and developmental stages. Such methods depend extensively on the accuracy with which genes have been mapped to the biochemical reactions in the plant metabolic pathways. Errors in these mappings lead to metabolic reconstructions with an inflated number of reactions and possible generation of unreliable metabolic phenotype predictions. In Chapter 8, Seaver and colleagues introduce a new evidence-based genome-scale metabolic reconstruction of maize, with significant improvements in the quality of the gene-reaction associations included within our model. They also present a new approach for applying our model to predict active metabolic genes based on transcriptome data. This method includes a minimal set of reactions associated with low expression genes to enable activity of a maximum number of reactions associated with high expression genes. The authors apply this method to construct an organ-specific model for the maize leaf, and tissue specific models for maize embryo and endosperm cells. They validate our models using fluxomics data for the endosperm and embryo, demonstrating an improved capacity of our models to fit the available fluxomics data. All models are publicly available via the DOE Systems Biology Knowledgebase and PlantSEED, and the new method is generally applicable for analysis transcript profiles from any plant, paving the way for further in silico studies with a wide variety of plant genomes.

Water deficit is one of the major constraints for soybean production in Vietnam. The soybean breeding research efforts conducted at the Agriculture Genetics Institute (AGI) of Vietnam resulted in the development of promising soybean genotypes, suitable for the drought-stressed areas in Vietnam and other countries. Such a variety, namely, DT2008, was recommended by AGI and widely used throughout the country. The aim of Sulieman and colleagues in Chapter 9 was to assess the growth of shoots, roots, and nodules of DT2008 versus Williams 82 (W82) in response to drought and subsequent rehydration in symbiotic association as a means to provide genetic resources for genomic research. Better shoot, root, and nodule growth and development were observed in the cultivar DT2008 under sufficient, water deficit, and recovery conditions. The results represent a good foundation for further comparison of DT2008 and W82 at molecular levels using high throughput omic technologies, which will provide huge amounts of data, enabling researchers to understand the genetic network involved in regulation of soybean responses to water deficit and increasing the chances of developing drought-tolerant cultivars.

Most of the lentil growing countries face a certain set of abiotic and biotic stresses causing substantial reduction in crop growth, yield, and production. Until-to date, lentil breeders have used conventional plant breeding techniques of selection-recombination-selection cycle to develop improved cultivars. These techniques have been successful in mainstreaming some of the easy-to-manage monogenic traits. However, in case of complex quantitative traits, these conventional techniques are less precise. As most of the economic traits are complex, quantitative, and often influenced by environments and genotype–environment interaction, the genetic improvement of these traits becomes difficult. Genomics assisted breeding is relatively powerful and fast approach to develop high yielding varieties more suitable to adverse environmental conditions. New tools such as molecular markers and bioinformatics are expected to generate new knowledge and improve our understanding on the genetics of complex traits. In the past, the limited availability of genomic resources in lentil could not allow breeders to employ these tools in mainstream breeding program. The recent application of the next generation sequencing and genotyping by sequencing technologies has facilitated to speed up the lentil genome sequencing project and large discovery of genome-wide single nucleotide polymorphism (SNP) markers. Currently, several linkage maps have been developed in lentil through the use of expressed sequenced tag (EST) derived simple sequence repeat (SSR) and SNP markers.These maps have emerged as useful genomic resources to identify quantitative trait loci imparting tolerance to biotic and abiotic stresses in lentil. In Chapter 10, Kumar and colleagues discuss, the current knowledge on available genomic resources and its application in lentil breeding program.

PART I

CROP GENOMICS
AND CLIMATE CHANGE

Genomics Strategies for Germplasm Characterization and the Development of Climate Resilient Crops

ROBERT J. HENRY

1.1 NEED TO ADAPT CROPS TO NEW AND CHANGING ENVIRONMENTS AND THE ROLE OF GENOMICS

Agriculture needs significant increases in productivity to satisfy the expected growth in demand for food in the next few decades. The impact of climate variability and climate change on agricultural productivity is likely to be a major constraint to achieving increased food production. This makes the development of crop genotypes with resilience to climate change an important strategy for food security. Innovations in crop improvement based upon application of advanced genomics tools may be a way to address this need. The delivery of these technologies will require significant efforts in coordinated development and delivery of improved germplasm (Lybbert et al., 2013). Genomics allows resources available for crop adaptation to environmental stress to be characterized and utilized (Bansal et al., 2013). An evolutionary perspective may assist in the effective application of the power of genomic tools to the development of climate resilient crops adapted to a changing environment.

1.2 GENOMIC ANALYSIS OF CROP EVOLUTION AND ADAPTATION TO CLIMATE CHANGE

Crop evolution has been relatively rapid under human selection over the last 10,000 years of agriculture. However, it is built on a very much longer period of evolution of wild crop relatives and the plant groups from which they are sourced. Understanding the processes and history of crop domestication and the evolution of related wild species provides critical knowledge to guide the development of crop varieties that are resilient to climate change in the future.

Analysis of wild plant populations provides evidence of factors contributing to success in periods of climate change. For example, hybridization between species may be an advantage in adapting to rapid climate change by providing new genetic combinations to cope with new environmental circumstances. Closely related species that can hybridize are more likely to survive than highly divergent species that cannot hybridize (Becker et al., 2013). Analysis of the genetics of populations growing across environmental gradients or from contrasting environments may be used to identify how plant populations adapt to climate under natural selection (Cronin et al., 2007). Sampling of populations at the same time over a long period of time can also be used to monitor adaptation to climate change but few sites have been sampled in the past in a way that allows this type of analysis to be conducted. Establishment of long term experiments of this type would be of great value. Recent dramatic improvements in genome analysis tools due to rapid advances in DNA sequencing technology make feasible research that should deliver much greater understanding of the relationships between wild and domesticated plant populations (Henry, 2012, 2013).

Recent fossil evidence suggests early diversification of groups of crop wild relatives such as the grasses (Prasad et al., 2011). The climate resilience of domesticated rice populations may be related to their evolutionary history. For example, expansion of the range of climates to which crops are adapted will require the transfer of genes from wild populations adapted to new environments or the use of novel genes. Crop species are derived for many different flowering plant groups but many are from a small number of families (e.g., Poaceae and Fabaceae). Crop plants have many traits that reflect the environments in which they evolved prior to domestication. Humans have collected plants for food for a long period of time prior to domestication of plants and the establishment of agriculture in the last 10,000 years. Pre-domestication use of plants by humans or natural variants that suit domestication (Ishii et al., 2013) may have also impacted upon some plant populations but domestication has usually resulted in significant genetic alteration of plants to suit human production in agriculture and food uses (Jin et al., 2008).

1.3 CHOICE OF SPECIES FOR CLIMATE RESILIENT AGRICULTURE

Domesticated crop species are few in number compared to the total number of land plant species (Henry, 2010). A small number of plant species that have been adapted to wide scale production account for a large part of the energy and protein in human diets. These have become the key crops contributing to global food security. A larger number of species have been domesticated for more limited local production in specific regions. Some of these could be considered for adaptation to a wider range of environments.

Genomics tools provide new options for accelerated domestication of new species to allow adaptation of agriculture to climate change (Shapter et al., 2013).

1.3.1 MONO-PHYLETIC AND POLYPHYLETIC DOMESTICATION

Domestication may have been a single genetic event with all the domesticated plants being descendent from the same wild parents or have involved a few or many independent domestication events with many wild plants contributing to the domesticated gene pool. This understanding may provide the opportunity to repeat the domestication of important crop species from a different or more diverse gene pool. Genome analysis may be used to guide this process.

1.3.2 CENTERS OF ORIGIN

The center of origin of a crop species is the region from which the species is believed to have been domesticated. These are the environments that the crop plant was originally best adapted to survive at the time of domestication. Domestication from a different population selected by genome analysis may provide an opportunity to develop genotypes adapted to a new environment.

1.3.3 CENTERS OF DIVERSITY

Genome analysis allows rapid identification of geographic centers of genome diversity. The center of diversity of a crop species is the region displaying the greatest genetic diversity of the crop species or its wild relatives. This may be distinct from the center of origin as plant species may have been domesticated in areas that are not those including the greatest diversity. Identification of these locations may provide new and diverse germplasm and define new environments for production of the crop now or in the future. Asian rice (*Oryza sativa*) was probably domesticated in China from wild *O. rufipogon*. The A genome clade of wild rice relatives is now considered to be most diverse further south with a center of diversity in New Guinea, Australia, and Indonesia. These locations may prove to be good sources of novel germplasm for rice improvement. Species from more temperate regions could be used to adapt rice to production in cooler climates.

1.3.4 PRIMARY, SECONDARY, AND TERTIARY GENE POOLS

The gene pools of crop species may be considered at several levels. Genomic analysis may have value at all of these levels. The primary gene pool is the gene pool of the plant found in domestication and usually the species from which the crop was domesticated. The primary gene pool includes those plants that are available for direct use in genetic improvement of the species. The secondary gene pool may include more diverse material from other species that can be accessed but with a greater degree of difficulty. This often includes other species in the same or a related genus (Dillon et al., 2007). The tertiary gene pool is a wider group of plants from which genes can be accessed but only with significant difficulty (e.g., plants in the family outside the genus that can be accessed as a source of new genes but only with technological intervention). Understanding the genetic basis of domestication and the issues associated with access of genes from more difficult (or distant) relatives facilitates their use in crop improvement and in the domestication of new species to adapt agriculture to climate change (Malory et al., 2011). These analyses are more powerful at the whole genome level.

1.4 ADVANCES IN GENOMICS OF CROPS

Advances in DNA sequencing in the last few years have resulted in genomic sequence data becoming more readily available (Edwards et al., 2012). Major efforts have been made to produce reference genome sequences for key species. This allows rapid analysis of sequence variation within species. However, de novo assembly of sequence data may be necessary to detect all differences and advances in sequencing technology to make this routinely possible with large plant genomes will be a significant advance.

Analysis of the genomes of plants growing along environmental gradients may provide a greater understanding of how plants adapt to climate variation under natural selection (Cronin et al., 2007; Fitzgerald et al., 2011; Shapter et al., 2012).

1.5 GENOMIC ANALYSIS OF GENETIC RESOURCES

Analysis of the genomes of plant genetic resources will become a key tool to enable their utilization in crop improvement for climate adaptation. Targeting of genetic resources from environments that match the one being breed for is an important strategy. Large scale sequencing of accessions in plant germplasm collections will provide a platform to enable these approaches (Henry, 2013).

Increased utilization of wild crop relatives will remain a major strategy for adaptation of crops to the environmental factors associated with climate change. Many crop wild relatives remain poorly collected and are not yet represent well in seed banks. Climate change and human development risk loss of this genetic diversity making accelerated collection of crop wild relatives urgent. Rice illustrates this challenge. The closest wild relatives of rice are those from the A genome clade from which rice was domesticated (Vaughan et al., 2006). Recent research has identified two possible new species in this group that represent important new sources of diversity for rice improvement (Sotowa et al., 2013). Rice wild relatives from some regions such as Africa (Wambugu et al., 2013) and Australia (Henry et al., 2010) are poorly known.

1.6 ANALYSIS OF NATURAL POPULATIONS AS A GUIDE TO IMPROVEMENT OF CROPS FOR AGRICULTURAL PRODUCTION

The analysis of populations of wild relatives of barley (Cronin et al., 2007; Fitzgerald et al., 2011) and rice (Fitzgerald et al., 2011; Shapter et al., 2012) indicate the potential value of genome analysis of these populations to support efforts to develop crop varieties adapted to new climates.

In these studies, wild plants were collected from diverse environments or along a sharp environmental gradient. Sampling of the same population over time as the climate changes could be simulated by this strategy. In only a few cases we can access samples that have been sampled from the same population over a significant period of time. Key findings were that adaptation to hotter or dryer environments was associated with increased diversity of biotic stress genes. Coping with abiotic stress may be confounded by overriding associated changes in the biotic environment (Fitzgerald et al., 2011).

1.7 REMOVING THE CONSTRAINT OF END USE QUALITY ON RAPID CROP ADAPTATION TO CLIMATE

Productivity gains in crop production require elimination of constraints to utilization of more diverse germplasm. In some species the requirements of end uses are a major limitation. Market requirements for specific food or processing attributes that are complex or not well understood at the genetic level can greatly hamper attempts to use diverse adapted germplasm. Genomics tools that allow these traits to be readily selected for in breeding will assist by removing these as constraints to rapid climate adaptation (Henry, 2014). Food quality traits are often associated with human selection in domestication. They are often relatively simply controlled genetically because of their relatively recent and brief evolution under human selection in the last 10,000 years or less. Improved understanding these genes can be targeted as achievable steps toward removing a major constraint on climate adaptation.

1.8 AVOIDING SELECTION THAT REDUCES CLIMATIC RESILIENCE

Human selection for quality may result in loss of environmental adaptation. Fragrance in rice is highly attractive to humans and adds significant value to rice. The sequencing of the rice genome allowed the identification of the genetic basis of this trait (Bradbury et al., 2005) due to the gene being flanked by closely linked known markers (Qingsheng et al., 2003). The gene responsible is an aldehyde dehydrogenase (Bradbury et al., 2008) the activity of which is lost in fragrant genotypes. The loss of the gene reduces the ability of the plant to cope with salt stress (Fitzgerald et al., 2010). Whole genome understanding of genes responsible for quality (Kharabian-Masouleh et al., 2012) will allow their relationship to abiotic stress tolerance genes to be carefully evaluated. Very attractive traits like fragrance may require strategies such as selection of compensating abiotic stress tolerance genes to counteract the deleterious effects of the quality gene.

1.9 DURABLE PEST AND DISEASE RESISTANCE IN A CHANGING CLIMATE

The breeding of crops to cope with new pests and diseases will be a key strategy to allow plants to cope with new climates. Genes from wild populations will continue to be a major option but this may need to be complemented by the use of novel transgenes or genetic modifications.

1.10 ROLE OF CONTINUING TECHNOLOGY ADVANCES

Technology advances will continue to be critical. Ultimately we need to be able to access whole genome information on all crop species and their wild relatives to be effective in rapid crop adaptation to climate. Ongoing developments in the chemistry of DNA sequencing and in information technology hardware and software will be required to allow these very large amounts of information to be captured and managed.

REFERENCES

1. Bansal, K. C., Lenaka, S. K., and Mondal, T. K. (2013). Genomic resources for breeding crops with enhanced abiotic stress tolerance. Plant Breed. doi: 10.1111/pbr.12117
2. Becker, M., Gruenheit, N., Steel, M., Voelckel, C., Deusch, O., Heenan, P. B., et al. (2013). Hybridization may facilitate in situ survival of endemic species through periods of climate change Nat. Clim. Chang. doi: 10.1038/nclimate2027
3. Bradbury, L. M. T., Fitzgerald, T. L., Henry, R. J., Jin, Q., and Waters, D. L. E. (2005). The gene for fragrance in rice. Plant Biotechnol. J. 3, 363–370. doi: 10.1111/j.1467-7652.2005.00131.x
4. Bradbury, L. M. E., Gillies, S. A., Brushett, D., Waters, D. L. E., and Henry, R. J. (2008). Inactivation of an aminoaldehyde dehydrogenase is responsible for fragrance in rice. Plant Mol. Biol. 68, 439–449. doi: 10.1007/s11103-008-9381-x
5. Cronin, J. K., Bundock, P. C., Henry, R. J., and Nevo, E. (2007). Adaptive climatic molecular evolution in wild barley at the Isa defense locus. Proc. Nat. Acad. Sci. U.S.A. 104, 2773–2778. doi: 10.1073/pnas.0611226104
6. Dillon, S. L., Shapter, F. M., Henry, R. J., Cordeiro, G., Izquierdo, L., and Lee, L. S. (2007). Domestication to crop improvement: genetic resources for Sorghum and Saccharum (Andropogoneae). Ann. Bot. 100: 975–989. doi: 10.1093/aob/mcm192.

7. Edwards, D., Henry, R. J., and Edwards, K. J. (2012). Advances in DNA sequencing accelerating plant biotechnology. Plant Biotechnol. J. 10, 621–622. dol: 10.1111/j.1467-7652.2012.00724.x

8. Fitzgerald, T. L., Shapter, F. M., McDonald, S., Waters, D. L. E., Chivers, I. H., Drenth, A., et al. (2011) Genome diversity in wild grasses under environmental stress. Proc. Nat. Acad. Sci. U.S.A. 108, 21139–21144. doi: 10.1073/pnas.1115203108

9. Fitzgerald, T. L., Waters, D. L. E., Brooks, L. O., and Henry, R. J. (2010). Fragrance in rice (Oryza sativa) is associated with reduced yield under salt treatment. Environ. Exp. Bot. 68, 292–300. doi: 10.1016/j.envexpbot.2010.01.001

10. Henry, R. J. (2010). Plant Resources for Food, Fuel and Conservation. London: Earthscan, 200.

11. Henry, R. J. (2012). Next generation sequencing for understanding and accelerating crop domestication. Brief. Funct. Genomics 11, 51–56. doi: 10.1093/bfgp/elr032

12. Henry, R. J. (2013). Sequencing crop wild relatives to support the conservation and utilization of plant genetic resources. Plant Genet. Resour. doi: 10.1017/S1479262113000439

13. Henry, R. J. (2014). Wheat genomics for grain quality improvement. Cereal foods world (in press).

14. Henry, R. J., Rice, N., Waters, D. L. E., Kasem, S., Ishikawa, R., Dillon, S. L., et al. (2010). Australian Oryza: utility and conservation. Rice 3, 235–241. doi: 10.1007/s12284-009-9034-y

15. Ishii, T., Numaguchi, K., Miura, K., Yoshida, K., Thien Thanh, P., Myint Htun, T., et al. (2013). OsLG1 regulates a closed panicle trait in domesticated rice. Nat. Genet. 45, 462–465. doi: 10.1038/ng.2567

16. Jin, J., Huang, W., Gao, J.-P., Yang, J., Shi, M., Zhu, M.-Z., et al. (2008). Genetic control of rice plant architecture under domestication. Nat. Genet. 40, 1365–1369. doi: 10.1038/ng.247

17. Kharabian-Masouleh, A., Waters, D. L. E., Reinke, R. F., Ward, R., and Henry, R. J. (2012). SNP in starch biosynthesis genes associated with nutritional and functional properties of rice. Sci. Rep. 2, 557, doi: 10.1038/srep00557

18. Lybbert, T., Skerritt, J. H., and Henry, R. J. (2013). "Facilitation of future research and extension through funding and networking support", in Genomics and Breeding for Climate-Resilient Crops, Vol. 1, Concepts and Strategies, ed. C. Kole (Heidelberg: Springer), 415-432.

19. Malory, S., Shapter, F. M., Elphinstone, M. S., Chivers, I. H., and Henry, R. J. (2011). Characterizing homologues of crop domestication genes in poorly described wild relatives by high-throughput sequencing of whole genomes. Plant Biotechnol. J. 9, 1131–1140. doi: 10.1111/j.1467-7652.2011.00640.x

20. Prasad, V., Stromberg, C. A. E., Leache, A. D., Samant, B., Patnaik, R., Tang, L., et al. (2011). Late Cretaceous origin of the rice tribe provides evidence for early diversification in Poaceae. Nat. Commun. 2, 480. doi: 10.1038/ncomms1482

21. Qingsheng, J., Waters, D. L. E., Cordeiro, G. M., Henry, R. J., and Reinke, R. F. (2003). A single nucleotide polymorphism (SNP) marker linked to fragrance in rice (Oryza sativa L.). Plant Sci. 165, 359–364. doi: 10.1016/S0168-9452(03)00195-X

22. Shapter, F. M., Cross, M., Ablett, G., Malory, S., Chivers, I. H., King, G. J., et al. (2013). High-throughput sequencing and mutagenesis to accelerate the domestica-

tion of Microlaena stipoides as a new food crop. PLoS ONE 8:e82641. doi: 10.1371/journal.pone.0082641

23. Shapter, F. M., Fitzgerald, T. L., Waters, D. L. E., McDonald, S., Chivers, I. H., and Henry, R. J. (2012). Analysis of adaptive ribosomal gene diversity in wild plant populations from contrasting climatic environments. Plant Signal. Behav. 7, 1–3. doi: 10.4161/psb.19938

24. Sotowa, M., Ootsuka, K., Kobayashi, Y., Hao, Y., Tanaka, K., Ichitani, K., et al. (2013). Molecular relationships between Australian annual wild rice, Oryza meridionalis, and two related perennial forms. Rice 6, 26. doi: 10.1186/1939-8433-6-26

25. Vaughan, D. A., Ge, S., Kaga, A., and Tomooka, N. (2006). "Phylogeny and Biogeography of the Genus Oryza," in Rice Biology in the Genomics Era, eds H.-Y. Hirano, Y. Sano, A. Hirai, and T. Sasaki (Berlin: Springer), 218–234.

26. Wambugu, P., Furtado, A., Waters, D., Nyamongo, D., and Henry, R. (2013). Conservation and utilization of African Oryza genetic resources. Rice 6, 29. doi: 10.1186/1939-8433-6-29

CHAPTER 2

Plant Responses to Elevated Temperatures: A Field Study on Phenological Sensitivity and Fitness Responses to Simulated Climate Warming

DAVID A. SPRINGATE AND PAULA X. KOVER

2.1 INTRODUCTION

Flowering time can affect many aspects of a plant's ecology and fitness (Rathcke & Lacey, 1985; Parra-Tabla & Vargas, 2004; Kover et al., 2009a; Amasino, 2010). Accordingly, natural variation in flowering time has been shown to be under selection (Le Corre et al., 2002; Franks et al., 2007; Korves et al., 2007; Anderson et al., 2011; Munguía-Rosas et al., 2011). Mean global temperatures have risen by around 0.8 °C in the last hundred years and further increases of 2–3 °C are expected by the end of the century (Minorsky, 2002; IPCC, 2007). Climate change is expected to have its strongest and most immediate effects on plant phenology (Forrest &

Plant Responses to Elevated Temperatures: A Field Study on Phenological Sensitivity and Fitness Responses to Simulated Climate Warming. © Springate DA and Kover PX. Global Change Biology 20,2 (2014). DOI: 10.1111/gcb.12430. Licensed under Creative Commons 3.0 Unported License, http://creativecommons.org/licenses/by/3.0/.

Miller-Rushing, 2010; Munguía-Rosas et al., 2011), and accelerated phenologies have already been observed in many species (Sparks et al., 2000; Abu-Asab et al., 2001; Menzel et al., 2006; Cleland et al., 2012). There is concern that accelerated phenologies may alter patterns of resource allocation, interactions with pollinators, the size and diversity of the soil seed bank and compromise species persistence (Visser & Holleman, 2001; Minorsky, 2002; Walther et al., 2002; Post & Pedersen, 2008; Hegland et al., 2009). Here, we use a climate manipulation experiment under field conditions, to investigate the consequences of elevated temperature to flowering time and fitness in the plant *Arabidopsis thaliana*.

The use of environmental cues to flower at the right time is critical to capitalize on the best environmental conditions to produce fruits. It has been suggested that species that do change their phenology in response to climate change (phenologically sensitive species) are better at tracking optimal environmental conditions, and therefore more likely to persist (Cleland et al., 2012). Accordingly, phenological sensitivity has been recently incorporated in species vulnerability assessments (Glick et al., 2011). Although phenological sensitivity at the population or species level must be connected to responses at the individual level, a connection between this proposed species-level phenomenon and a mechanistic understanding at the population and genotypic level has not been investigated (Forrest & Miller-Rushing, 2010). 'Phenological sensitivity' is a concept analogous to 'flowering plasticity', which is the difference in flowering time expressed by the same genotype under different environmental conditions. Although phenotypic plasticity is very common in plants, its adaptive value and its role in facilitating plants coping with environmental change is much debated (Ghalambor et al., 2007; Valladares et al., 2007; Nicotra et al., 2010). Here, we investigate the genetic architecture of plastic responses, and test specifically whether genotypes that are more phenologically responsive to temperature (show higher plasticity in flowering time) are better adapted to changes in climate.

Life-history theory predicts that early flowering genotypes (i.e. genotypes that flower earlier in relation to planting date) will transition into reproduction at smaller vegetative size, reducing their reproductive output (Mitchell-Olds, 1996). Therefore, if earlier flowering in response to climate warming is achieved by earlier onset of reproductive development;

we would also expect a reduction in both vegetative size and fitness. However, if in response to climate change, plants transition to flowering earlier due to a general increase in the rate of vegetative development (because climate change improves the quality of the environment), plants can both flower earlier and be larger (therefore increasing their fitness). A better understanding of the mechanism through which flowering is accelerated in response to elevated temperature would clarify whether there is a need for concern. Yet, studies that combine responses to temperature and life-history trade-offs are rare (Metcalf & Mitchell-Olds, 2009).

Flowering time is a complex trait that is affected by a range of environmental factors such as photoperiod, ambient temperature, vernalization and plant size (Boss et al., 2004; Cockram et al., 2007; Colasanti & Coneva, 2009). Many of these studies were carried out in the model plant *Arabidopsis thaliana* because this species is easily manipulated experimentally, amenable to the construction of inbred lines and genetically well characterized (e.g. Johanson et al., 2000; Lempe et al., 2005; Wilczek et al., 2010). More than 60 genes have been identified to affect flowering time in *A. thaliana*, including genes that affect the thermosensory pathway (Blazquez et al., 2003; Balasubramanian, 2006) under laboratory conditions. However, little is known about the importance of small changes in average temperature under field conditions, or the genetic basis of responses to such changes. The few studies on *A. thaliana* performed under field conditions (e.g. Weinig et al., 2002; Brachi et al., 2010) suggest that plants in the field can respond very differently from laboratory experiments because they simultaneously experience a larger number and a wider range of environmental cues (including variations in light quality, temperature and photoperiod). Because under field conditions, plants in different locations will experience different photoperiods and temperatures, photo-thermal models that incorporate both photoperiod and temperature data have been proposed to translate flowering time measured in days into photothermal units (taking into account local environmental conditions). It is hypothesized that plants integrate photoperiod and thermal cues to transition into reproduction once a genetically determined threshold of accumulated photothermal units has been reached. These models have been used successfully to study the importance of different mutants and genetic pathways in *A. thaliana* (Wilczek et al., 2009), and they suggest that for a given site (where the photoperiod is a constant), flowering time should be a linear function of temperature.

Field studies on the effect of climate warming on phenology typically compare populations or species over long periods of time or across environments (e.g. Fitter & Fitter, 2002; Wilczek et al., 2010; Ågren & Schemske, 2012). However, because flowering is affected by both temperature and photoperiod, among other variables, comparisons across sites and time periods cannot clearly separate the effect of temperature. Here, we used surface-level heating cables (as championed by Grime et al., 2000, 2008) to investigate the effect of temperature on a set of *A. thaliana* recombinant inbred lines. Such climate manipulation under field conditions is particularly powerful because it allows plants to be exposed to small elevations in temperatures without losing information about daily variation in temperature, day length, or other environmental cues, which will be equal across treatments. The use of *A. thaliana* mapping lines allows us to also explore the underlying genetic basis and the genetic variation in response to elevated temperature. Using this approach, we ask the following questions:

1. Are small changes in temperature sufficient to change the phenology and fitness of *A. thaliana* given all other environmental cues in the field?
2. What is the genetic architecture of the plastic response in flowering, vegetative size and fitness to elevated temperature?
3. Does acceleration in phenology in response to elevated temperatures cause smaller vegetative size and compromise fitness?
4. Do more phenologically responsive genotypes have higher fitness under elevated temperatures?

2.2 MATERIALS AND METHODS

2.2.1 EXPERIMENTAL DESIGN

We used a set of 320 Multiparental Advanced Genetic InterCross (MAGIC) *A. thaliana* lines (Kover et al., 2009b). These nearly isogenic lines are derived from an outbred population composed of 19 natural accessions of *A. thaliana* that have been genotyped with 1260 single nucleotide polymorphisms (SNPs). Thus, they can be used for QTL mapping as well as to estimate response to the environment. That is because replicates of each MAGIC line are nearly genetically identical, allowing an estimation of the average difference in phenotype under different environments, i.e. their 'plasticity'. For each MAGIC line, we prepared 10 microtubes, each containing five seeds in a 0.2% agar solution. All tubes were cold stratified for 7 days to promote synchronized germination. All seeds from each tube were directly planted into the soil in April 2009 with the help of a pipette into one of 10 plots set-up at the Botanical research station of the University of Manchester (UK). Multiple seeds were used to ensure we have at least one successful germinant per planting.

Each plot was 320 by 100 cm, and contained a single replicate of each line. Seeds and agar mixture were planted into a 5 mm diameter plastic ring pressed into the surface of the soil at 10 cm intervals to mark planting positions. The position of each line within a plot was randomly assigned. An 8 by 12 m fruit cage was erected over the 10 plots to protect the experiment from herbivores. Plots were arranged in two rows of five plots, and every other plot was assigned to a warming treatment (forming a reticulate pattern). Warming cables (600W; Thermoforce HQ, Cockermouth, Cumbria, UK) were connected to differential thermostats which maintained the surface temperature a constant 2–3 °C above ambient. Thus, the elevated temperature plots experienced the same variation in temperature, day length, light quality and humidity that the control plots experienced, the only difference being that the temperature was constantly elevated by 2–3 °C. Cables were placed in the surface of the plot, in between each row of plantings. Data loggers (Hobo U2 Temperature data loggers) were set up to record temperatures in both treatments. The plots were treated with Roundup™ herbicide (Scotts HQ, Marysville, OH, USA) 3 weeks prior to planting then tilled and levelled 1 week later.

Plants were inspected daily, and seedlings within a single planting ring were thinned down to a single one 10 days after planting. Any seed that germinated after this could not have been detected and thinned because the plant density prevented close inspection without damage to the growing plants. Flowering time was recorded as the first day an open flower was visible. At flowering time, we estimated the plant's vegetative size by measuring the rosette diameter. The diameter of each plant's rosette was measured across two perpendicular axes, and the estimated diameter was calculated as the average of the two measures. The same approximate axes were used for all plants, by orienting a ruler at a 45° and 135° angle from the plant label. After senescence, all plants were harvested and the number of fruits on each plant (larger than 1 cm) was counted in the laboratory to estimate fitness. When more than one seedling was present in a single planting ring, we phenotyped the largest plant because the smaller plants were more likely to have been the later germinants.

Because we were interested in the effect of temperature on phenology in a complex environment with multiple environmental cues, we analysed flowering time both in terms of number of days from planting to flowering, and in terms of 'photothermal time' or PTT, which was calculated as:

$$PTT = \sum_{p=i}^{ft} \lambda i (\mu i - \mu b)$$

(Brachi et al., 2010), where p is the planting date; ft is the flowering date; μi is the mean daily temperature during daylight (calculated from the data collected from the data loggers); μb is the base temperature for development (estimated to be equal to 3 °C, Granier et al., 2002); and λi is the daily photoperiod as a proportion of 24 h (photoperiod data were extracted from www.timeanddate.com). The PTTs represent the threshold number of photothermal units (PTUs) needed for a genotype to flower. This threshold is expected to be determined genetically, and independently of the environment, in which a genotype is grown, if the model is correct. Here, as the only difference between the two treatments is temperature, no difference in flowering time when measured in PTUs is expected across treatments if time to flower is a linear function of temperature, and temperature does not alter the threshold.

2.2.2 DATA ANALYSIS

All statistical analyses were performed in R version 2.13.0. We initially tested for significant effects of temperature on germination and surviv-ability to flowering using logistic regression using all data available. As temperature had no significant effect on germination or survivability to flowering ("Results"), the subsequent analyses were performed on 278 lines for which we have a complete data set for three or more replicates. All replicates of these 278 lines that died after germinating without pro-ducing fruits were assigned a fitness of zero and included in the analysis.

To determine the effect of elevated temperatures on flowering time, rosette size, and fruit number; as well as genetic variation and geno-type-by-environment interactions for these traits, we fitted the follow-ing mixed effects model using the R package *lme4*(Bates & Maechler, 2009): Trait = Treatment + MAGIC line +Treatment × MAGIC line + Density + Edge + Plot(Treatment) + error. In this model, treatment was set as a fixed effect, whereas genotype was set as a random effect. Plot was set as a random effect nested within treatment. Each replicate was scored as being on the edge of a plot or not, and edge effects were con-trolled for as a fixed effect. We controlled for density effects for each data point in all models by setting density (number of plants within a planting ring) as a fixed effect. MCMC p-values (10000 MCMC sam-ples) were calculated for the fixed effects using pvals.fnc from the R package languageR(Baayen, 2008). Significance of random effect vari-ance components were determined by likelihood ratio tests. To provide the final unbiased variance components, the model was then re-fitted using restricted maximum likelihood (REML).

To estimate whether genotypic trait values under ambient temperatures could predict responses in elevated temperatures, we calculated cross-en-vironment genetic correlations as the Pearson correlation coefficients of the best linear unbiased predictors (BLUPs) for the MAGIC line random effects of each trait. Standard errors for genetic correlations were obtained by jackknife.

To determine whether phenological sensitivity (i.e. the magnitude and direction of the plastic response in flowering time to elevated temperature) was positively associated with fitness under elevated temperature, we estimated the average phenological response of each MAGIC line as its flowering plasticity (equal to the mean flowering time in the control, minus in the elevated temperature plots). We then regressed the mean fruit production under elevated temperature for each line against its flowering time under elevated temperature and its phenological response (as in Weinig et al., 2006 and Springate et al., 2011). The choice of regressing fruit production under the elevated temperature is because we are specifically interested in testing whether phenological responsiveness is adaptive under climate change. We also included the mean flowering value under elevated temperature in the model, to separate the effect of the plastic response from the trait mean value. To determine if any association was a general association with more plastic genotypes, or specific to the phenological response, we also performed the same analysis using plasticity in rosette diameter.

2.2.3 QTL MAPPING

We used a quantitative trait loci approach to determine if the genetic architecture underlying flowering time and fitness was affected by elevated temperatures. We used the BLUPs for the MAGIC lines from a REML mixed effect model to give estimates of the line effects for each trait. The use of BLUPs was chosen due to the small effects detected for edges, plot and density (see results). The model used to generate the BLUPs was: Trait = MAGIC line + plot + Edge + density, with MAGIC line set as a random effect, and the others as fixed effects. The models were run for all traits in both ambient and elevated temperature treatments separately. To determine if there were QTL affecting the plastic response of traits independent of QTL that directly affected the trait value, we also mapped QTLs for the magnitude of the plastic response. QTL were mapped using the method described in Kover et al. (2009b), where a probabilistic reconstruction of the haplotype mosaic of each MAGIC line is initially calculated, and the genome is scanned for evidence of a QTL in each SNP interval using a fixed effects model. The genome-wide evidence in favour of a QTL was evaluated by resampling the data 500 times and fitting multiple QTL models.

2.3 RESULTS

In the first week of the experiment, daytime temperatures averaged 14.9 °C and night time temperatures averaged 10.7 °C. At the end of the experiment, 3 months later, the average temperatures were 16.2 °C and 14.8 °C respectively. Day length varied during the experiment from 14 to 17 h long. During the whole experiment elevated plots were on average 2.6 °C warmer.

Germination was high in both control and elevated plots (87.5% and 87.9% of the plantings respectively). Mortality was overall quite sporadic, with 96% of the plantings that have germinated surviving to flowering in the control plots and 98.5% in the elevated plots. There was no significant effect of temperature treatment on either germination ($\chi^2 = 3.1$, $P_{(df = 1)} = 0.08$), or survival of seedlings to flowering ($\chi^2 = 1.0$, $P_{(df = 1)} = 0.31$).

Elevated temperature significantly affected days to flowering time, rosette diameter and number of fruits: Plants under elevated temperature, on average, flowered ca. 4 days earlier, had rosettes ca. 13 mm larger and produced ca. 200 more fruits (Table 1). However, there was no significant effect on flowering time in terms of photothermal units, suggesting that temperature effects on flowering time is consistent with photothermal models. The effect of plant density, plot and edge effects were significant (Table S1) but explained a small amount of variation ($R^2 < 5\%$). Significant genetic variance was observed for all traits as indicated by the between MAGIC line variance (V_g) (Table 2). No significant interaction between line and treatment was observed for rosette diameter ($P = 0.254$), suggesting that most lines respond similar to elevated temperatures. In contrast, a significant line by treatment interaction was observed for number of fruits ($P = 0.019$), and a marginally significant effect was detected for flowering time (days to flowering $P = 0.071$; PTT $P = 0.072$). Norms of reaction for all four traits is shown in Fig. 1.

TABLE 1: Mean trait values (and SE in parenthesis) under ambient and elevated temperature treatments. Mean Squares (MS), F statistic (F) and probability (P) values are for the effect of the elevated temperature treatment. Results for all variables included in the model are shown in Table S1. P values were calculated using MCMC resampling (df = 1)

Trait	Control	Elevated	MS	F	P
Rosette diameter (mm)	36.69 (2.54)	49.92 (3.41)	4219.6	13.2	0.004
Flowering time					
Days	53.15 (0.77)	49.02 (1.04)	255.5	14.5	0.003
PTT	540.51 (9.45)	522.9 (12.69)	4067.8	1.5	0.205
Number of fruits	786.5 (52.81)	989.73 (68.78)	3053558.0	9.8	0.023

Bold values indicate significant at $P < 0.05$.

TABLE 2: Variance components [genetic (V_g,) and genotype-by-environment variance (V_gX_e,)], heritabilities (H^2) and cross-environment genetic correlations (R_g) for MAGIC lines grown under ambient and elevated temperature treatments

Trait	V_g	V_gX_e	Ambient H^2 (SE)	Elevated H^2 (SE)	R_g (SE)
Rosette diameter	84.7 **	6.21	0.15 (0.03)	0.28 (0.04)	0.45 (0.39)
Flowering time (days)	11.1 **	0.71	0.45 (0.04)	0.37 (0.04)	0.64 (0.04)
Flowering time (PTT)	1677.43 **	113.19	0.46 (0.04)	0.35 (0.05)	0.64 (0.04)
Number of fruits	30678.1 **	18703.1 *	0.10 (0.4)	0.16 (0.04)	0.23 (0.05)

Significance levels of genetic variances (determined by likelihood ratio tests) are indicated by asterisks (*$P < 0.05$, **$P < 0.0001$). Components in bold have $P < 0.05$.

Pairwise genetic correlations among traits (based on genotype mean trait values) for ambient and elevated temperature treatments are shown in Fig. 2 (PTT is not included because it is perfectly correlated with flowering time within treatments). Rosette diameter is significantly correlated with number of fruits in both treatments, meaning that larger plants also have more fruits. In contrast, flowering time is not significantly correlated with number of fruits, even though flowering time is positively correlated with rosette diameter—confirming that within treatment plants that flower earlier tend to have smaller vegetative size.

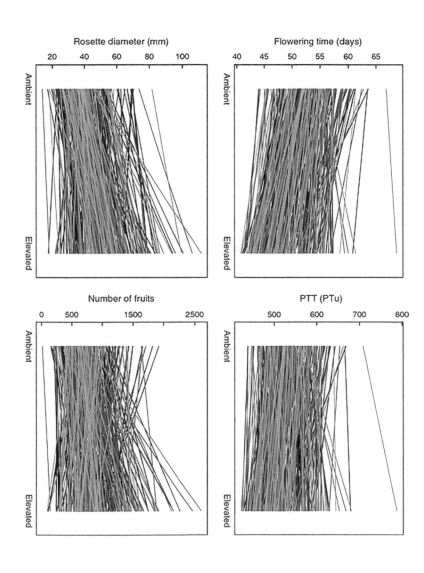

FIGURE 1: Mean reaction norms between ambient and elevated temperature treatments for four traits in *A. thaliana* MAGIC lines.

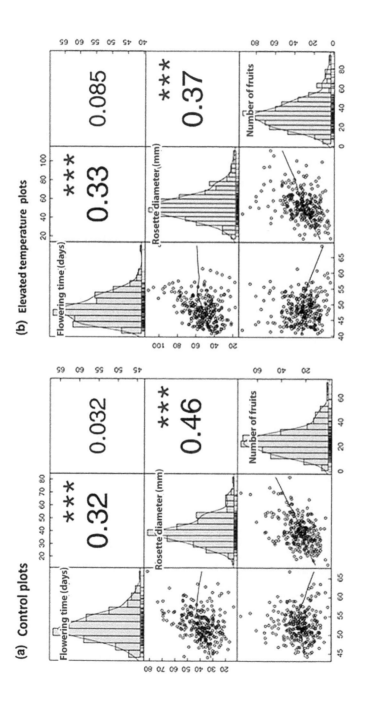

FIGURE 2: Correlations between mean trait values within treatment. Panel 2a shows values for the ambient treatment, and panel 2b for elevated temperature. Histograms and kernel density plots of the univariate distributions are shown on the diagonal. Pairwise Pearson correlations with starred significance levels are shown on the right of the diagonal. Scatter plots of the correlations with LOESS smoothers are shown left of the diagonal.

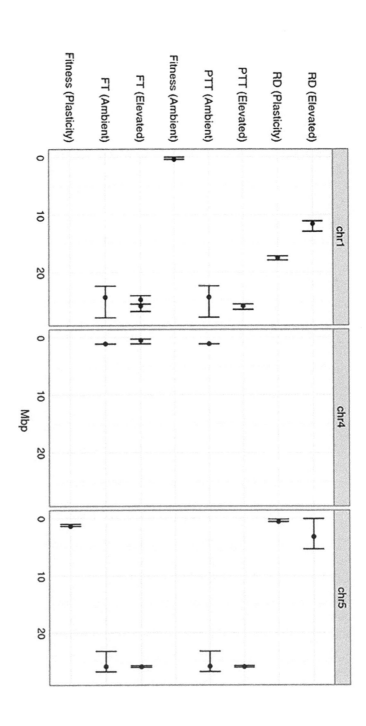

FIGURE 3: Positions of each QTL identified in this study. Only chromosomes 1, 4 and 5 (where QTL were identified) are shown. Further details about each of these QTL can be found on supplementary materials (Table S2).

A significant correlation between fruit number under elevated temperatures and flowering plasticity shows that lines that are more phenologically responsive also tended to have higher fitness under elevated temperature (Table 3). It is interesting to notice that flowering time under elevated temperatures did not have a significant direct effect on fruit production. In contrast, rosette diameter had a significant direct effect on fruit production under elevated temperature, but plasticity in rosette diameter did not (Table 3).

TABLE 3: Regression of fruit number (fitness) under elevated temperature on mean trait value and its plasticity. Plasticity was measured as the difference between elevated and control means, and represents the response to changes in the temperature of the same genotype. Analyses were done twice using flowering time (measured in days) and rosette diameter

	Flowering (days)			Rosette diameter (mm)		
	b	t	P	b	t	P
Mean in elevated plots	-8.0 ± 6.8	1.2	0.237	10.1 ± 2.1	4.8	3.0×10^{-06}
Plasticity	-40.9 ± 9.8	4.2	3.9×10^{-05}	3.9 ± 2.4	1.6	0.102

Bold values indicate significant at P <0.05.

Full genome LOD scans identified a total of 18 QTL for traits in ambient and elevated treatments, as well as for plasticity (Fig. 3, Table S1 in the supporting material). Three QTL were detected for days to flower in ambient temperatures in chromosomes 1, 4 and 5. While under elevated temperatures 4 QTL were observed, the only difference is that in the elevated temperature treatment 2 distinct QTLs are detected on chromosome 1 (where a broader peak in the same location is observed in the controls). No QTL for plasticity in flowering time was detected. All QTLs for PTT overlapped with those for flowering time (Fig. 3). Two QTL for rosette size were detected in the elevated temperature treatment on chromosomes 1 and 5. No QTLs for rosette size were detected in the control plants, but two QTLs for rosette size plasticity were detected on chromosomes 1 and 5. The QTL for rosette size plasticity on chromosome 1 did not colocate with a QTL for rosette diameter itself, suggesting that there is some separate genetic control for trait value and its plasticity in rosette size. One QTL was found for number of fruits on chromosome 1 under control conditions, but no colocated QTL for fruit number is seen under elevated temperatures (Fig. 3). A QTL for fruit number plasticity is observed in chromosome 5, suggesting again separate genetic control from trait value.

2.4 DISCUSSION

We used a climate warming simulation approach and a set of RILs to investigate the effect of small increases in mean temperature on the phenology and fitness of *A. thaliana*, in a complex environmental background. We found that plants that experienced elevated temperatures flower significantly earlier than plants in control plots, in agreement with previous studies that compared flowering time across years or different locations (Fitter & Fitter, 2002; Menzel et al., 2006; Anderson et al., 2012). It is noteworthy that in this study, the changes in average temperature were quite small (<3 °C), and many other cues could be used to trigger flowering (such as daily variations in temperature, seasonal increase in day length, precipitation, etc.). Although plants flower significantly earlier, we found that, on average, plants flower at similar thresholds of PTUs, suggesting that photothermal models (e.g. Brachi et al., 2010; Chew et al., 2012) provide a good approximation of the effect of temperature on flowering under natural conditions. This is important because it indicates that the effect of temperature in climate change models can be efficiently modelled by a predictable linear relationship. Although the average change is well captured by the photothermal model, it is important to notice that some genotypes show clear deviation from the expected reduction in days to flowering time (Fig. 1), and the test for GXE effects approaches significance (Table S1). Thus, it should be interesting in the future to pursue the genetic basis of the few lines that were identified to have uncharacteristic norms of reaction.

Plasticity is the result of differential phenotypic expression of the same genotype in response to different environments. Here, we detected significant differences in the flowering time, rosette size and fruit production of

MAGIC lines under elevated temperature. Understanding the genetic basis of variation in plasticity could be useful in developing genetic models of plant response to climate change (Chevin et al., 2010; Nicotra et al., 2010). The different phenotypic expression may be the result of a different set of QTL underlying the trait under different environments (i.e. some genes are expressed only under some circumstances, or some gene function is only relevant under some environments) (Gardner & Latta, 2006; Lacaze et al., 2008). Alternatively, it may be due to the existence of a 'plasticity gene' that affects the magnitude of the plastic response independent of the QTL that affect variation in trait value (Schlichting & Pigliucci, 1993). Here, we found that plasticity may have different genetic causes depending on the trait. A distinct QTL was observed for rosette size and fruit number plasticity that does not correspond to the QTL identified for trait mean value, suggesting the existence of genetic factors that affect the plasticity of a trait somewhat independently of genes that affect the trait value. In contrast, the same QTL for flowering time were identified under both treatments, whether flowering was measured in terms of days or in PTUs. These results suggest that a common genetic pathway regulates flowering time under ambient and elevated temperature, and that temperature affects flowering in a linear and predictable way. The plastic response in term of days to flowering appears to be a simple consequence of an accelerated accumulation of PTUs with higher temperature without significant changes in the genetically determined threshold.

There has been concern that accelerated phenology could reduce yield and fitness because an earlier trigger in the transition to flowering could constrain vegetative development, leading to smaller rosettes and lower seed production (Lobell & Field, 2007; Craufurd & Wheeler, 2009). However, although we observe a positive relationship between rosette size and flowering time within treatments (confirming life-history theory predictions), across treatments this pattern is not upheld. Plants growing under elevated temperature flower earlier but at a larger vegetative size than plants under ambient temperature, leading to an increase in fitness. This may be explained by faster development due to improvement of growing conditions with the increase in temperature during spring, so that flowering occurred earlier without compromising vegetative growth. A similar effect of higher temperatures on the rate of leaf production has been previously reported in laboratory experiments (Granier et al., 2002; Hoffmann et al., 2005). It is possible that the same effect would not have been observed if the experiment was run later in the summer or in a different geographical region where the increase in temperature would put averages above ideal temperatures for *A. thaliana* growth in late spring. Further experiments should investigate the effect of season and geographical location on the generality of our results.

Early flowering in *A. thaliana* has been proposed to be favoured in spring (Korves et al., 2007), so the increase in fitness under elevated temperature could have simply resulted from earlier flowering. However, we found that flowering time was only weakly correlated with fitness in either treatment ($R < 0.1$). In contrast, we found that rosette diameter is more strongly correlated with number of fruits, and that elevated temperatures cause an increase in rosette size. Thus, the increase in fruit production (and therefore, fitness) due to elevated temperature is most likely mediated by the increase in vegetative size due to faster development.

Although on average there is an increase in fitness under elevated temperature, this response was not consistent across genotypes, as we detect a significant line by treatment interaction for fruit production. Most of this interaction is due to changes in the rank order of genotypes (Fig. 1). The low genetic correlation across treatment in fitness means that the current distribution of fitness is a poor predictor of future fitness ranking in responses to climate warming. This conclusion is further supported by the fact the QTL for fitness identified in ambient temperatures is not observed under elevated temperatures. It follows that current populations would likely be maladapted if the environment change by an average of approximately 3 °C, and populations are unable to respond through genetic adaptation. For plant species with small generation times and large populations, such as *A. thaliana* and many other spring annuals, a slow increase in temperature is likely to result in significant changes in population genetic composition.

There is concern that mismatch in responses to climate change between plants and their pollinators may cause severe fitness decrease for some species (Hegland et al., 2009). Because *A. thaliana* can self-fertilize, the observed fitness response to temperature in this study is unconstrained by the need to coordinate its flowering times with other mating plants or pollinators. In addition, the fact that our experiment was carried under a fruit cage, minimize any impact herbivores could have had. Thus, it is important to notice that in our experiment, any impact of a mismatch between *A. thaliana* and other members of the community on fitness was minimized, and the effect of elevated temperature on fitness might be overestimated. Although recent work suggest that the dangers of asynchrony might be compensated by a diverse community of pollinators and herbivores (the biodiversity insurance hypothesis, e.g. Bartomeus et al., 2013), broad generalization of our results will require further studies using plants with other mating characteristics and different phenological schedule.

Identification of genetic mechanisms that help plants cope with environmental changes will improve the likelihood that populations would persist long enough for the necessary genetic adaptation to take place. Plastic responses might provide a buffer for the effect of climate change and help populations persist (Chevin et al., 2010; Nicotra et al., 2010). Here, we identified a QTL for plasticity in fruit number, where plants with an allele from the Ws-0 accession at this location produce ca. 500 more fruits under elevated temperatures, whereas plants with an allele from the Rsch-4 accession produces ca. 100 fewer fruits (Table S2). Although a previous study has observed environment-specific QTL for fruit production in the field (Weinig et al., 2003), we are not aware of any previous studies that identified QTL for the magnitude of the plastic response in fitness. Further studies on the nature of this QTL in *A. thaliana* may provide significant insight in mechanisms that mediate fitness response to environmental changes. Investigation of currently annotated genes under the QTL peak identifies DREB2 (At5G05410) as a possible candidate gene. This gene is known to respond to temperature and to affect growth (Yoshida et al., 2011), but has not been previously associated with fitness under elevated temperatures.

In summary, the novel combination of temperature manipulation in the field with the use of MAGIC lines in this experiment has allowed us to determine that plasticity may have an important role in a plant's ability to cope with climate change. While on average plants under elevated temperature had higher fitness, not all genotypes responded equally, and some clearly declined in fitness. We found evidence for natural genetic variation for adaptive plasticity that significantly affects fitness ranking. Thus, significant changes in the genetic composition of populations are likely in response to imminent climate changes. The magnitude of this change, and whether populations will find themselves maladapted, will depend on the population size and the species' generation time. Annual plants such as *A. thaliana* may be able to respond quickly given their large progeny and short generation time, but plants with longer generation or smaller populations may experience significant reduction in fitness. These results suggest that when plasticity is included in models of species persistence under climate change, it is important to also consider that there is genetic variation in plastic responses. Future work should further investigate the genetic mechanisms underlying plastic responses that increase fitness under climate change, since depending on pleiotropic effects of these genetic factors, the phenological and ecological characteristics of the populations can be significantly changed. Nevertheless our results do support the idea that genotypes that are more phenologically responsive tend to be more successful under elevated temperatures, suggesting that phenological sensitivity can be a useful indicator of genotypes to be used in restoration and conservation projects.

REFERENCES

1. Abu-Asab M, Peterson P, Shetler S, Orli S (2001) Earlier plant flowering in spring as a response to global warming in the Washington, DC area. Biodiversity and Conservation, 10, 597–612.
2. Ågren J, Schemske DW (2012) Reciprocal transplants demonstrate strong adaptive differentiation of the model organism *Arabidopsis thaliana* in its native range. New Phytologist, 194, 1112–1122.
3. Amasino R (2010) Seasonal and developmental timing of flowering. The Plant Journal, 61, 1001–1013.
4. Anderson JT, Lee C-R, Mitchell-Olds T (2011) Life-history QTLs and natural selection on flowering time in Boechera stricta, a perennial relative of Arabidopsis. Evolution, 65, 771–787.
5. Anderson JT, Inouye DW, Mckinney AM, Colautti RI, Mitchell-Olds T (2012) Phenotypic plasticity and adaptive evolution contribute to advancing flowering phenology in response to climate change. Proceedings of the Royal Society B: Biological Sciences, 279, 3843–3852.
6. Baayen R (2008) Analyzing Linguistic Data: A Practical Introduction to Statistics. Cambridge Univesrity Press, Cambridge.

7. Balasubramanian S (2006) Potent induction of *Arabidopsis thaliana* flowering by elevated growth temperature. PLoS Genetics, 2, e106.
8. Bartomeus I, Park MG, Gibbs J, Danforth BN, Lakso AN, Winfree R (2013) Biodiversity ensures plant–pollinator phenological synchrony against climate change. Ecology Letters, 11, 1331–1338.
9. Bates D., Maechler M (2009) lme4: Linear mixed-effects models using S4 classes. R package version 0.999375-31.
10. Blazquez M, Ahn J, Weigel D (2003) A thermosensory pathway controlling flowering time in *Arabidopsis thaliana*. Nature Genetics, 33, 168–171.
11. Boss PK, Bastow RM, Mylne JS, Dean C (2004) Multiple pathways in the decision to flower: enabling, promoting, and resetting. The Plant Cell, 16, S18–S31.
12. Brachi B, Faure N, Horton M et al. (2010) Linkage and association mapping of *Arabidopsis thaliana* flowering time in nature. PLoS Genetics, 6, e1000940.
13. Chevin L-M, Lande R, Mace GM (2010) Adaptation, plasticity, and extinction in a changing environment: towards a predictive theory. PLoS Biology, 8, e1000357.
14. Chew Y, Wilczek AM, Williams M, Welch ME, Schmitt J, Halliday KJ (2012) An augmented Arabidopsis phenology model reveals seasonal temperature control of flowering time. New Phytologist, 194, 654–665.
15. Cleland EE, Allen JM, Crimmins TM et al. (2012) Phenological tracking enables positive species responses to climate change. Ecology, 93, 1765–1771.
16. Cockram J, Jones H, Leigh FJ, O'Sullivan D, Powell W, Laurie DA, Greenland AJ. (2007) Control of flowering time in temperate cereals: genes, domestication, and sustainable productivity. Journal of Experimental Botany, 58, 1231–1244.
17. Colasanti J, Coneva V (2009) Mechanisms of floral induction in grasses: something borrowed, something new. Plant Physiology, 149, 56–62.
18. Craufurd PQ, Wheeler TR (2009) Climate change and the flowering time of annual crops. Journal of Experimental Botany, 60, 2529–2539.
19. Fitter AH, Fitter RSR (2002) Rapid changes in flowering time in British plants. Science, 296, 1689–1691.
20. Forrest J, Miller-Rushing AJ (2010) Toward a synthetic understanding of the role of phenology in ecology and evolution. Philosophical Transactions of the Royal Society B: Biological Sciences, 365, 3101–3112.
21. Franks SJ, Sim S, Weis AE (2007) Rapid evolution of flowering time by an annual plant in response to a climate fluctuation. Proceedings of the National Academy of Sciences, 104, 1278–1282.
22. Gardner KM, Latta RG (2006) Identifying loci under selection across contrasting environments in Avena barbata using quantitative trait locus mapping. Molecular Ecology, 15, 1321–1333.
23. Ghalambor CK, Mckay JK, Carroll SP, Reznick DN (2007) Adaptive versus non-adaptive phenotypic plasticity and the potential for contemporary adaptation in new environments. Functional Ecology, 21, 394–407.
24. Glick P, Stein B, Na E (2011) Scanning the Conservation Horizon: A Guide to Climate Change Vulnerability Assessment. National Wildlife Federation, Washington DC.

25. Granier C, Massonnet C, Turc O, Muller B, Chenu K, Tardieu F (2002) Individual leaf development in *Arabidopsis thaliana*: a stable thermal-time-based programme. Annals of Botany, 89, 595–604.

26. Grime JP, Brown VK, Thompson K et al. (2000) The response of two contrasting limestone grasslands to simulated climate change. Science, 289, 762–765.

27. Grime JP, Fridley JD, Askew AP, Thompson K, Hodgson JG, Bennett CR (2008) Long-term resistance to simulated climate change in an infertile grassland. Proceedings of the National Academy of Sciences, USA, 105, 10028–10032.

28. Hegland SJ, Nielsen A, Lázaro A, Bjerknes A-L, Totland Ø (2009) How does climate warming affect plant-pollinator interactions? Ecology Letters, 12, 184–195.

29. Hoffmann MH, Tomiuk J, Schmuths H, Koch C, Bachmann K (2005) Phenological and morphological responses to different temperature treatments differ among a world-wide sample of accessions of *Arabidopsis thaliana*. Acta Oecologica, 28, 181–187.

30. IPCC (2007) Climate Change 2007: Impacts, Adaptation and Vulnerability. Cambridge University Press, Cambridge.

31. Johanson U, West J, Lister C, Michaels S, Amasino R, Dean C (2000) Molecular analysis of FRIGIDA, a major determinant of natural variation in Arabidopsis flowering time. Science, 290, 344–347.

32. Korves TM, Schmid KJ, Caicedo AL, Mays C, Stinchcombe JR, Purugganan MD, Schmitt J (2007) Fitness effects associated with the major flowering time gene FRIGIDA in *Arabidopsis thaliana* in the field. The American Naturalist, 169, E141–E157.

33. Kover PX, Rowntree JK, Scarcelli N, Savriama Y, Eldridge T, Schaal BA (2009a) Pleiotropic effects of environment-specific adaptation in *Arabidopsis thaliana*. New Phytologist, 183, 816–825.

34. Kover PX, Valdar W, Trakalo J et al. (2009b) A multiparent advanced generation intercross to fine-map quantitative traits in *Arabidopsis thaliana*. PLoS Genetics, 5, e1000551.

35. Lacaze X, Hayes PM, Korol A (2008) Genetics of phenotypic plasticity: QTL analysis in barley, Hordeum vulgare. Heredity, 102, 163–173.

36. Le Corre V, Roux F, Reboud X (2002) DNA polymorphism at the FRIGIDA gene in *Arabidopsis thaliana*: extensive nonsynonymous variation is consistent with local selection for flowering time. Molecular Biology and Evolution, 19, 1261–1271.

37. Lempe J, Balasubramanian S, Sureshkumar S, Singh A, Schmid M, Weigel D (2005) Diversity of flowering responses in wild *Arabidopsis thaliana* strains. PLoS Genetics, 1, e6.

38. Lobell D, Field C (2007) Global scale climate-crop yield relationships and the impacts of recent warming. Environmental Research Letters, 2, p014002.

39. Menzel A, Sparks TH, Estrella N et al. (2006) European phenological response to climate change matches the warming pattern. Global Change Biology, 12, 1969–1976.

40. Metcalf JE, Mitchell-Olds T (2009) Life history in a model system: opening the black box with *Arabidopsis thaliana*. Ecology Letters, 12, 593–600.

41. Minorsky PV (2002) Global warming - effects on plants. Plant Physiology, 129, 1421–1422.

42. Mitchell-Olds T (1996) Genetic constraints on life-history evolution: quantitative-trait loci influencing growth and flowering in *Arabidopsis thaliana*. Evolution, 50, 140–145.

43. Munguía-Rosas MA, Ollerton J, Parra-Tabla V, De-Nova JA (2011) Meta-analysis of phenotypic selection on flowering phenology suggests that early flowering plants are favoured. Ecology Letters, 14, 511–521.

44. Nicotra AB, Atkin OK, Bonser SP et al. (2010) Plant phenotypic plasticity in a changing climate. Trends in Plant Science, 15, 684–692.

45. Parra-Tabla V, Vargas CF (2004) Phenology and phenotypic natural selection on the flowering time of a deceit-pollinated tropical orchid, Myrmecophila christinae. Annals of Botany, 94, 243–250.

46. Post E, Pedersen C (2008) Opposing plant community responses to warming with and without herbivores. Proceedings of the National Academy of Sciences, 105, 12353–12358.

47. Rathcke B, Lacey EP (1985) Phenological patterns of terrestrial plants. Annual Review of Ecology and Systematics, 16, 179–214.

48. Schlichting CD, Pigliucci M (1993) Control of phenotypic plasticity via regulatory genes. American Naturalist, 142, 366–370.

49. Sparks TH, Jeffree EP, Jeffree CE (2000) An examination of the relationship between flowering times and temperature at the national scale using long-term phenological records from the UK. International Journal of Biometeorology, 44, 82–87.

50. Springate DA, Scarcelli N, Rowntree J, Kover PX (2011) Correlated response in plasticity to selection for early flowering in *Arabidopsis thaliana*. Journal of Evolutionary Biology, 24, 2280–2288.

51. Valladares F, Gianoli E, Gómez JM (2007) Ecological limits to plant phenotypic plasticity. New Phytologist, 176, 749–763.

52. Visser ME, Holleman LJM (2001) Warmer springs disrupt the synchrony of oak and winter moth phenology. Proceedings of the Royal Society of London. Series B: Biological Sciences, 268, 289–294.

53. Walther G-R, Post E, Convey P et al. (2002) Ecological responses to recent climate change. Nature, 416, 389–395.

54. Weinig C, Ungerer MC, Dorn LA et al. (2002) Novel loci control variation in reproductive timing in *Arabidopsis thaliana* in natural environments. Genetics, 162, 1875–1884.

55. Weinig C, Dorn LA, Kane NC et al. (2003) Heterogeneous selection at specific loci in natural environments in *Arabidopsis thaliana*. Genetics, 165, 321–329.

56. Weinig C, Johnston J, German ZM, Demink LM (2006) Local and global costs of adaptive plasticity to density in *Arabidopsis thaliana*. The American Naturalist, 167, 826–836.

57. Wilczek AM, Roe JL, Knapp MC et al. (2009) Effects of genetic perturbation on seasonal life history plasticity. Science, 323, 930–934.

58. Wilczek AM, Burghardt LT, Cobb AR, Cooper MD, Welch SM, Schmitt J (2010) Genetic and physiological bases for phenological responses to current and predicted climates. Philosophical Transactions of the Royal Society B: Biological Sciences, 365, 3129–3147.

59. Yoshida T, Ohama N, Nakajima J et al. (2011) Arabidopsis HsfA1 transcription factors function as the main positive regulators in heat shock-responsive gene expression. Molecular Genetics and Genomics, 286, 321–332.

There are several supplemental files that are not available in this version of the article. To view this additional information, please use the citation on the first page of this chapter.

CHAPTER 3

Biological Invasions, Climate Change, and Genomics

STEVEN L. CHOWN, KATHRYN A. HODGINS, PHILIPPA C. GRIFFIN, JOHN G. OAKESHOTT, MARGARET BYRNE, AND ARY A. HOFFMANN

3.1 INTRODUCTION

Biological invasions constitute a major environmental change driver, affecting conservation, agriculture and human health. Invasive alien species (IAS) impacts include, alterations of ecosystem features such as hydrology, fire regimes, food webs, soil nutrients and nutrient cycling (Hänel and Chown 1998; Asner and Vitousek 2005; Van Wilgen 2009; Veldtman et al. 2011); negative effects on populations (Blackburn et al. 2004; Butchart et al. 2010; Ziska et al. 2011); facilitation of further invasion and its associated impacts (O'Dowd et al. 2003); changes to evolutionary trajectories, such as by hybridization (McDonald et al. 2008); and evolutionary shifts in species responding to the new introductions (Strauss et al. 2006). Much

Biological Invasions, Climate Change, and Genomics © *Chown SL, Hodgins KA, Griffin PC, Oakeshott JG, Byrne M, and Hoffmann AA.* Evolutionary Applications *8,1 (2015). DOI: 10.1111/eva.12234.* *Licensed under a Creative Commons Attribution 4.0 International License, https://creativecommons. org/licenses/by/4.0/.*

attention has therefore been given to understanding the invasion process. A broadly accepted perspective on biological invasions now exists that distinguishes the stages of invasion and the mechanisms that are involved in each stage (Blackburn et al. 2011).

Although the field has grown rapidly (Mooney and Hobbs 2000; Sax et al. 2005; Richardson 2011; McGeoch et al. 2012; Spear et al. 2013), several significant challenges remain. Foremost among these is limited understanding of the mechanisms underlying each of the key transitions in the invasion process, reflected by periodic calls for additional work in these areas (e.g. Puth and Post 2005; Hulme et al. 2008). Most recently, attention has turned to the quantification and forecasting of IAS impacts. Several reviews have concluded that current knowledge and capability in this area are inadequate, rendering management responses either ineffective or inefficient (Pyšek et al. 2012; Hulme et al. 2013; Ricciardi et al. 2013; Simberloff et al. 2013).

Further challenges are presented by the simultaneous increase in the impacts of other environmental change drivers. Biological invasions (e.g. Lambdon et al. 2008; Chown et al. 2012; Richardson and Ricciardi 2013) are not the only forms of change increasing in frequency and extent. Climate change, as a consequence of anthropogenic greenhouse gas emissions, is continuing, with indications that the rate of change may be increasing owing to a rise in CO_2 emissions (Rignot et al. 2011). Similarly, landscapes continue to change through human interventions (Barnosky et al. 2012). Exploitation is ongoing with apparently little abatement, resulting in substantial population declines, even within protected areas (Chown 2010; Craigie et al. 2010; Jackson 2010; Laurance et al. 2012). While some forms of pollution have declined notably, others have come to replace these (Sutherland et al. 2010, 2012).

Although the form of the interactions between biological invasions and these other environmental change drivers has yet to be fully generalized (Darling and Côté 2008; Bradley et al. 2010; Treasure and Chown 2014), the current view is that interactions are likely most often to be synergistic (Brook et al. 2008; Walther et al. 2009). Moreover, the simultaneous effects of environmental change drivers on indigenous species, and human responses to counter at least some of the effects of environmental change, such as through managed relocation (Schwartz et al. 2012), are likely to make management decisions about biological invasions much more difficult than in the past (Webber and Scott 2012).

The field of invasion science has recognized many of these challenges and research directions are clearly changing to meet them (McGeoch et al. 2012; Ricciardi et al. 2013; Simberloff et al. 2013). Much of this new work remains ecological in nature (at the organismal, population, species or ecosystem levels), including work that aims to guide management interventions to address the impacts of biological invasions (Richardson and Pyšek 2012; Blackburn et al. 2014). Nonetheless, from early on in the development of the field, several other approaches have been considered. Among these, the utility of genetics and the significance of evolutionary processes have long been recognized (e.g. Baker 1974). However, it is only recently that they have seen a resurgence of interest (Huey et al. 2000; Callaway and Maron 2006; Dormontt et al. 2011; Lawson Handley et al. 2011). Indeed, the significance of genetic approaches and an evolutionary perspective are now recognized as important not only for understanding the ability of species to progress along the stages of invasion (e.g. Pandit et al. 2011; Richardson and Pyšek 2012), but also for improving management interventions that might reduce the rates and impacts of invasions (Lawson Handley et al. 2011; Prentis and Pavasovic 2013). Examples include better genotypic matching of biocontrol agents and their hosts, and understanding the coevolutionary dynamics of invasive species and those with which they interact (Phillips and Shine 2006; McDonald et al. 2008; Gaskin et al. 2011). Nonetheless, additional scope exists for modern genetic approaches, and notably genomics, to contribute to the understanding and forecasting that is required to limit the rates and impacts of biological invasions, especially given the expectation of interactions with other global change drivers.

Here, we focus on the use of genomics to understand the changing form of invasion with climate change. We briefly review the contributions that genetic and evolutionary approaches have already made, discuss recent technological developments and then highlight further potential of this approach. We base our overview on the unified framework for biological invasions (Blackburn et al. 2011), recognizing that each of the stages in the invasion process can be mitigated by an accompanying management response (Fig. 1). In addition, although various views exist about how invasive species should be defined (Pyšek et al. 2004; Valéry et al. 2008), here, to avoid uncertainty (McGeoch et al. 2012), we adopt the definition provided by Richardson et al. (2011). In consequence, we are concerned with the processes by which species are transported outside their natural range through direct or indirect anthropogenic intervention and become established, proliferate and interact with the biota in their new ranges.

3.2 GENETIC AND EVOLUTIONARY STUDIES OF INVASIVE ORGANISMS

An increasing number of studies have documented evolutionary changes in invasive populations, typically over ecological timescales (Lee 2002; Prentis et al. 2008; Whitney and Gabler 2008). These studies have shown that evolutionary adaptation can be rapid in a broad range of invasive organisms. Examples include copepods adapting to different salinity levels (Lee and Petersen 2002), soapberry bugs adapting to new host plants (Carroll et al. 2001), and invasive weeds adapting through mimicry to evade eradication (Barrett 1983). Moreover, several of the documented examples of rapid evolutionary adaptation involve responses to climate change over a few decades. These include the evolution of *Drosophila* body size and inversion frequencies following invasion across continents with latitudinal climate variation (Huey et al. 2000), the evolution of rabbits along a heat gradient following invasion into Australia (Williams and Moore 1989), and changes in flowering time in *Solidago* goldenrods following introduction into Europe (Weber and Schmid 1998).

Recent work has continued to reinforce the notion that evolutionary genetic changes in invasive species can be rapid and often involve responses to climate. For instance, purple loosestrife, *Lythrum salicaria*, has invaded wetlands across an introduced range that covers a latitudinal gradient extending in eastern North America from 39°N to 48.5°N. Along this gradient, Colautti and Barrett (2013) showed an adaptive cline in flowering time, with earlier flowering at the northern invasion front, where it increases survival and fruit set, but with selection against earlier flowering in the south where it decreases vegetative growth. Climate adaptation involving changes in flowering traits has been recorded in several other invasive plant species (Moran and Alexander 2014). Evolution of traits in response to climate in invasive species has also been found for a range of animals. In *Drosophila subobscura*, introduced into Chile in the late 1970s, evolutionary divergence in thermal preferences among populations from different climates has weakened clines in chromosomal inversion (Castañeda et al. 2013). Likewise, physiological tolerances appear to have diverged in response to climate in the mite *Halotydeus destructor* (Hill et al. 2013). Several researchers have argued that evolution in invasive species can provide general answers to questions of whether species will be able to adapt to rapid climate change (Moran and Alexander 2014), including those about the speed of evolutionary change, the specific nature of the change that has occurred and factors that might limit evolutionary change.

Apart from their use to investigate adaptive changes, genetic tools have also been used to understand the dynamic nature of invasive processes. In particular, genetic markers have been widely used to detect invasive organisms, investigate patterns of historical movement in invasive species and examine the role of genetic variation in ensuring the success of invasions (Armstrong and Ball 2005; Lawson Handley et al. 2011; Blanchet 2012). Most past studies have focussed on markers thought to be selectively neutral, initially allozymes and then later microsatellites, AFLPs and other DNA markers. These markers can indicate the status of invasive species, origin of invasive populations and the levels of genetic variation in invading populations when compared to the populations of origin. Recent examples include tracking invasions of the winter annual *Geranium carolinianum* in China from multiple origins (Shirk et al. 2014), identifying a key translocation event in the invasion by a coregonid fish of a major Scandinavian watershed (Præbel et al. 2013), the use of mtDNA markers and microsatellites to trace the invasion history of eastern mosquitofish (*Gambusia holbrooki*) into Australia (Ayres et al. 2012) and determining multiple origins of invasive acacias in South Africa and South Australia (Le Roux et al. 2011; Millar et al. 2012). Genetic markers have also provided much information on the biology of invasive species, such as modes of reproduction (Weeks et al. 1995; Ali et al. 2014; Molins et al. 2014), movement patterns and rates (Bronnenhuber et al. 2011; Kirk et al. 2011; Berthouly-Salazar et al. 2013) and predator–prey relationships (Kasper et al. 2004; Blanchet 2012; Valdez-Moreno et al. 2012).

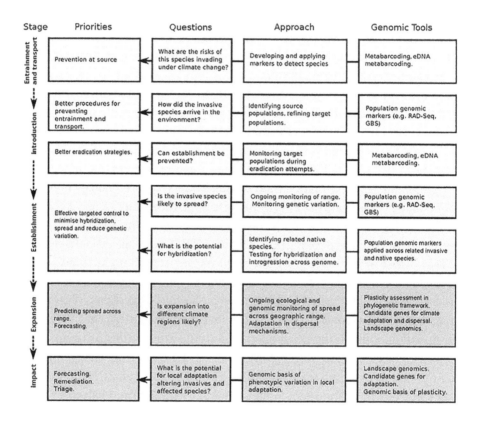

FIGURE 1: Genomic tools that can add to capacity for understanding risk and monitoring management actions at each stage of the invasion process. Tools can assist in the initial detection of an invasion, monitoring its spread after the initial detection and understanding the potential for invasions to expand and impact native species following evolutionary adaptation. Management priorities and questions relevant to each stage are also shown.

BOX 1: SEQUENCING APPROACHES
FOR ECOLOGICAL GENOMICS

Whole-genome resequencing is cost-prohibitive and unnecessary in many cases. As an alternative, several methods of genome reduction make it possible for many individuals to be combined in one sequencing run while maintaining high coverage. This results in the reliable identification of SNPs and genotypes for each individual at a subset of regions throughout the genome.

Restriction-enzyme-based reduced-representation sequencing: Genomic DNA is digested with specific restriction enzymes, followed by ligation of adaptors, amplification and sequencing. Several methods employ this strategy, including genotyping-by-sequencing (GBS; Elshire et al. 2011) and RAD-Seq (Hohenlohe et al. 2010). Repetitive regions of genomes can be avoided using methylation-sensitive enzymes (Elshire et al. 2011). This approach is relatively inexpensive and easy to use in nonmodel organisms, but can suffer from biases such as those associated with the loss of restriction sites in some individuals (Gautier et al. 2013).

RNA-Seq: Sequencing mRNA from the genes that are expressed in the whole organism or specific sampled tissues at a particular point in time and development and under specific conditions (the 'transcriptome' of a whole organism or specific tissue; Wang et al. 2009). This method targets the expressed portion of the genome (protein-coding sequences and untranslated regions of genes). Gene expression and sequence information is simultaneously identified, as it is possible to compare counts of reads that map to genes and identify variants given sufficient expression of that gene in each individual.

Targeted sequence capture by hybridization: This approach uses oligoprobes to enrich specific regions of the genome for subsequent next-generation sequencing. The scale of the capture can range from a handful to several thousand targeted loci, making it appropriate for wide range of projects. However, prior genomic knowledge is essential for the development of the probes (e.g. through RAD-Seq or RNA-Seq projects). The development of noncommercial protocols and PCR-based probes (Peñalba et al. 2014) is making this approach more attractive for smaller-scale sequencing projects in nonmodel species. Several alternative methods are available that enrich target regions prior to sequencing (for review, see Mamanova et al. 2010).

Gene-space sequencing: Many genomes are replete with repetitive regions along with genes present in high copy number (e.g. chloroplast genes), and whole-genome sequencing without reducing their quantity can be a substantial waste of sequencing space. Consequently, some approaches seek to deplete these regions rather than enrich specific portions of the genome (e.g. Matvienko et al. 2013).

GLOSSARY OF GENETIC TERMS

Genetic architecture
The genetic underpinnings of phenotypic traits, including the number loci, their effect sizes and location in the genome.

Linkage disequilibrium
The nonrandom association between alleles at two or more loci.

Quantitative trait locus (QTL)
A genomic area associated with variation in a quantitative trait in the progeny of a genetic cross.

Standing genetic variation
Existing variation in a population as opposed to variation that results from new mutations.

Wright fixation index (FST)
The proportion of the total genetic variability that occurs among populations. It is a measure of the level of population genetic differentiation.

These types of studies are now being supplemented with a range of genomic and transcriptomic approaches that are providing answers to questions about the origins of invasive populations at a new level of resolution, providing new ways of monitoring for invasive organisms, and leading to an understanding of the nature of the changes that underlie adaptive shifts. We consider the promise provided by these studies within the context of detecting and managing biological invasions as climate change proceeds, and within a framework of the different steps involved in detecting and responding to an invader, understanding its source, and then tracking its impact on the surroundings (Fig. 1).

3.3 TECHNOLOGY

3.3.1 MODEL AND NONMODEL ORGANISMS

The genomes of several invasive species have now been sequenced and assembled as well as annotated to differing extents. These include the tunicate *Ciona intestinalis* (Dehal et al. 2002), dengue mosquito vector *Aedes aegypti* (Nene et al. 2007), argentine ant *Linepithema humile* (Smith et al. 2011), fire ant *Solenopsis invicta*, (Wurm et al. 2011), diamondback moth *Plutella xylostella* (You et al. 2013), domestic cat *Felis catus* (Pontius et al. 2007) and mallard *Anas platyrhynchos* (Huang et al. 2013). Then, there are the genomes of model species that are also invasive, which include *Drosophila* species such as *D. melanogaster* and *D. suzukii* (Adams et al. 2000; Chiu et al. 2013), rats and mice (Waterston et al. 2002; Gibbs et al. 2004), and some weedy plants, such as the castor bean (Chan et al. 2010). Some of these species are useful for climatic studies, based on evidence for geographic variation in climate responses in species such as the drosophilids (Hoffmann et al. 2003) and *C. intestinalis* (Dybern 1965).

Many genomic approaches can be applied to population genetic studies of nonmodel species that lack a reference genome assembly. The use of sequence tagging and 'Pool-Seq' technologies, together with relatively inexpensive RNA-Seq, genotyping-by-sequencing (GBS) and other reduced-representation library methods, means that data on many thousands of single nucleotide polymorphisms (SNPs) can be collected at moderate cost on multiple samples from a population without a reference genome (Box 1) (e.g. Hohenlohe et al. 2010; Elshire et al. 2011; Bi et al. 2013; Neves et al. 2013; Yeaman et al. 2014). However, even draft reference genomes or genomes from related species can be useful when implementing these approaches to identify variants. The expansion of reference genomes across a wider array of species will facilitate the study of invasion genomics. Several impediments to genome construction still exist (Ellegren 2014), although new technologies are emerging to overcome them and the cost of short-read sequencing is declining. Nonetheless, many plants and animals have large genomes, and the common occurrence of higher ploidy, especially in invasive plants, adds significant difficulty and expense. Higher ploidy also complicates other approaches such as SNP identification and analyses (see below).

An important feature of invasive species for evolutionary studies is that the introduction and spread of the invader is often documented (Baker 1974), and in many cases, this includes historic collections. Genomic studies are benefiting from increased progress in obtaining information from specimens stored in herbariums, museum collections or recovered from natural repositories such as sediment layers in lake bottoms (Krehenwinkel and Tautz 2013; Martin et al. 2014). Such studies are particularly applicable to invasive organisms that are often represented well in collections with longitudinal sampling. These specimen records often cover periods over which populations have been exposed to relatively rapid climate change (such as between 1970 and 2000) (Hansen et al. 2013). More formal approaches are also now being applied to secure the preservation of material across time, such as the resurrection initiative for plants (Franks et al. 2008).

In future, methods can be employed that maximize the genomic information obtained from such limited material, rather than focusing on a few anonymous markers. Whole-genome resequencing has been used successfully for historic samples (Rowe et al. 2011), although contamination by nontarget DNA can make this expensive proposition even more cost-prohibitive. Sequence capture, which utilizes probes to enrich target sequences (e.g. Bi et al. 2013) (Box 1), offers a compromise between minimizing sequencing costs and maximizing sequence information from target regions of the study organism. Pooling individual samples together prior to either whole-genome (e.g. Futschik and Schlotterer 2010) or reduced-representation (e.g. Vandepitte et al. 2012) sequencing is another approach that can greatly reduce costs, while still providing useful information about the genomic extent and location of variation and divergence. Genome-wide analysis of historic DNA in invasive species will provide exciting opportunities to assess the temporal dynamics of evolutionary change during colonization and spread, and over periods of rapid climate change. It can also provide evidence of past changes in population size across time through coalescence methods (do Amaral et al. 2013) and indicate parts of the genome that have been affected by selection and population processes (Excoffier et al. 2009). Such data will improve inference of invasion history and aid in the identification of the genetic basis of recent adaptation.

3.3.2 TOP-DOWN AND BOTTOM-UP APPROACHES

Ecological and landscape genomics are approaches that incorporate information about the phenotype, genotype and the local environment to make inferences about the loci involved in local adaptation. 'Bottom-up' approaches use the genomic signatures of selection to identify loci likely important for adaptation and do not rely on phenotypic information (Wright and Gaut 2005). Several different methods are used, such as tests examining the site frequency spectrum or the extent of linkage disequilibrium, which can indicate the location of a recent selective sweep (Smith and Haigh 2007; Thornton et al. 2007). The popular FST outlier method identifies regions of the genome under divergent selection by determining which sites have strong differentiation among populations relative to the entire genome or a set of putatively neutrally evolving loci (Lewontin and Krakauer 1973; Beaumont and Nichols 1996; Foll and Gaggiotti 2008; Steane et al. 2014). Another approach involves identifying associations between alleles and the local environment, while controlling for population structure, to make inferences about local adaptation (e.g. Coop et al. 2010; Günther and Coop 2013). Many of these tests of selection could be applied to invasive species to identify the genes associated with adaptation during invasion and in responses to climate change. These methods are appealing because genes under selection during invasion can be identified without a priori identification of specific functional traits and do not require common garden experiments. However, the loci identified may only be in linkage disequilibrium with the functional site(s). Moreover, population structure can hinder the ability of these tests to correctly identify adaptive loci, and certain demographic scenarios, particularly those involving nonequilibrium conditions, have been shown to generate a large number of false positives (Lotterhos and Whitlock 2014). A critical assessment of how complicated invasion histories, including founder events and genetic admixture, might impact the ability of these methods to detect loci under very recent divergent selection has not yet been conducted, but these processes could represent an impediment to the successful implementation of these tests in some invasive species.

'Top-down' methods offer a complementary approach for identifying the genetic basis of adaptation (Wright and Gaut 2005; Barrett and Hoekstra 2011). Here, the genetic basis of specific traits under selection during invasion can be dissected by looking for the cosegregation of these traits and genetic markers. Quantitative trait loci (QTL) mapping uses pedigreed populations descended from known parents. Linkage disequilibrium between genetic markers and causal loci enables the identification of genomic regions segregating with the phenotypes of interest. This method limits the identification of loci to those present in the initial cross, and the genomic resolution is low due to high linkage disequilibrium in the mapping population. Association mapping studies use unpedigreed mapping to look for associations between traits and markers, either using a candidate gene approach or genome-wide markers (GWAS studies) (Box 1). Such studies can be prone to false positives through spurious associations brought about by population structure, although there are statistical means to control for these effects (Tian et al. 2008). The approaches offer greater resolution than QTL studies due to the lower linkage disequilibrium found in natural populations relative to pedigreed mapping populations.

Top-down and bottom-up approaches each have their own advantages and disadvantages (Wright and Gaut 2005; Barrett and Hoekstra 2011; Le Corre and Kremer 2012; Sork et al. 2013), suggesting that tackling the problem from both ends will often yield the most useful results. However, there are biases common to all these methods that may obscure the genetic architecture of adaptation during invasion, or to climate change signals (Moran and Alexander 2014). For example, detecting small effect loci underlying polygenic traits has been a major challenge regardless of the method applied (Pritchard et al. 2010; Le Corre and Kremer 2012). Well-powered QTL and GWAS studies appear to be an effective solution to this problem (Bloom et al. 2013), and new analytical methods that attempt to integrate information from GWAS with allele frequency data from many populations are providing a powerful way to identify the signal of local adaptation in polygenic traits (Turchin et al. 2012; Berg and Coop 2014).

Although there have been few studies that have applied these approaches to identify genes important for adaptation during invasion, the growth of genomic data in nonmodel organisms suggest that these methods will soon be commonplace for invasive species. As strong selection will be required to see evolutionary changes over short time scales, theory suggests that large-effect loci (Yeaman and Whitlock 2011) or perhaps clusters of adaptive loci may be more likely to underlie a rapid response to selection. However, genetic changes from standing variation as opposed to new mutations are likely to be essential for adaptation during invasion, due to the waiting time required for new mutations (Barrett and Schluter 2008; Prentis et al. 2008). This suggests that polygenic adaptation from standing variation present in the original introduction is likely, so adaptation through multiple loci of small effect may also play an important role in the invasion. Both empirical and theoretical studies are needed to determine which features of the genetic architecture of rapid adaptation, if any, might distinguish it from longer-term adaptive evolution and if differences in genetic architecture could impact the propensity of organisms to become invasive and/or rapidly adapt to continuing climate change.

The identification of candidate 'invasion' genes using these top-down and bottom-up approaches is only the first step. Comparisons across species will be essential for determining whether there are similar functional groups, genes or genetic pathways that frequently evolve during invasion and in responses to changing environments further precipitated by local climate change. This might be expected, for example, if invading species commonly evolve along similar fitness trade-offs in response to changes in the biotic and abiotic environment in the introduced range (Blossey and Notzold 1995; Bossdorf et al. 2005; He et al. 2010). The context dependence and sometimes idiosyncratic nature of invasion and adaptation may, however, make such generalities unlikely (though see also Richardson and Pyšek 2006, 2012). Experimental studies, where possible, will be important to assess fitness effects of these candidate loci in nature, while controlling for genetic background (Barrett and Hoekstra 2011). The development of more model invasive species amenable to reverse genetics will be key for dissecting the functional role of candidate loci and will be a step forward for invasion genomics (Stewart et al. 2009).

3.4 THE INITIAL STAGES: ENTRAINMENT RISK AND TRANSPORT

The introduction of alien species is either intentional, usually for economic benefit of some form, or unintentional, owing to the fact that species move around in all kinds of ways and include human vectors and human-mediated pathways in their dispersal portfolio (Hulme et al. 2008). For intentional introductions, understanding entrainment risk amounts to investigating the socio-political motivations for introducing new species. These are many and diverse and, through time, have ranged from agriculture and horticulture, which are ongoing (Dehnen-Schmutz 2011), to the activities of acclimatization societies (Cassey et al. 2004). In the nineteenth century, acclimatization societies were established to 'improve' what were considered impoverished biotas in newly colonized regions (e.g. North America, New Zealand and Australia) through the importation of plants, birds and mammals, typically from Europe, but also from other regions. Investigations of survival during transport belong in the realm either of plant propagation and animal husbandry, or maintenance of biocontrol agents (e.g. Teshler et al. 2004). While these areas may have much to inform the investigation of biological invasions and their evolutionary dynamics (e.g. Vorsino et al. 2012), we will not consider them here. Rather, our emphasis is on unintentional introductions.

Much of the understanding required to reduce the rates of unintentional introductions is concerned with improving knowledge of the pathways and vectors of invasion (Hulme 2009). Preventing an invasion is the most cost-effective way of dealing with it (Simberloff et al. 2013). These pathways and vectors are now becoming much better understood across a range of activities in different areas, and for different organisms (e.g. Drake and Lodge 2004; Tatem and Hay 2007; Huiskes et al. 2014). Changing human traffic patterns in tandem with changing climates (including their variability) is likely to change substantially the sites of entrainment risk for any given location (Tatem 2009). In consequence, although something of an invasion cycle exists, where disturbed entrainment areas tend to harbour similar suites of species (Lee and Chown 2009), forecasting which species and areas to prioritize for surveillance may be difficult because of rapidly changing environments and species distributions. Ongoing invasion of source areas (see e.g. Roy et al. 2011) may likewise mean changing risk profiles. Genomic approaches offer a powerful tool to help manage these risks.

At their most straightforward, DNA barcoding approaches can identify alien species at a given location (Darling and Blum 2007). They are now proving useful for a range of taxa. Mostly, the aim has been to characterize the diversity of an assemblage and identify alien species among its members in a given recipient area, or to verify whether given populations are from species thought to be nonindigenous (e.g. Scheffer et al. 2006; Smith and Fisher 2009; Porco et al. 2012, 2013; Fernández-Álvarez and Machordom 2013; Zhang et al. 2013a). Such approaches nonetheless are subject to the same kinds of problems as those facing DNA barcoding generally (see review by Taylor and Harris 2012). These include substantial error rates in some taxonomic groups (e.g. Hickerson et al. 2006; Meier et al. 2006; Zhang et al. 2013b) and lack of clarity about the questions being posed or rigorous independent assessment of the hypotheses being tested (Collins and Cruickshank 2013).

As DNA barcoding approaches were developed, it has now become possible to screen simultaneously samples of many species. Such metabarcoding provides relatively reliable estimates of community composition and turnover (Yoccoz et al. 2012; Yu et al. 2012; Ji et al. 2013). The approach has also been broadened to include what may be termed environmental metabarcoding or eDNA metabarcoding (Taberlet et al. 2012a; Bohmann et al. 2014). Here, rather than the organisms being sequenced directly, DNA is typically extracted from soil or water (Collins et al. 2013; Porco et al. 2013; Piaggio et al. 2014), but may also come from other sources such as the gut contents of flies (Calvignac-Spencer et al.

2013). The terms metagenomics and ecogenomics have also been applied to this approach, although typically metagenomics involves analysis of functional characteristics and assembly of whole genomes for microbes (Taberlet et al. 2012a). Considerable success has been had in detecting invasive alien species using this approach, with the method surpassing traditional survey approaches both in terms of sampling effort and sensitivity. Notable demonstrations thereof are for carp in waterways linking to the Laurentian Great Lakes (Jerde et al. 2011), the American bullfrog in France (Ficetola et al. 2008; Dejean et al. 2012) and several other aquatic species (Bohmann et al. 2014; Rees et al. 2014). Much interest exists in the application of eDNA barcoding to surveillance of ballast water in ships, a substantial source of aquatic invasions, with novel tools being developed to enable rapid screening by nonspecialists (Mahon et al. 2012).

A further area where eDNA metabarcoding is proving useful is in understanding the extent, ecology and ecosystem functioning consequences of microbial invasions (Litchman 2010; Shade et al. 2012). Microbial invasions are significant across the globe with substantial impacts on a range of ecosystems and on human and animal health (Van der Putten et al. 2007; Jones et al. 2008; Vellinga et al. 2009; Keesing et al. 2010; Cowan et al. 2011; Carroll et al. 2014). Because metagenomics has deep roots in microbiology, the approach enables sophisticated portraits to be painted of the identity of the players involved in a given system, the extent to which invasions are significant and the functional roles that key members of the community play (Litchman 2010; Cowan et al. 2011; Shade et al. 2012; Adriaennsens and Cowan 2014; Ramirez et al. 2014).

Metabarcoding and eDNA metabarcoding approaches are not without their problems, including challenges to do both with resolving taxa (see above and Taberlet et al. 2012b; Brodin et al. 2013; Pyšek et al. 2013), and with the skills, methods and databases required to ensure that these approaches deliver their full potential (Yoccoz 2012). Nonetheless, the field is developing rapidly (e.g. Liu et al. 2013; Zhang et al. 2013b; Bohmann et al. 2014), and management risks posed by differentiating among increasing numbers of species introductions and range shifts precipitated by changing climates (Walther et al. 2009), could readily be reduced using these approaches.

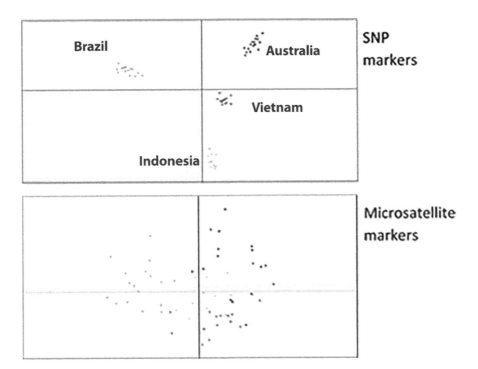

FIGURE 2: Comparison of population structure of *Aedes aegypti* mosquitoes as determined from microsatellite markers versus SNP markers. Based on a discriminant analysis, the SNP markers provide a far higher level of resolution of populations including the invaded (non-Asian) range of this species (from Rasic et al. 2014).

Although barcoding and metabarcoding approaches are increasingly being applied to the detection of invasives and have been used in several successful detections (e.g. Dejean et al. 2012; Collins et al. 2013; Egan et al. 2013; Takahara et al. 2013), they could also be used to reduce risks of introduction by understanding the composition of species in source hot spots. While doing so might at first seem impracticable, reductions in costs of the technical methods (Lemmon et al. 2012; Rocha et al. 2013), and routine surveillance of areas for incoming species (e.g. Bourlat et al. 2013), which are often the same sources for outgoing ones (Lee and Chown 2009), mean that the data may become increasingly available for species detection. The main barriers to overcome are making sure that sequence information, biological records and distributional data are available in linked and readily useable forms [although their downstream use also faces a variety of challenges (Amano and Sutherland 2013)], achieving interoperability among the various approaches and information databases, and estimating the extent of errors of omission and commission for various taxa (Collins and Cruickshank 2013). While these barriers are high, they are being reduced by increasing discussions among major data providers and ongoing work to improve interoperability. Although this provider approach is important, good practice by investigators (such as coherence in data submitted to major providers) is probably the most significant way to reduce barriers. DNA barcoding approaches may be especially helpful where changing climates encourage change in human agricultural and business practices, creating rapidly changing invasion scenarios.

3.5 IDENTIFYING SOURCE POPULATIONS

Although genetic markers have been used for some time and are now used almost routinely for identifying source populations of species, genomic markers promise a much higher level of resolution and they are being increasingly applied. A recent example is provided by the dengue vector mosquito, *Aedes aegypti*, whose source populations have traditionally been identified using mtDNA markers coupled with microsatellites (Endersby et al. 2011; Brown et al. 2014). These provide some indication of population sources and are currently used in countries such as Australia to locate source populations. However, a much higher level of resolution for identifying source populations can be obtained from SNP nuclear markers that can now be scored in the thousands; this was demonstrated in the extent to which geographic samples of *Ae. aegypti* could be completely resolved based on 18 000 SNP markers compared to 10–15 other markers (Rasic et al. 2014) (Fig. 2).

SNP markers are now being developed for other invasive species including invasive plants (De Kort et al. 2014) and mammals (White et al. 2013) as well as insects. These markers promise to revolutionize risk assessment and quarantine control because they can reveal both contemporary and historical population processes. The potential of this approach in detecting and understanding past patterns based on whole-genome sequencing is evident from recent human studies on invasions and contractions of different cultural groups that provide new insights into patterns of relatedness and the spread of cultural practices (Sikora et al. 2014).

Understanding source populations and spread dynamics may benefit from using nuclear and extranuclear genomes. For instance, in the Russian wheat aphid, Zhang et al. (2014) tested patterns of invasion by characterizing more than 500 clones for genetic markers in nuclear DNA, mtDNA and endosymbionts. They were able to show that Turkey and Syria were the most likely sources of invasion to Kenya and South Africa respectively, as well as establishing patterns of invasion into the New World. Having the different sources of genomic information was essential to this study as the recent introduction limited the genetic resolution of the aphid markers used. Combining nuclear DNA and organellar markers has traditionally improved the understanding of invasive processes, and it is becoming clear that including vertically transmitted endosymbiont or parasitic organisms can provide new insight into the origins of invasive species as well as the diseases they carry (Gupta et al. 2014; Mobegi et al. 2014). The addition of mtDNA provides information on introgression patterns between diverged populations or subspecies because mtDNA genomes track maternal lineages (McCormick et al. 2014).

In the future, genomic information will likely prove useful in identifying source risk and predicted outcomes of invasions when they are detected. For instance, in ragweed, it appears, based on microsatellite and chloroplast markers, that invasive populations have developed in the native range of this species associated with human habitation, and these populations are then likely to form the basis of invasive populations into other areas (Martin et al. 2014). Predictions are that invasions will grow in number and be increasingly difficult to distinguish from range shifts (Walther et al. 2009). Genomic techniques that can facilitate rapid identification of source populations will prove especially useful for addressing this problem (for an application in identifying the basis of range expansion, see Krehenwinkel and Tautz 2013).

3.6 COLONIZATION AND ESTABLISHMENT: GENETIC DIVERSITY

Multiple founders can facilitate population establishment in two ways. First, propagule pressure mitigates the impact of demographic stochasticity; second, increased genetic and phenotypic variation improves the chances of successful colonization (Colautti et al. 2006; Simberloff 2009; Rius and Darling 2014). A recent meta-analysis, which controlled for the effects of population size, found strong evidence that higher levels of genotypic and phenotypic diversity in founder groups increased establishment success (Forsman 2014). Genetic variation can improve the likelihood of establishment for many reasons (Zhu et al. 2000; Willi et al. 2006; Gamfeldt and Källström 2007; Dlugosch and Parker 2008; Prentis et al. 2008; Thompson et al. 2012; Berthouly-Salazar et al. 2013), and evidence is mounting that a large proportion of invasive species have experienced multiple introductions (e.g. Bossdorf et al. 2005; Roman and Darling 2007; Thompson et al. 2012; Rius and Darling 2014).

The role of genetic variation in successful invasions has proven contentious though, with authors often taking opposing views depending on the nature of the system that they study (Rius and Darling 2014). Indeed, many species have spread successfully with low genetic diversity (Myburgh et al. 2007; Darling et al. 2008a; Richards et al. 2012). The success of invaders with low genetic variability may depend on the nature of the environment into which they invade; weedy species moving into agricultural landscapes may require less genetic variability because they enter a relatively homogeneous landscape, whereas for species invading natural environments, a high level of variability may be required (Moran and Alexander 2014).

Genomic approaches provide a means to help resolve the apparent paradox. For example, success with limited genetic diversity may be due to the expression of phenotypic plasticity, aided by epigenetic responses to the environment (Richards et al. 2012) that can be identified with genomic approaches. Epigenetic mechanisms may be particularly important in these situations because they generate heritable changes in response to environmental cues, although their ecological significance remains unclear (Bossdorf et al. 2008). Alternatively, functional traits may remain variable even when genetic variation is low, as the loss of rare alleles will not greatly affect the quantitative trait distribution, and nonadditive genetic variance may either aid in fitness directly, or be converted to additive variance due to allele frequency shifts (Dlugosch and Parker 2008; Dawson et al. 2012; Le Roux et al. 2013). Genomic tools provide a way of identifying variation in functional traits (see next section). They also provide new means to understand the nature of introductions of invasive species. As mentioned in the previous section, genomic markers provide a much higher level of resolution when identifying source populations, and they can therefore clearly indicate the number of introductions from different environmental regions involved in invasions.

3.7 COLONIZATION AND ESTABLISHMENT: ADAPTIVE LOCI

Identifying risk factors associated with establishment would be assisted by more information on the genetic basis of adaptive changes in colonizing species. If these changes were known and repeatable, it might help predict when colonizations occur. One of the best examples is the evolutionary changes involved in the repeated independent colonization of freshwater environments by sticklebacks (summary in Bell and Aguirre 2013). These colonization events have been associated with repeatable changes in lateral plate morphology across populations involving the ectodysplasin (EDA) locus. Many loci in different genomic regions also appear to be involved, but nevertheless there appear to be changes across the same set of genomic regions in different, isolated freshwater populations of sticklebacks following invasion (Jones et al. 2012). The stickleback situation may be unusual because there has been a repeated process of invasions that have been occurring over 10 million years (Bell and Aguirre 2013), but more cases need to be examined. Because invasive species often occupy similar habitats in the introduced range to those in the native range (see Guisan et al. 2014 for a recent overview and perspective), there is an opportunity to determine whether the same genomic regions contribute to adaptations in the native and introduced ranges across many invasive species.

While there is abundant evidence for changes in neutral markers following invasion, characterization of adaptive molecular differences, such as those established in stickleback populations, remains relatively scarce. Transcriptome comparisons of plants in common garden experiments suggest that invasive populations of ragweed show altered patterns of gene expression that are distinct from most populations from the native range, and genes showing altered patterns of expression appear to be linked to stress and biotic responses (Hodgins et al. 2013). A similar pattern has since been identified in Canada thistle (Guggisberg et al. 2013), but whether changes to expression during invasion primarily involve certain functional groups, as indicated by these studies, will require more examples across a broader array of organisms.

One of the challenges in finding the genetic basis of adaptive change is that causal connections between genetic changes and traits can be difficult to identify. Often researchers are faced with the problem of locating small but crucial changes within a mass of genomic data. However, the implementation of population genomic analysis of SNP variation across the genome has enabled the identification of putatively selected loci (or those linked to the causal loci). This approach has already proven useful in several cases. Pyrenean rocket (*Sisymbrium austriacum* subsp. *chrysanthum*) is a small colonizing herb of rocky soils that has recently invaded Belgium and the Netherlands, and Vandepitte et al. (2014) used a RAD-Seq approach to identify outlier loci in comparisons between invaded and native populations. Multiple native and invaded populations were investigated including those represented by herbarium specimens covering ca. 100 years in the invasive range. Several outlier loci that could underlie adaptive differences were discovered, and the sampling enabled a comparison of outliers across space as well as time. Some of the SNPs were located in genes known to affect flowering time in the closely related *Arabidopsis thaliana*. These findings point to likely adaptive shifts rather than changes associated with demographic processes. This study demonstrates the utility of phenotype–genotype approaches for identifying candidate adaptive loci in invasive species coupled with functional information from model species. As the genomes of invasive species become better defined and genetic tools are developed, it should be increasingly possible to also pursue bottom-up approaches in a range of invasive species particularly because these can often be relatively easily grown under controlled conditions.

3.8 COLONIZATION AND ESTABLISHMENT: HYBRIDIZATION AND POLYPLOIDY

Hybridization can act as a stimulus to invasion (Ellstrand and Schierenbeck 2000; Schierenbeck and Ellstrand 2009) and occurs on a continuum from the combining of largely reproductively isolated genomes to the mixing of distinct populations within a species. Hybridization offers many potential advantages during invasion. It may increase adaptive potential through the injection of genetic variation and the formation of beneficial gene combinations (Anderson 1948). Hybridization is known to result in transgressive segregation, where phenotypic variation in the hybrids exceeds that of the parents (Rieseberg et al. 2003). In sunflowers and Louisiana irises, there are particularly compelling cases for hybridization facilitating the colonization of novel habitats (Rieseberg et al. 2003, 2007; Arnold et al. 2012).

Over the short term, heterosis (improved performance of hybrid offspring) could be important for overcoming demographic stochasticity associated with initial establishment (Drake 2006; Keller and Taylor 2010; Rius and Darling 2014). Although heterosis is transitory in most cases (Hochholdinger and Hoecker 2007), fixed heterozygosity can be maintained across generations in allopolyploids (assuming disomic inheritance occurs) and asexually reproducing species (Ellstrand and Schierenbeck 2000; Rieseberg et al. 2007). In addition, if dominance is the primary cause of heterosis, the advantage of hybridization may continue in later generations and facilitate the purging of genetic load (Burke and Arnold 2001). Alternative mildly deleterious alleles can become fixed within diverging populations due to the effects of genetic drift. Hybridization between these lineages could produce descendants with an intrinsic fitness advantage over their parents that is maintained if the combined effects of recombination and natural selection erode the frequency of these deleterious alleles (Ellstrand and Schierenbeck 2000; Burke and Arnold 2001).

Genomic tools can help in identifying the consequences of these processes because they can be followed across different parts of the genome. This applies particularly to plants where invasive taxa often arise as hybrids (Moody and Les 2007; Ainouche et al. 2009; Schierenbeck and Ellstrand 2009). One famous case is of the damaging riparian invasive *Tamarix* species in the USA. The main invasive lineage arose from hybridization between two introduced species, *T. chinensis* and *T. ramosissima*, which have largely separate ranges with some overlap in Asia, but do not appear to hybridize there (Gaskin and Schaal 2002). Genome-wide neutral markers revealed that hybrids dominate the invasive range, and the level of introgression is strongly correlated with latitude (Gaskin and Kazmer 2009). A corresponding latitudinal cline in cold hardiness suggests that hybridization may have contributed to the rapid adaptation of *Tamarix* to climatic extremes in North America (Friedman et al. 2008). In reed canarygrass, multiple introductions into North America of genetic material from different European regions have mitigated the impact of genetic bottlenecks (Lavergne and Molofsky 2007). This species has higher genetic diversity and heritable phenotypic variation in its invasive range relative to its native range. Comparisons of neutral differentiation (F_{ST}) to differentiation in quantitative traits among populations (Q_{ST}) provide evidence for rapid selection of genotypes with greater vegetative colonization ability and phenotypic plasticity in the introduced range.

Hybridization is expected to be important from the perspective of invasiveness under climate change for two reasons. First, evidence is growing that climate change is increasing rates of hybridization, as species that were previously geographically isolated come into contact with each other (Garroway et al. 2010; Muhlfeld et al. 2014). Second, evidence from genomic studies is revealing that genes associated with climate change adaptation can be of hybrid origin (Becker et al. 2013; De La Torre et al. 2014). Thus, when hybrids are formed, they may contribute to rapid adaptation under climate change. Although most information on hybridization comes from plants, hybridization is also likely to play an important role in the adaptation of animal lineages to climate change. Genomic comparisons have indicated introgression of invasive genes across lineage boundaries affected by climate in birds (Taylor et al. 2014) and stick insects (Nosil et al. 2012). Introgressed genotypes can be favoured under changing climatic conditions as evident in hybridization events in swallowtail butterflies (Scriber 2011) and in Anopheles mosquitoes (Besansky et al. 2003). Genome scans across multiple related species along climate gradients are likely to provide many other examples of candidate genomic regions associated with climate adaptation and exchanged through hybridization and introgression.

Polyploidy has occurred frequently in the evolutionary history of angiosperms, often associated with hybridization events (Wood et al. 2009). There are several examples of recently derived invasive polyploids (Ainouche et al. 2009; Schierenbeck and Ellstrand 2009; te Beest et al. 2012), and polyploidy is over-represented among alien plant taxa (e.g. Pandit et al. 2011). Polyploidy is associated with several advantages (reviewed in Otto and Whitton 2000; Comai 2005; Otto 2007; te Beest et al. 2012) that could aid in invasion, including pre-adapting species to the environmental conditions in the new range (e.g. Henery et al. 2010), masking deleterious alleles (Otto and Whitton 2000; Otto 2007), restoring fertility through chromosome doubling (e.g. Ainouche et al. 2009) or through changes to the reproductive system to that contribute to colonization ability (e.g. Robertson et al. 2011). In addition, allopolyploids benefit from fixed heterozygosity due to the combining of divergent parental genomes. This feature may have contributed to the evolutionary success of polyploids over diploids in the Arctic enabling them to survive dramatic shifts in climate (Brochmann et al. 2004).

One of the best examples of polyploidy contributing to invasion success is tetraploid spotted knapweed, *Centaurea stoebe*. Although both diploids and tetraploids were introduced into the USA from Europe, invasive populations are dominated by tetraploids (Treier et al. 2009). Tetraploids are pre-adapted to a wider range of climates compared to diploids, and following introduction, they may have further adapted to their introduced environment (Treier et al. 2009; Henery et al. 2010). A classic example of hybridization and polyploidy coinciding with invasion is *Spartina anglica*. This species is a recently formed allopolyploid ($12\times$) that is highly invasive compared to its parental species, is able to colonize a broader range of habitats and exhibits substantial phenotypic plasticity (Thompson 1991; Ainouche et al. 2009), despite a strong genetic bottleneck that occurred during its formation (Baumel et al. 2001). Neo-allopolyploids often undergo extensive alterations to their genome and transcriptome, and genomic studies of S. anglica are dissecting the genetic, epigenetic and expression changes following hybridization and genome duplication (Salmon et al. 2005; Ainouche et al. 2009; Chelaifa et al. 2010). However, the mechanism by which S. anglica has become invasive remains elusive, and it is not known whether the success of this species reflects fixed heterosis and restored fertility of the hybrid following genome duplication, or phenotypic novelty due to epigenetic and transcriptome remodelling.

Whole-genome and reduced-representation genomic approaches are invaluable for detecting hybridization, and more powerful than any other method to quantify the extent of introgression of genomic material and (potentially) functional genes involved in environmental tolerance and invasiveness. They have been used to estimate rates and genomic extent of hybridization (Larson et al. 2013; Parchman et al. 2013) and identify the potential for genetic swamping, or hybrid incompatibilities that may be deleterious to native species (Fitzpatrick et al. 2010). Applying genetic and genomic methods to recent polyploids poses extra challenges due to the presence of multiple genome and gene copies; this can complicate short-read alignment and de novo assembly of genomes, but can also be useful in establishing patterns of relatedness among species and populations (Griffin et al. 2011). Reference genomes are especially valuable tools in these cases and are helping to reveal within-species diversity and hybrid origins of polyploids (e.g. Evans et al. 2014; Marcussen et al. 2014), and reduced-representation library methods can now be applied even without a reference genome (Lu et al. 2013). Genomic studies of several plant groups are pointing to porous species barriers across thousands of years where ongoing hybridization and gene flow between species is present despite species maintaining their morphological integrity (e.g. De La Torre et al. 2014; Griffin and Hoffmann 2014). Such a dynamic process of hybridization and polyploidy may also occur among different chromosomal races within a species as in the case of buffel grass, which is weedy and invasive in some areas of its current range and consists of different polyploids that vary in their environmental distributions and invasiveness (Kharrat-Souissi et al. 2014).

3.9 COLONIZATION AND ESTABLISHMENT: PHENOTYPIC PLASTICITY

Although the link between phenotypic plasticity and distribution has not been comprehensively supported (Dawson et al. 2012), phenotypic plasticity clearly contributes to environmental tolerance in various ways (e.g. Ghalambor et al. 2007; Des Marais et al. 2013; Schilthuizen and Kellermann 2014). Moreover, at least under conditions of resource abundance, it appears that invasive species do show greater plasticity than their indigenous counterparts (Davidson et al. 2011), although the way in which such comparisons should be made has been the subject of much recent discussion (Van Kleunen et al. 2010; Leffler et al. 2014).

Phenotypic plasticity plays a significant role in the responses of both plants and animals to environmental change (De Witt and Scheiner 2004; Ellers and Stuefer 2010; Nicotra et al. 2010). The extent to which plasticity promotes longer-term adaptation may vary, however, ranging from cases where it promotes adaptation through to those where it might inhibit adaptation. The outcome depends on a range of circumstances, including the extent and form of environmental predictability, the extent of dispersal and the genetic variance in new environments (Chown and Terblanche 2007; Ghalambor et al. 2007; Scoville and Pfrender 2010; Chevin et al. 2013). Nonetheless, currently, it appears that phenotypic plasticity can play a major role in species responses to climate change, such as in changing the timing of breeding in birds (Gienapp et al. 2008) and responses of plants along climate gradients (Matesanz and Valladares 2014; McLean et al. 2014). Some evidence also exists suggesting that invasive species have advantages over indigenous ones as a consequence either of the extent or form of their plasticity (Stachowicz et al. 2002; Chown et al. 2007; Kleynhans et al. 2014). Given that at least some evidence supports the idea that phenotypic plasticity may enhance the success of nonindigenous species as climates change, an understanding of the genomic basis of plasticity may improve forecasts of where such advantage should be expected and what form it should take.

The genomic and transcriptomic basis of variation in phenotypic plasticity is still poorly understood. Several gene expression studies have now been undertaken across related species or populations exposed to different environmental conditions to assess whether these might provide clues about the mechanistic basis of variability in plastic responses (Smith et al. 2013; Dunning et al. 2014; Meier et al. 2014). A few of them have considered populations from native and invaded parts of their range to assess whether there are differences in plastic responses to the different environmental conditions encountered in invasive situations. For instance, in sticklebacks, the invasion of freshwater environments by marine populations has involved an increase in expression plasticity, including some genes that are thought to be involved in adaptation to these new habitats (Morris et al. 2014), while in D. melanogaster populations from warm environments, there was an enrichment of down-regulated genes when flies were raised in conditions that were not commonly encountered in their native range (Levine et al. 2011).

A challenge is to link these expression changes to adaptive variation in plasticity; population differences in gene expression under different conditions may indicate adaptive changes in plasticity, but expression changes have not yet been clearly linked mechanistically to such changes. One problem in making such connections is that there is a level of complexity in organism's transcriptome that largely remains to be explored even in model organisms. For instance, in D. melanogaster, the transcriptome can be modified qualitatively in a variety of ways, such as through alternative splicing and expression of alternative isoforms, that may ultimately have more impact on plastic responses than quantitative changes in gene expression. These sources of variation for climate adaptation are only just starting to be explored (Telonis-Scott et al. 2013).

Another challenge is to consider plastic changes across generations as well as within them. Epigenetic modifications such as DNA methylation and histone modifications can be induced by environment and influence gene expression and transposable element activity, with direct effects on the phenotype (Glastad et al. 2011; Herrera et al. 2012; Zhang et al. 2013c). In plants, methylation can also be inherited through multiple generations, thus partly acting like heritable genetic variation. There are several examples of invasive species where epigenetic variation far exceeds standing genetic variation (Richards et al. 2012; Liebl et al. 2013). Genomics is integral to the field of ecological epigenetics. Whole-genome bisulphite sequencing can reconstruct entire methylation profiles (Schmitz et al. 2013; Wang et al. 2013) while reduced-representation genomic methods and genetic methods such as MS-AFLP can enable quantification of the extent of epigenetic compared to genetic variation within and among populations (Pérez et al. 2006). The hypothesis that epigenetic variation is important to invasion in some cases is intriguing, and a handful of studies exist directly linking epigenotype to phenotypic means, plasticities and environmental tolerance (Herrera et al. 2012; Zhang et al. 2013c), but in general, few data have been collected that examine adaptive and heritable epigenetic effects (Furrow and Feldman 2014).

3.10 SPREAD

Understanding the spread/expansion phase of an invasion has similarities with species range projections in climate change biology (Caplat et al. 2013). To expand its range into new environments, a species requires appropriate dispersal traits and tolerance of a range of environments or capacity to experience the new environment as consistently receptive. Dispersal traits are a key characteristic of invasive species, and dispersal occurs through direct movement (in the case of animals) or indirect movement, by seed or vegetative dispersal (in the case of plants), dispersal through soil or water (e.g. the plant pathogen *Phytophthora cinnamomi*) or human-mediated dispersal (e.g. weedy species along roadsides, cane toads in transport vehicles and vertebrates on ships).

Genomic studies on populations can be used to test ecological hypotheses about dispersal patterns in invasive species once these have become established in a new area and started to spread, including the widely documented lag phase before populations disperse widely (Richardson et al. 2011). For instance, Rohfritsch et al. (2013) used genomic data to map the spread of an invasive oyster across Europe following its introduction from Asia. This approach is particularly powerful when combined with ecological models that predict the extent to which populations are expected to occupy new space across time or maintain refugia under a changing climate, both of which should be reflected in patterns of genetic variation (Fordham et al. 2014). For example, genomic approaches have been used to understand patterns of spread in Australian *Acacia* species that have subsequently invaded southern Africa. These include estimating the impacts of human-assisted dispersal in the native range prior to introduction (Le Roux et al. 2013) and better forecasting of the effects of climate change by accounting for genotypic variation (see Millar et al. 2011; Thompson et al. 2011).

Population genomic data can also be used to understand the impact of spreading invaders through hybridization and introgression. As discussed above, invaders spreading into non-native habitat, particularly under climate change, are expected to hybridize with local species where reproductive isolation is incomplete. This can lead to the incursion of genes from the invaders into native species. As a consequence, local species can become genetically 'polluted', resulting in the loss of species integrity. Genomic markers provide a way of detecting such pollution (e.g. Yamazaki et al. 2005; Sampson and Byrne 2008; Millar et al. 2012).

Mechanisms of dispersal can evolve, as in the case of leg length in the cane toad (Phillips et al. 2006), and seed dispersal in *Abronia umbellata* (Darling et al. 2008b). In these cases, genomic studies on quantitative traits can provide information on the genes and pathways involved. Dispersal evolution may involve traits directly involved in movement, such as cane toad leg length, or more subtle changes that nevertheless enhance dispersal potential, such as altered timing of reproduction that increases seed dispersal. Genomic data can identify the nature of these evolved changes in mechanisms that alter the potential of populations to spread. For instance, QTLs have been isolated that control the formation of rhizomes in perennial wild rye that represents an important trait facilitating colonization in weedy ryegrass species (Yun et al. 2014). Understanding the genomic basis of traits involved in non-human-mediated dispersal may aid in predicting the invasive potential of suspect species before introduction or at the establishment stage (Pérez et al. 2006), as well as projecting their future spread in cases where dispersal traits can evolve.

3.11 MANAGEMENT IMPLICATIONS

The broad consensus in the literature is that climate change is likely to interact synergistically with biological invasions, leading to increasing economic and biodiversity impact (Walther et al. 2009; Scholes et al. 2014). Much evidence exists for recent changes in pest and disease distributions (Bebber et al. 2013), although substantial variation exists among taxa and geographic regions in the extent to which an overall increase in the burden of invasives under climate change might be expected (Bellard et al. 2013). A complicating factor is that human activity in the form of trade, transport and adaptation to climate change is likely to increase too. This will complicate decisions about range shifts relative to anthropogenic introductions and thus potentially confounding management decisions that are required to limit the overall burden of invasive species impacts. Even in relatively remote areas, these problems are playing out (Chown et al. 2012). Overall, the prospects are for increased impacts of invasion as climate change proceeds (Scholes et al. 2014).

From a management perspective, prevention, usually involving some form of risk assessment, is clearly the most cost-effective solution to invasions (Simberloff et al. 2013). Therefore, considerable focus should be on prevention including the identification of the source and receiving areas for introductions (such as ports). Internationally, risk assessment is accepted as an essential policy instrument for managing biological risk. Risk is quantified by examining the invasiveness, impacts and potential distribution of species, areas which are all being improved by the adoption of genomic approaches. Thereafter (i.e. in a postborder context), substantial focus on interventions across the stages of invasion is critical for management success both in relation to typical population processes (such as those involved in emergence from lag phases) (Kueffer et al. 2013; Ricciardi et al. 2013; Simberloff et al. 2013) and to changes in the environment associated with climate change and human adaptation to such change. Likewise, understanding when invasion success may be driven by factors other than climate change, such as local adaptation to climates, evolution of increased competitive ability and enemy release (Blossey and Notzold 1995; Chun et al. 2010; Van Kleunen et al. 2010; Hill et al. 2013), is also important for managing invasions. Prioritization of management actions to species of greatest risk (Randall et al. 2008; Byrne et al. 2011; Forsyth et al. 2012) is critical given the requirement for effective outcomes with investment of limited resources. As hybridization has been demonstrated to facilitate invasiveness, assessment of genetic risk is an important management tool

(Byrne et al. 2011), particularly in the context of managed relocation as a climate change adaptation strategy (Schwartz et al. 2012).

Throughout this review, we have demonstrated how genomics approaches can improve understanding, and in many cases forecasting, at each stage of invasion (Fig. 1). These include better understanding of source areas and the identity of introduced species. In this sense, genomic approaches are typically improvements on genetic tools that have been available for sometime; they allow source populations to be identified more accurately, provide a clearer picture regarding multiple interactions, including the role of hybridization, and lead to better estimates of population size, gene flow and other processes. NGS techniques can provide ways of detecting invasive organisms that have previously not been available, and identifications become more accurate as additional sequence information is added. Microsatellite and mtDNA markers have been widely used to understand the population dynamics of agricultural pests and disease vectors in the past and thereby assisting in programs aimed at controlling pests and/or limiting their impact. However, the new genomic tools provide an unprecedented view of past and present population processes (e.g. Brown et al. 2014; Rasic et al. 2014). With invasion frequency increasing under climate change, these are all important tools in improving an understanding of invasion processes.

Where control of invasive species is being considered, genomic approaches can help in defining units for eradication and risks of reinvasion once eradication is achieved (Fig. 1). At present, markers such as microsatellites are often used for defining movement patterns among populations of invasive pest mammals (e.g. Hampton et al. 2004; Abdelkrim et al. 2005; Berry et al. 2012; Adams et al. 2014). With the use of genomics approaches, managers should become much more confident of likely movement patterns and reinvasion potential as local movement patterns can be tracked much more accurately than in the past.

However, genomics tools promise much more than ways of simply improving detection and an understanding of spread. High-density markers provide a way of tracking changes in different parts of the genome, essential for understanding hybridization and introgression as well as adaptation under climate change including the application of quantitative genomic approaches to identify regions with candidate genes (Fig. 1). This can be used to understand the roles of various evolutionary processes at key steps of the invasion pathway, and the ways in which climate change will interact with the latter. Key areas include investigation of evolutionary change underpinning increased competitive ability and enemy release, especially in a biological control context. Genomics methods can also help redirect resources by improving understanding of the burden of invasion, an especially important role as that burden increases (e.g. Pfenninger et al. 2014). They can assist with rehabilitation in the case where substantial invasive impacts result in widespread ecosystem changes, by understanding patterns of resistance and identifying resistant genotypes that can persist into the future. For example, where trees are affected by introduced pathogens, there is the possibility of identifying tree genotypes that survive outbreaks (Stukely and Crane 1994; McKinney et al. 2014). As information becomes available on adaptive changes in traits controlling the spread and climate change adaptation, it should also be possible to refine models about the non-native range of an invasive species. For instance, Kearney et al. (2009) modelled the expected range of *Aedes aegypti* in northern Australia based on anticipated evolutionary changes in egg desiccation tolerance under climate change and showed the potential threat posed by this species around a population centre.

In summary, genomic tools are proving to be extremely useful for managing invasive species. While they may be seen as costly for less well-resourced nations, several studies have demonstrated that they are more cost-effective than traditional surveillance tools and other methods. Moreover, technology is now advancing to the stage where users do not need to be familiar with its intricacies (in much the same way, most mobile phone users could not build one), and substantial programmes of technology assistance are in place through various international agencies (such as the Consultative Group on International Agricultural Research, www.cgiar.org). In consequence, the myriad ways genomic tools can be utilized will enable society to contain the economic cost of biological invasions accentuated by climate change, and prevent a world dominated by weedy species.

REFERENCES

1. Abdelkrim, J., M. Pascal, C. Calmet, and S. Samadi 2005. Importance of assessing population genetic structure before eradication of invasive species: examples from insular Norway rat populations. Conservation Biology 19:1509–1518.
2. Adams, M. D., S. E. Celniker, R. A. Holt, C. A. Evans, J. D. Gocayne, P. G. Amanatides, S. E. Scherer et al. 2000. The genome sequence of Drosophila melanogaster. Science 287:2185–2195.
3. Adams, A. L., Y. van Heezik, K. J. M. Dickinson, and B. C. Robertson 2014. Identifying eradication units in an invasive mammalian pest species. Biological Invasions 16:1481–1496.
4. Adriaennsens, E. M., and D. A. Cowan 2014. Using signature genes as tools to address environmental viral ecology and diversity. Applied and Environmental Microbiology. doi:10.1128/AEM.00878-14.
5. Ainouche, M. L., P. M. Fortune, A. Salmon, C. Parisod, M. A. Grandbastien, K. Fukunaga, M. Ricou, and M. T. Misset 2009. Hybridization, polyploidy and invasion: lessons from Spartina (Poaceae). Biological Invasions 11:1159–1173.
6. Ali, S., P. Gladieux, H. Rahman, M. S. Saqib, M. Fiaz, H. Ahmad, M. Leconte et al. 2014. Inferring the contribution of sexual reproduction, migration and off-season survival to the temporal maintenance of microbial populations: a case study on the wheat fungal pathogen Puccinia striiformis f.sp tritici. Molecular Ecology 23:603–617.
7. Amano, T., and W. J. Sutherland 2013. Four barriers to the global understanding of biodiversity conservation: wealth, language, geographical location and security. Proceedings of the Royal Society B 280:20122649.
8. do Amaral, F. R., P. K. Albers, S. V. Edwards, and C. Y. Miyaki. 2013. Multilocus tests of Pleistocene refugia and ancient divergence in a pair of Atlantic Forest antbirds (Myrmeciza). Molecular Ecology 22:3996–4013.
9. Anderson, E. 1948. Hybridization of the habitat. Evolution 2:1–9.
10. Armstrong, K. F., and S. L. Ball 2005. DNA barcodes for biosecurity: invasive species identification. Philosophical Transactions of the Royal Society B 360:1813–1823.
11. Arnold, M. L., E. S. Ballerini, and A. N. Brothers 2012. Hybrid fitness, adaptation and evolutionary diversification: lessons learned from Louisiana Irises. Heredity 108:159–166.
12. Asner, G. P., and P. M. Vitousek. 2005. Remote analysis of biological invasion and biogeochemical change. Proceedings of the National Academy of Sciences of the USA 102:4383–4386.
13. Ayres, R. M., V. J. Pettigrove, and A. A. Hoffmann 2012. Genetic structure and diversity of introduced eastern mosquitofish (Gambusia holbrooki) in south-eastern Australia. Marine and Freshwater Research 63:1206–1214.
14. Baker, H. G. 1974. The evolution of weeds. Annual Review of Ecology and Systematics 5:1–24.

15. Barnosky, A. D., E. A. Hadly, J. Bascompte, E. L. Berlow, J. H. Brown, M. Forte-lius, W. M. Getz et al. 2012. Approaching a state shift in Earth's biosphere. Nature 486:52–58.
16. Barrett, S. C. H. 1983. Crop mimicry in weeds. Economic Botany 37:255–282.
17. Barrett, R. D. H., and H. E. Hoekstra 2011. Molecular spandrels: tests of adaptation at the genetic level. Nature Reviews Genetics 12:767–780.
18. Barrett, R. D. H., and D. Schluter 2008. Adaptation from standing genetic variation. Trends in Ecology and Evolution 23:38–44.
19. Baumel, A., M. L. Ainouche, and J. E. Levasseur 2001. Molecular investigations in populations of Spartina anglica C.E. Hubbard (Poaceae) invading coastal Brittany (France. Molecular Ecology 10:1689–1701.
20. Beaumont, M. A., and R. A. Nichols 1996. Evaluating loci for use in the genetic analysis of population structure. Proceedings of the Royal Society B 263:1619–1626.
21. Bebber, D. P., M. A. T. Ramatowski, and S. J. Gurr 2013. Crop pests and pathogens move polewards in a warming world. Nature Climate Change 3:985–988.
22. Becker, M., N. Gruenheit, M. Steel, C. Voelckel, O. Deusch, P. B. Heenan, P. A. McLenachan et al. 2013. Hybridization may facilitate in situ survival of endemic species through periods of climate change. Nature Climate Change 3:1039–1043.
23. te Beest, M., J. J. Le Roux, D. M. Richardson, A. K. Brysting, J. Suda, M. Kubesova, and P. Pyšek 2012. The more the better? The role of polyploidy in facilitating plant invasions. Annals of Botany 109:19–45.
24. Bell, M. A., and W. E. Aguirre 2013. Contemporary evolution, allelic recycling, and adaptive radiation of the threespine stickleback. Evolutionary Ecology Research 15:377–411.
25. Bellard, C., W. Thuiller, B. Leroy, P. Genovesi, M. Bakkenes, and F. Courchamp 2013. Will climate change promote future invasions? Global Change Biology 19:3740–3748.
26. Berg, J., and G. Coop 2014. The population genetic signature of polygenic local adaptation. Plos Genetics 8:e1004412.
27. Berry, O., D. Algar, J. Angus, N. Hamilton, S. Hilmer, and D. Sutherland 2012. Genetic tagging reveals a significant impact of poison baiting on an invasive species. Journal of Wildlife Management 76:729–739.
28. Berthouly-Salazar, C., C. Hui, T. M. Blackburn, B. Gaboriaud, B. J. van Rensburg, B. J. van Vuuren, and J. J. Le Roux 2013. Long-distance dispersal maximizes evolutionary potential during rapid geographic range expansion. Molecular Ecology 22:5793–5804.
29. Besansky, N. J., J. Krzywinski, T. Lehmann, F. Simard, M. Kern, O. Mukabayire, D. Fontenille et al. 2003. Semipermeable species boundaries between Anopheles gambiae and Anopheles arabiensis: evidence from multilocus DNA sequence variation. Proceedings of the National Academy of Sciences of the USA 100:10818–10823.
30. Bi, K., T. Linderoth, D. Vanderpool, J. M. Good, R. Nielsen, and C. Moritz 2013. Unlocking the vault: next-generation museum population genomics. Molecular Ecology 22:6018–6032.
31. Blackburn, T. M., P. Cassey, R. P. Duncan, K. L. Evans, and K. J. Gaston 2004. Avian extinction and mammalian introductions on oceanic islands. Science 305:1955–1958.

32. Blackburn, T. M., P. Pyšek, S. Bacher, J. T. Carlton, R. P. Duncan, V. Jarošík, J. R. U. Wilson, and D. M. Richardson 2011. A proposed unified framework for biological invasions. Trends in Ecology and Evolution 26:333–339.

33. Blackburn, T. M., F. Essl, T. A. Evans, P. E. Hulme, J. M. Jeschke, I. Kühn, S. Kumschick et al. 2014. A unified classification of alien species based on the magnitude of their environmental impacts. PLoS Biology 12:e1001850.

34. Blanchet, S. 2012. The use of molecular tools in invasion biology: an emphasis on freshwater ecosystems. Fisheries Management and Ecology 19:120–132.

35. Bloom, J. S., I. M. Ehrenreich, W. T. Loo, L. Thuy-Lan Vo, and L. Kruglyak 2013. Finding the sources of missing heritability in a yeast cross. Nature 494:234–237.

36. Blossey, B., and R. Notzold 1995. Evolution of increased competitive ability in invasive nonindigenous plants - a hypothesis. Journal of Ecology 83:887–889.

37. Bohmann, K., A. Evans, M. T. Gilbert, G. R. Carvalho, S. Creer, M. Knapp, D. W. Yu et al. 2014. Environmental DNA for wildlife biology and biodiversity monitoring. Trends in Ecology and Evolution 29:358–367.

38. Bossdorf, O., H. Auge, L. Lafuma, W. E. Rogers, E. Siemann, and D. Prati 2005. Phenotypic and genetic differentiation between native and introduced plant populations. Oecologia 144:1–11.

39. Bossdorf, O., C. L. Richards, and M. Pigliucci 2008. Epigenetics for ecologists. Ecology Letters 11:106–115.

40. Bourlat, S. J., A. Borja, J. Gilbert, M. I. Taylor, N. Davies, S. B. Weisberg, J. F. Griffith et al. 2013. Genomics in marine monitoring: new opportunities for assessing marine health status. Marine Pollution Bulletin 74:19–31.

41. Bradley, B. A., D. M. Blumenthal, D. S. Wilcove, and L. H. Ziska 2010. Predicting plant invasions in an era of global change. Trends in Ecology and Evolution 25:310–318.

42. Brochmann, C., A. K. Brysting, I. G. Alsos, L. Borgen, H. H. Grundt, A. C. Scheen, and R. Elven 2004. Polyploidy in arctic plants. Biological Journal of the Linnean Society 82:521–536.

43. Brodin, Y., G. Ejdung, J. Strandberg, and T. Lyrholm 2013. Improving environmental and biodiversity monitoring in the Baltic Sea using DNA barcoding of Chironomidae (Diptera). Molecular Ecology Resources 13:996–1004.

44. Bronnenhuber, J. E., B. A. Dufour, D. M. Higgs, and D. D. Heath 2011. Dispersal strategies, secondary range expansion and invasion genetics of the nonindigenous round goby, Neogobius melanostomus, in Great Lakes tributaries. Molecular Ecology 20:1845–1859.

45. Brook, B. W., N. S. Sodhi, and C. J. A. Bradshaw 2008. Synergies among extinction drivers under global change. Trends in Ecology and Evolution 23:453–460.

46. Brown, J. E., B. R. Evans, W. Zheng, V. Obas, L. Barrera-Martinez, A. Egizi, H. Y. Zhao, A. Caccone, and J. R. Powell 2014. Human impacts have shaped historical and recent evolution in Aedes aegypti, the dengue and yellow fever mosquito. Evolution 68:514–525.

47. Burke, J. M., and M. L. Arnold 2001. Genetics and the fitness of hybrids. Annual Review of Genetics 35:31–52.

48. Butchart, S. H. M., M. Walpole, B. Collen, A. van Strien, J. P. W. Scharlemann, R. E. A. Almond, J. E. M. Baillie et al. 2010. Global biodiversity: indicators of recent declines. Science 328:1164–1168.

49. Byrne, M., L. Stone, and M. A. Millar 2011. Assessing genetic risk in revegetation. Journal of Applied Ecology 48:1365–1373.

50. Callaway, R. M., and J. L. Maron 2006. What have exotic plant invasions taught us over the past 20 years? Trends in Ecology and Evolution 21:369–374.

51. Calvignac-Spencer, S., K. Merkel, N. Kutzner, H. Kuhl, C. Boesch, P. M. Kappeler, S. Metzger et al. 2013. Carrion fly-derived DNA as a tool for comprehensive and cost-effective assessment of mammalian biodiversity. Molecular Ecology 22:915–924.

52. Caplat, P., P. O. Cheptou, J. Diez, A. Guisan, B. M. H. Larson, A. S. Macdougall, D. A. Peltzer et al. 2013. Movement, impacts and management of plant distributions in response to climate change: insights from invasions. Oikos 122:1265–1274.

53. Carroll, S. P., H. Dingle, T. R. Famula, and C. W. Fox 2001. Genetic architecture of adaptive differentiation in evolving host races of the soapberry bug, Jadera haematoloma. Genetica 112:257–272.

54. Carroll, S. P., P. S. Jørgensen, M. T. Kinnison, C. T. Bergstrom, R. F. Denison, P. Gluckman, T. B. Smith et al. 2014. Applying evolutionary biology to address global challenges. Science. doi:10.1126/science.1245993.

55. Cassey, P., T. M. Blackburn, D. Sol, R. P. Duncan, and J. L. Lockwood 2004. Global patterns of introduction effort and establishment success in birds. Proceedings of the Royal Society B 271(Suppl 6):S405–S408.

56. Castañeda, L. E., J. Balanyá, E. L. Rezende, and M. Santos 2013. Vanishing chromosomal inversion clines in Drosophila subobscura from Chile: is behavioral thermoregulation to blame? American Naturalist 182:249–259.

57. Chan, A. P., J. Crabtree, Q. Zhao, H. Lorenzi, J. Orvis, D. Puiu, A. Melake-Berhan et al. 2010. Draft genome sequence of the oilseed species Ricinus communis. Nature Biotechnology 28:951–956.

58. Chelaifa, H., A. Monnier, and M. Ainouche 2010. Transcriptomic changes following recent natural hybridization and allopolyploidy in the salt marsh species Spartina x townsendii and Spartina anglica (Poaceae). New Phytologist 186:161–174.

59. Chevin, L.-M., S. Collins, and F. Lefèvre 2013. Phenotypic plasticity and evolutionary demographic responses to climate change: taking theory out to the field. Functional Ecology 27:967–979.

60. Chiu, J. C., X. T. Jiang, L. Zhao, C. A. Hamm, J. M. Cridland, P. Saelao, K. A. Hamby et al. 2013. Genome of Drosophila suzukii, the Spotted Wing Drosophila. G3-Genes Genomes. Genetics 3:2257–2271.

61. Chown, S. L. 2010. Temporal biodiversity change in transformed landscapes: a southern African perspective. Philosophical Transactions of the Royal Society B 365:3729–3742.

62. Chown, S. L., and J. S. Terblanche 2007. Physiological diversity in insects: ecological and evolutionary contexts. Advances in Insect Physiology 33:50–152.

63. Chown, S. L., S. Slabber, M. A. McGeoch, C. Janion, and H. P. Leinaas 2007. Phenotypic plasticity mediates climate change responses among invasive and indigenous arthropods. Proceedings of the Royal Society B 274:2661–2667.

64. Chown, S. L., A. H. L. Huiskes, N. J. M. Gremmen, J. E. Lee, A. Terauds, K. Crosbie, Y. Frenot et al. 2012. Continent-wide risk assessment for the establishment of nonindigenous species in Antarctica. Proceedings of the National Academy of Sciences of the USA 109:4938–4943.
65. Chun, Y. J., M. van Kleunen, and W. Dawson 2010. The role of enemy release, tolerance and resistance in plant invasions: linking damage to performance. Ecology Letters 13:937–946.
66. Colautti, R. I., and S. C. H. Barrett 2013. Rapid adaptation to climate facilitates range expansion of an invasive plant. Science 342:364–366.
67. Colautti, R. I., I. A. Grigorovich, and H. J. MacIsaac 2006. Propagule pressure: A null model for biological invasions. Biological Invasions 8:1023–1037.
68. Collins, R. A., and R. H. Cruickshank 2013. The seven deadly sins of DNA barcoding. Molecular Ecology Resources 13:969–975.
69. Collins, R. A., K. F. Armstrong, A. J. Holyoake, and S. Keeling 2013. Something in the water: biosecurity monitoring of ornamental fish imports using environmental DNA. Biological Invasions 15:1209–1215.
70. Comai, L. 2005. The advantages and disadvantages of being polyploid. Nature Reviews Genetics 6:836–846.
71. Coop, G., D. Witonsky, A. Di Rienzo, and J. K. Pritchard 2010. Using environmental correlations to identify loci underlying local adaptation. Genetics 185:1411–1423.
72. Cowan, D., S. L. Chown, P. Convey, I. M. Tuffin, K. E. Hughes, S. Pointing, and W. F. Vincent 2011. Non-indigenous microorganisms in the Antarctic: assessing the risks. Trends in Microbiology 19:540–548.
73. Craigie, I. D., J. E. M. Baillie, A. Balmford, C. Carbone, B. Collen, R. E. Green, and J. M. Hutton 2010. Large mammal population declines in Africa's protected areas. Biological Conservation 143:2221–2228.
74. Darling, J. A., and M. J. Blum 2007. DNA-based methods for monitoring invasive species: a review and prospectus. Biological Invasions 9:751–765.
75. Darling, E. S., and I. M. Côté 2008. Quantifying the evidence for ecological synergies. Ecology Letters 11:1278–1286.
76. Darling, J. A., M. J. Bagley, J. Roman, C. K. Tepolt, and J. B. Geller 2008a. Genetic patterns across multiple introductions of the globally invasive crab genus Carcinus. Molecular Ecology 17:4992–5007.
77. Darling, E., K. E. Samis, and C. G. Eckert 2008b. Increased seed dispersal potential towards geographic range limits in a Pacific coast dune plant. New Phytologist 178:424–435.
78. Davidson, A. M., M. Jennions, and A. B. Nicotra 2011. Do invasive species show higher phenotypic plasticity than native species and, if so, is it adaptive? A meta-analysis. Ecology Letters 14:419–431.
79. Dawson, W., R. P. Rohr, M. van Kleunen, and M. Fischer 2012. Alien plant species with a wider global distribution are better able to capitalize on increased resource availability. New Phytologist 194:859–867.
80. De Kort, H., K. Vandepitte, J. Mergeay, and O. Honnay 2014. Isolation, characterization and genotyping of single nucleotide polymorphisms in the non-model tree species Frangula alnus (Rhamnaceae). Conservation Genetics Resources 6:267–269.

81. De La Torre, A. R., D. R. Roberts, and S. N. Aitken 2014. Genome-wide admixture and ecological niche modelling reveal the maintenance of species boundaries despite long history of interspecific gene flow. Molecular Ecology 23:2046–2059.

82. De Witt, T. J., and S. M. Scheiner 2004. Phenotypic Plasticity. Oxford University Press, Functional and Conceptual Approaches. Oxford.

83. Dehal, P., Y. Satou, R. K. Campbell, J. Chapman, B. Degnan, A. De Tomaso, B. Davidson et al. 2002. The draft genome of Ciona intestinalis: Insights into chordate and vertebrate origins. Science 298:2157–2167.

84. Dehnen-Schmutz, K. 2011. Determining non-invasiveness in ornamental plants to build green lists. Journal of Applied Ecology 48:1374–1380.

85. Dejean, T., A. Valentini, C. Miquel, P. Taberlet, E. Bellemain, and C. Miaud 2012. Improved detection of an alien invasive species through environmental DNA barcoding: the example of the American bullfrog Lithobates catesbeianus. Journal of Applied Ecology 49:953–959.

86. Des Marais, D. L., K. M. Hernandez, and T. E. Juenger 2013. Genotype-by-environment interaction and plasticity: exploring genomic responses of plants to the abiotic environment. Annual Review of Ecology, Evolution, and Systematics 44:5–29.

87. Dlugosch, K. M., and I. M. Parker 2008. Founding events in species invasions: genetic variation, adaptive evolution, and the role of multiple introductions. Molecular Ecology 17:431–449.

88. Dormontt, E. E., A. J. Lowe, and P. J. Prentis 2011. Is rapid evolution important in successful invasions? In D. M. Richardson, ed. Fifty years of invasion ecology. The legacy of Charles Elton, pp. 175–193. Wiley-Blackwell, Oxford.

89. Drake, J. M. 2006. Heterosis, the catapult effect and establishment success of a colonizing bird. Biology Letters 2:304–307.

90. Drake, J. M., and D. M. Lodge 2004. Global hot spots of biological invasions: evaluating options for ballast-water management. Proceedings of the Royal Society of London B 271:575–580.

91. Dunning, L. T., A. B. Dennis, B. J. Sinclair, R. D. Newcomb, and T. R. Buckley 2014. Divergent transcriptional responses to low temperature among populations of alpine and lowland species of New Zealand stick insects (Micrarchus). Molecular Ecology 23:2712–2726.

92. Dybern, B. I. 1965. The life cycle of Ciona intestinalis (L.) f. typica in the relation to environmental temperature. Oikos 16:109–131.

93. Egan, S. P., M. A. Barnes, C.-T. Hwang, A. R. Mahon, J. L. Feder, S. T. Ruggiero, C. E. Tanner, and D. M. Lodge 2013. Rapid invasive species detection by combining environmental DNA with light transmission spectroscopy. Conservation Letters 6:402–409.

94. Ellegren, H. 2014. Genome sequencing and population genomics in non-model organisms. Trends in Ecology and Evolution 29:51–63.

95. Ellers, J., and J. F. Stuefer 2010. Frontiers in phenotypic plasticity research: new questions about mechanisms, induced responses and ecological impacts. Evolutionary Ecology 24:523–526.

96. Ellstrand, N. C., and K. A. Schierenbeck. 2000. Hybridization as a stimulus for the evolution of invasiveness in plants? Proceedings of the National Academy of Sciences of the USA 97:7043–7050.

97. Elshire, R. J., J. C. Glaubitz, Q. Sun, J. A. Poland, K. Kawamoto, E. S. Buckler, and S. E. Mitchell 2011. A robust, simple genotyping-by-sequencing (GBS) approach for high diversity species. PLoS ONE 6:e19379.

98. Endersby, N. M., A. A. Hoffmann, V. L. White, S. A. Ritchie, P. H. Johnson, and A. R. Weeks 2011. Changes in the genetic structure of Aedes aegypti (Diptera: Culicidae) populations in Queensland, Australia, across two seasons: implications for potential mosquito releases. Journal of Medical Entomology 48:999–1007.

99. Evans, J., J. Kim, K. L. Childs, B. Vaillancourt, E. Crisovan, A. Nandety, D. J. Gerhardt et al. 2014. Nucleotide polymorphism and copy number variant detection using exome capture and next-generation sequencing in the polyploid grass Panicum virgatum. The Plant Journal 79:993–1008.

100. Excoffier, L., T. Hofer, and M. Foll 2009. Detecting loci under selection in a hierarchically structured population. Heredity 103:285–298.

101. Fernández-Álvarez, F. A., and A. Machordom 2013. DNA barcoding reveals a cryptic nemertean invasion in Atlantic and Mediterranean waters. Helgoland Marine Research 67:599–605.

102. Ficetola, G. F., F. Miaud, F. Pompanon, and P. Taberlet 2008. Species detection using environmental DNA from water samples. Biology Letters 4:423–425.

103. Fitzpatrick, B. M., J. R. Johnson, D. K. Kump, J. J. Smith, S. R. Voss, and H. B. Shaffer 2010. Rapid spread of invasive genes into a threatened native species. Proceedings of the National Academy of Sciences of the USA 107:3606–3610.

104. Foll, M., and O. Gaggiotti 2008. A genome-scan method to identify selected loci appropriate for both dominant and codominant markers: a Bayesian perspective. Genetics 180:977–993.

105. Fordham, D. A., B. W. Brook, C. Moritz, and D. Nogues-Bravo 2014. Better forecasts of range dynamics using genetic data. Trends in Ecology and Evolution 29:436–443.

106. Forsman, A.. 2014. Effects of genotypic and phenotypic variation on establishment are important for conservation, invasion, and infection biology. Proceedings of the National Academy of Sciences of the USA 111:302–307.

107. Forsyth, G. G., D. C. Le Maitre, P. J. O'Farrell, and B. W. van Wilgen 2012. The prioritisation of invasive alien plant control projects using a multi-criteria decision model informed by stakeholder input and spatial data. Journal of Environmental Management 103:51–57.

108. Franks, S. J., J. C. Avise, W. E. Bradshaw, J. K. Conner, J. R. Etterson, S. J. Mazer, R. G. Shaw, and A. E. Weis 2008. The Resurrection Initiative: storing ancestral genotypes to capture evolution in action. BioScience 58:870–873.

109. Friedman, J. M., J. E. Roelle, J. F. Gaskin, A. E. Pepper, and J. R. Manhart 2008. Latitudinal variation in cold hardiness in introduced Tamarix and native Populus. Evolutionary Applications 1:598–607.

110. Furrow, R. E., and M. W. Feldman 2014. Genetic variation and the evolution of epigenetic regulation. Evolution 68:673–683.

111. Futschik, A., and C. Schlotterer 2010. The next generation of molecular markers from massively parallel sequencing of pooled DNA samples. Genetics 186:207–218.

112. Gamfeldt, L., and B. Källström 2007. Increasing intraspecific diversity increases predictability in population survival in the face of perturbations. Oikos 116:700–705.

113. Garroway, C. J., J. Bowman, T. J. Cascaden, G. L. Holloway, C. G. Mahan, J. R. Malcolm, M. A. Steele et al. 2010. Climate change induced hybridization in flying squirrels. Global Change Biology 16:113–121.

114. Gaskin, J. F., and D. J. Kazmer 2009. Introgression between invasive saltcedars (Tamarix chinensis and T. ramosissima) in the USA. Biological Invasions 11:1121–1130.

115. Gaskin, J. F., and B. A. Schaal. 2002. Hybrid Tamarix widespread in US invasion and undetected in native Asian range. Proceedings of the National Academy of Sciences of the USA 99:11256–11259.

116. Gaskin, J. F., M.-C. Bon, M. J. W. Cock, M. Cristofaro, A. D. Biase, R. De Clerck-Floate, C. A. Ellison et al. 2011. Applying molecular-based approaches to classical biological control of weeds. Biological Control 58:1–21.

117. Gautier, M., K. Gharbi, T. Cezard, J. Foucaud, C. Kerdelhue, P. Pudlo, J.-M. Cornuet, and A. Estoup 2013. The effect of RAD allele dropout on the estimation of genetic variation within and between populations. Molecular Ecology 22:3165–3178.

118. Ghalambor, C. K., J. K. McKay, S. P. Carroll, and D. N. Reznick 2007. Adaptive versus non-adaptive phenotypic plasticity and the potential for contemporary adaptation in new environments. Functional Ecology 21:394–407.

119. Gibbs, R. A., G. M. Weinstock, M. L. Metzker, D. M. Muzny, E. J. Sodergren, S. Scherer, G. Scott et al. 2004. Genome sequence of the Brown Norway rat yields insights into mammalian evolution. Nature 428:493–521.

120. Gienapp, P., C. Teplitsky, J. S. Alho, J. A. Mills, and J. Merila 2008. Climate change and evolution: disentangling environmental and genetic responses. Molecular Ecology 17:167–178.

121. Glastad, K. M., B. G. Hunt, S. V. Yi, and M. A. D. Goodisman 2011. DNA methylation in insects: on the brink of the epigenomic era. Insect Molecular Biology 20:553–565.

122. Griffin, P. C., and A. A. Hoffmann 2014. Limited genetic divergence among Australian alpine Poa tussock grasses coupled with regional structuring points to ongoing gene flow and taxonomic challenges. Annals of Botany 113:953–965.

123. Griffin, P. C., C. Robin, and A. A. Hoffmann 2011. A next-generation sequencing method for overcoming the multiple gene copy problem in polyploid phylogenetics, applied to Poa grasses. BMC Biology 9:18.

124. Guggisberg, A., Z. Lai, J. Huang, and L. H. Rieseberg 2013. Transcriptome divergence between introduced and native populations of Canada thistle, Cirsium arvense. New Phytologist 199:595–608.

125. Guisan, A., B. Petitpierre, O. Broennimann, C. Daehler, and C. Kuefferz 2014. Unifying niche shift studies: insights from biological invasions. Trends in Ecology and Evolution 29:260–269.

126. Günther, T., and G. Coop 2013. Robust identification of local adaptation from allele frequencies. Genetics 195:205–220.

127. Gupta, S. K., S. Singh, A. Nischal, K. K. Pant, and P. K. Seth 2014. Molecular-based identification and phylogeny of genomic and proteomic sequences of mosquito-borne flavivirus. Genes & Genomics 36:31–43.

128. Hampton, J. O., P. B. S. Spencer, D. L. Alpers, L. E. Twigg, A. P. Woolnough, J. Doust, T. Higgs, and J. Pluske 2004. Molecular techniques, wildlife management

and the importance of genetic population structure and dispersal: a case study with feral pigs. Journal of Applied Ecology 41:735–743.

129. Hänel, C., and S. L. Chown 1998. The impact of a small, alien invertebrate on a sub-Antarctic terrestrial ecosystem: Limnophyes minimus (Diptera, Chironomidae) at Marion Island. Polar Biology 20:99–106.

130. Hansen, J., P. Kharecha, M. Sato, V. Masson-Delmotte, F. Ackerman, D. J. Beerling, P. J. Hearty et al. 2013. Assessing "dangerous climate change": required reduction of carbon emissions to protect young people, future generations and nature. PLoS ONE 8:e81648.

131. He, W.-M., G. C. Thelen, W. M. Ridenour, and R. M. Callaway 2010. Is there a risk to living large? Large size correlates with reduced growth when stressed for knapweed populations. Biological Invasions 12:3591–3598.

132. Henery, M. L., G. Bowman, P. Mraz, U. A. Treier, E. Gex-Fabry, U. Schaffner, and H. Mueller-Schaerer 2010. Evidence for a combination of pre-adapted traits and rapid adaptive change in the invasive plant Centaurea stoebe. Journal of Ecology 98:800–813.

133. Herrera, C. M., M. I. Pozo, and P. Bazaga 2012. Jack of all nectars, master of most: DNA methylation and the epigenetic basis of niche width in a flower-living yeast. Molecular Ecology 21:2602–2616.

134. Hickerson, M. J., C. P. Meyer, and C. Moritz 2006. DNA barcoding will often fail to discover new animal species over broad parameter space. Systematic Biology 55:729–739.

135. Hill, M. P., S. L. Chown, and A. A. Hoffmann 2013. A predicted niche shift corresponds with increased thermal resistance in an invasive mite, Halotydeus destructor. Global Ecology and Biogeography 22:942–951.

136. Hochholdinger, F., and N. Hoecker 2007. Towards the molecular basis of heterosis. Trends in Plant Science 12:427–432.

137. Hodgins, K. A., Z. Lai, K. Nurkowski, J. Huang, and L. H. Rieseberg 2013. The molecular basis of invasiveness: differences in gene expression of native and introduced common ragweed (Ambrosia artemisiifolia) in stressful and benign environments. Molecular Ecology 22:2496–2510.

138. Hoffmann, A. A., J. G. Sørensen, and V. Loeschcke 2003. Adaptation of Drosophila to temperature extremes: bringing together quantitative and molecular approaches. Journal of Thermal Biology 28:175–216.

139. Hohenlohe, P. A., S. Bassham, P. D. Etter, N. Stiffler, E. A. Johnson, and W. A. Cresko 2010. Population genomics of parallel adaptation in threespine stickleback using Sequenced RAD Tags. PloS Genetics 6:e1000862.

140. Huang, Y., Y. Li, D. W. Burt, H. Chen, Y. Zhang, W. Qian, H. Kim et al. 2013. The duck genome and transcriptome provide insight into an avian influenza virus reservoir species. Nature Genetics 45:776–784.

141. Huey, R. B., G. W. Gilchrist, M. L. Carlson, D. Berrigan, and L. Serra 2000. Rapid evolution of a geographic cline in size in an introduced fly. Science 287:308–309.

142. Huiskes, A. H. L., N. J. M. Gremmen, D. M. Bergstrom, Y. Frenot, K. A. Hughes, S. Imura, K. Kiefer et al. 2014. Aliens in Antarctica: assessing transfer of plant propagules by human visitors to reduce invasion risk. Biological Conservation 171:278–284.

143. Hulme, P. E. 2009. Trade, transport and trouble: managing invasive species pathways in an era of globalization. Journal of Applied Ecology 46:10–18.
144. Hulme, P. E., S. Bacher, M. Kenis, S. Klotz, I. Kühn, D. Minchin, W. Nentwig et al. 2008. Grasping at the routes of biological invasions: a framework for integrating pathways into policy. Journal of Applied Ecology 45:403–414.
145. Hulme, P. E., P. Pyšek, V. Jarošik, J. Pergl, U. Schaffner, and M. Vilà 2013. Bias and error in understanding plant invasion impacts. Trends in Ecology and Evolution 28:212–218.
146. Jackson, J. B. 2010. The future of the oceans past. Philosophical Transactions of the Royal Society B 365:3765–3778.
147. Jerde, J. L., A. R. Mahon, W. L. Chadderton, and D. M. Lodge 2011. "Sight-unseen" detection of rare aquatic species using environmental DNA. Conservation Letters 4:150–157.
148. Ji, Y., L. Ashton, S. M. Pedley, D. P. Edwards, Y. Tang, A. Nakamura, R. Kitching et al. 2013. Reliable, verifiable and efficient monitoring of biodiversity via metabarcoding. Ecology Letters 16:1245–1257.
149. Jones, K. E., N. G. Patel, M. A. Levy, A. Storeygard, D. Balk, J. L. Gittleman, and P. Daszak 2008. Global trends in emerging infectious diseases. Nature 451:990–993.
150. Jones, F. C., Y. F. Chan, J. Schmutz, J. Grimwood, S. D. Brady, A. M. Southwick, D. M. Absher et al. 2012. A Genome-wide SNP genotyping array reveals patterns of global and repeated species-pair divergence in sticklebacks. Current Biology 22:83–90.
151. Kasper, M. L., A. F. Reeson, S. J. Cooper, K. D. Perry, and A. D. Austin 2004. Assessment of prey overlap between a native (Polistes humilis) and an introduced (Vespula germanica) social wasp using morphology and phylogenetic analyses of 16S rDNA. Molecular Ecology 13:2037–2048.
152. Kearney, M., W. P. Porter, C. Williams, S. Ritchie et al. 2009. Integrating biophysical models and evolutionary theory to predict climatic impacts on species' ranges: the dengue mosquito Aedes aegypti in Australia. Functional Ecology 23:528–538.
153. Keesing, F., L. K. Belden, P. Daszak, A. P. Dobson, C. D. Harvell, R. D. Holt, P. Hudson et al. 2010. Impacts of biodiversity on the emergence and transmission of infectious diseases. Nature 468:647–652.
154. Keller, S. R., and D. R. Taylor 2010. Genomic admixture increases fitness during a biological invasion. Journal of Evolutionary Biology 23:1720–1731.
155. Kharrat-Souissi, A., S. Siljak-Yakovlev, S. C. Brown, A. Baumel, F. Torre, and M. Chaieb 2014. The polyploid nature of Cenchrus ciliaris L. (Poaceae) has been overlooked: new insights for the conservation and invasion biology of this species - a review. Rangeland Journal 36:11–23.
156. Kirk, H., J. Paul, J. Straka, and J. R. Freeland 2011. Long-distance dispersal and high genetic diversity are implicated in the invasive spread of the common reed, Phragmites australis (Poaceae), in northeastern North America. American Journal of Botany 98:1180–1190.
157. Kleynhans, E., K. A. Mitchell, D. E. Conlong, and J. S. Terblanche 2014. Evolved variation in cold tolerance among populations of Eldana saccharina (Lepidoptera: Pyralidae) in South Africa. Journal of Evolutionary Biology 27:1149–1159.

158. Krehenwinkel, H., and D. Tautz 2013. Northern range expansion of European popu-
     lations of the wasp spider Argiope bruennichi is associated with global warming-
     correlated genetic admixture and population-specific temperature adaptations. Mo-
     lecular Ecology 22:2232–2248.
159. Kueffer, C., P. Pyšek, and D. M. Richardson 2013. Integrative invasion science:
     model systems, multi-site studies, focused meta-analysis and invasion syndromes.
     New Phytologist 200:615–633.
160. Lambdon, P. W., P. Pyšek, C. Basnou, M. Hejda, M. Arianoutsou, F. Essl, V. Jarošík
     et al. 2008. Alien flora of Europe: species diversity, temporal trends, geographical
     patterns and research needs. Preslia 80:101–149.
161. Larson, E. L., J. A. Andres, S. M. Bogdanowicz, and R. G. Harrison 2013. Differen-
     tial introgression in a mosaic hybrid zone reveals candidate barrier genes. Evolution
     67:3653–3661.
162. Laurance, W. F., D. C. Useche, J. Rendeiro, M. Kalka, C. J. Bradshaw, S. P. Sloan,
     S. G. Laurance et al. 2012. Averting biodiversity collapse in tropical forest protected
     areas. Nature 489:290–294.
163. Lavergne, S., and J. Molofsky. 2007. Increased genetic variation and evolutionary
     potential drive the success of an invasive grass. Proceedings of the National Acad-
     emy of Sciences of the USA 104:3883–3888.
164. Lawson Handley, L. J., A. Estoup, D. M. Evans, C. E. Thomas, E. Lombaert, B.
     Facon, A. Aebi, and H. E. Roy 2011. Ecological genetics of invasive alien species.
     BioControl 56:409–428.
165. Le Corre, V., and A. Kremer 2012. The genetic differentiation at quantitative trait
     loci under local adaptation. Molecular Ecology 21:1548–1566.
166. Le Roux, J. J., G. K. Brown, M. Byrne, J. Ndlovu, D. M. Richardson, G. D. Thomp-
     son, and J. R. U. Wilson 2011. Phylogeographic consequences of different introduc-
     tion histories of invasive Australian Acacia species and Paraserianthes lophantha
     (Fabaceae) in South Africa. Diversity and Distributions 17:861–871.
167. Le Roux, J. J., D. M. Richardson, J. R. U. Wilson, and J. Ndlovu 2013. Human usage
     in the native range may determine future genetic structure of an invasion: insights
     from Acacia pycnantha. BMC Ecology 13:8.
168. Lee, C. E. 2002. Evolutionary genetics of invasive species. Trends in Ecology &
     Evolution 17:386–391.
169. Lee, J. E., and S. L. Chown 2009. Breaching the dispersal barrier to invasion: quan-
     tification and management. Ecological Applications 19:1944–1959.
170. Lee, C. E., and C. H. Petersen 2002. Genotype-by-environment interaction for salin-
     ity tolerance in the freshwater-invading copepod Eurytemora affinis. Physiological
     and Biochemical Zoology 75:335–344.
171. Leffler, J. A., J. J. James, T. A. Monaco, and R. L. Sheley 2014. A new perspective
     on trait differences between native and invasive exotic plants. Ecology 95:298–305.
172. Lemmon, A. R., S. A. Emme, and E. M. Lemmon 2012. Anchored hybrid enrichment
     for massively high-throughput phylogenomics. Systematic Biology 61:727–744.
173. Levine, M. T., M. L. Eckert, and D. J. Begun 2011. Whole-genome expression plas-
     ticity across tropical and temperate Drosophila melanogaster populations from east-
     ern Australia. Molecular Biology and Evolution 28:249–256.

174. Lewontin, R. C., and J. Krakauer 1973. Distribution of gene frequency as a test of theory of selective neutrality of polymorphisms. Genetics 74:175–195.
175. Liebl, A. L., A. W. Schrey, C. L. Richards, and L. B. Martin 2013. Patterns of DNA methylation throughout a range expansion of an introduced songbird. Integrative and Comparative Biology 53:351–358.
176. Litchman, E. 2010. Invisible invaders: non-pathogenic invasive microbes in aquatic and terrestrial systems. Ecology Letters 13:1560–1572.
177. Liu, S., Y. Li, J. Lu, X. Su, M. Tang, R. Zhang, L. Zhou et al. 2013. SOAPBarcode: revealing arthropod biodiversity through assembly of Illumina shotgun sequences of PCR amplicons. Methods in Ecology and Evolution 4:1142–1150.
178. Lotterhos, K. E., and M. C. Whitlock 2014. Evaluation of demographic history and neutral parameterization on the performance of F-ST outlier tests. Molecular Ecology 23:2178–2192.
179. Lu, F., A. E. Lipka, J. Glaubitz, R. Elshire, J. H. Cherney, M. D. Casler, E. S. Buckler, and D. E. Costich 2013. Switchgrass genomic diversity, ploidy, and evolution: novel insights from a network-based SNP discovery protocol. PloS Genetics 9:e1003215.
180. Mahon, A. R., M. A. Barnes, F. Li, S. P. Egan, C. E. Tanner, S. T. Ruggiero, J. L. Feder et al. 2012. DNA-based species detection capabilities using laser transmission spectroscopy. Journal of the Royal Society Interface 10:20120637.
181. Mamanova, L., A. J. Coffey, C. E. Scott, I. Kozarewa, E. H. Turner, A. Kumar, E. Howard et al. 2010. Target-enrichment strategies for next-generation sequencing. Nature Methods 7:111–118.
182. Marcussen, T., S. R. Sandve, L. Heier, M. Spannagl, and M. Pfeifer, The International Wheat Genome Sequencing Consortium, K. S. Jakobsen et al. 2014. Ancient hybridizations among the ancestral genomes of bread wheat. Science 345:1251788.
183. Martin, M. D., E. A. Zimmer, M. T. Olsen, A. D. Foote, M. T. P. Gilbert, and G. S. Brush 2014. Herbarium specimens reveal a historical shift in phylogeographic structure of common ragweed during native range disturbance. Molecular Ecology 23:1701–1716.
184. Matesanz, S., and F. Valladares 2014. Ecological and evolutionary responses of Mediterranean plants to global change. Environmental and Experimental Botany 103:53–67.
185. Matvienko, M., A. Kozik, L. Froenicke, D. Lavelle, B. Martineau, B. Perroud, and R. Michelmore 2013. Consequences of normalizing transcriptomic and genomic libraries of plant genomes using a duplex-specific nuclease and tetramethylammonium chloride. PLoS ONE 8:e55913.
186. McCormick, H., R. Cursons, R. J. Wilkins, and C. M. King 2014. Location of a contact zone between Mus musculus domesticus and M. m. domesticus with M. m. castaneus mtDNA in southern New Zealand. Mammalian Biology 79:297–305.
187. McDonald, D. B., T. L. Parchman, M. R. Bower, W. A. Hubert, and F. J. Rahel 2008. An introduced and a native vertebrate hybridize to form a genetic bridge to a second native species. Proceedings of the National Academy of Sciences of the USA 105:10837–10842.
188. McGeoch, M. A., D. Spear, E. J. Kleynhans, and E. Marais 2012. Uncertainty in invasive alien species listing. Ecological Applications 22:959–971.

189. McKinney, L. V., L. R. Nielsen, D. B. Collinge, I. M. Thomsen, J. K. Hansen, and E. D. Kjaer 2014. The ash dieback crisis: genetic variation in resistance can prove a long-term solution. Plant Pathology 63:485–499.

190. McLean, E. H., S. M. Prober, W. D. Stock, D. A. Steane, B. M. Potts, R. E. Vaillancourt, and M. Byrne 2014. Plasticity of functional traits varies clinally along a rainfall gradient in Eucalyptus tricarpa. Plant Cell and Environment 37:1440–1451.

191. Meier, R., K. Shiyang, G. Vaidya, and P. K. Ng 2006. DNA barcoding and taxonomy in Diptera: a tale of high intraspecific variability and low identification success. Systematic Biology 55:715–728.

192. Meier, K., M. M. Hansen, E. Normandeau, K. L. D. Mensberg, J. Frydenberg, P. F. Larsen, D. Bekkevold, and L. Bernatchez 2014. Local adaptation at the transcriptome level in brown trout: evidence from early life history temperature genomic reaction norms. PLoS ONE 9:13.

193. Millar, M. A., M. Byrne, and W. O'Sullivan 2011. Defining entities in the Acacia saligna (Fabaceae) species complex using a population genetics approach. Australian Journal of Botany 59:137–148.

194. Millar, M. A., M. Byrne, I. K. Nuberg, and M. Sedgley 2012. High levels of genetic contamination in remnant populations of Acacia saligna from a genetically divergent planted stand. Restoration Ecology 20:260–267.

195. Mobegi, V. A., C. W. Duffy, A. Amambua-Ngwa, K. M. Loua, E. Laman, D. C. Nwakanma, B. MacInnis et al. 2014. Genome-wide analysis of selection on the malaria parasite Plasmodium falciparum in West African populations of differing infection endemicity. Molecular Biology and Evolution 31:1490–1499.

196. Molins, M. P., J. M. Corral, O. M. Aliyu, M. A. Koch, A. Betzin, J. L. Maron, and T. F. Sharbel 2014. Biogeographic variation in genetic variability, apomixis expression and ploidy of St. John's wort (Hypericum perforatum) across its native and introduced range. Annals of Botany 113:417–427.

197. Moody, M. L., and D. H. Les 2007. Geographic distribution and genotypic composition of invasive hybrid watermilfoil (Myriophyllum spicatum x M. sibiricum) populations in North America. Biological Invasions 9:559–570.

198. Mooney, H. A., and R. J. Hobbs 2000. Invasive Species in a Changing World. Island Press, Washington, D. C..

199. Moran, E. V., and J. M. Alexander 2014. Evolutionary responses to global change: lessons from invasive species. Ecology Letters 17:637–649.

200. Morris, M. R. J., R. Richard, E. H. Leder, R. D. H. Barrett, N. Aubin-Horth, and S. M. Rogers 2014. Gene expression plasticity evolves in response to colonization of freshwater lakes in threespine stickleback. Molecular Ecology 23:3226–3240.

201. Muhlfeld, C. C., R. P. Kovach, L. A. Jones, R. Al-Chokhachy, M. C. Boyer, R. F. Leary, W. H. Lowe et al. 2014. Invasive hybridization in a threatened species is accelerated by climate change. Nature Climate Change 4:620–624.

202. Myburgh, M., S. L. Chown, S. R. Daniels, and B. Jansen van Vuuren 2007. Population structure, propagule pressure, and conservation biogeography in the sub-Antarctic: lessons from indigenous and invasive springtails. Diversity and Distributions 13:143–154.

203. Nene, V., J. R. Wortman, D. Lawson, B. Haas, C. Kodira, Z. Tu, B. Loftus et al. 2007. Genome sequence of Aedes aegypti, a major arbovirus vector. Science 316:1718–1723.

204. Neves, L. G., J. M. Davis, W. B. Barbazuk, and M. Kirst 2013. Whole-exome targeted sequencing of the uncharacterized pine genome. Plant Journal 75:146–156.

205. Nicotra, A. B., O. K. Atkin, S. P. Bonser, A. M. Davidson, E. J. Finnegan, U. Mathesius, P. Poot et al. 2010. Plant phenotypic plasticity in a changing climate. Trends in Plant Science 15:684–692.

206. Nosil, P., Z. Gompert, T. E. Farkas, A. A. Comeault, J. L. Feder, C. A. Buerkle, and T. L. Parchman 2012. Genomic consequences of multiple speciation processes in a stick insect. Proceedings of the Royal Society B-Biological Sciences 279:5058–5065.

207. O'Dowd, D. J., P. T. Green, and P. S. Lake 2003. Invasional 'meltdown' on an oceanic island. Ecology Letters 6:812–817.

208. Otto, S. P. 2007. The evolutionary consequences of polyploidy. Cell 131:452–462.

209. Otto, S. P., and J. Whitton 2000. Polyploid incidence and evolution. Annual Review of Genetics 34:401–437.

210. Pandit, M. K., M. J. O. Pocock, and W. E. Kunin 2011. Ploidy influences rarity and invasiveness in plants. Journal of Ecology 99:1108–1115.

211. Parchman, T. L., Z. Gompert, M. J. Braun, R. T. Brumfield, D. B. McDonald, J. A. C. Uy, G. Zhang et al. 2013. The genomic consequences of adaptive divergence and reproductive isolation between species of manakins. Molecular Ecology 22:3304–3317.

212. Peñalba, J. V., L. L. Smith, M. A. Tonione, C. Sass, S. M. Hykin, P. L. Skipwith, J. A. McGuire et al. 2014. Sequence capture using PCR-generated probes: a cost-effective method of targeted high-throughput sequencing for nonmodel organisms. Molecular Ecology Resources 14:1000–1010.

213. Pérez, J. E., M. Nirchio, C. Alfonsi, and C. Muñoz 2006. The biology of invasions: The genetic adaptation paradox. Biological Invasions 8:1115–1121.

214. Pfenninger, M., A. Weigand, M. Balint, and A. Klussmann-Kolb 2014. Misperceived invasion: the Lusitanian slug (Arion lusitanicus auct. non-Mabille or Arion vulgaris Moquin-Tandon 1855) is native to Central Europe. Evolutionary Applications 7:702–713.

215. Phillips, B. L., and R. Shine 2006. An invasive species induces rapid adaptive change in a native predator: cane toads and black snakes in Australia. Proceedings of the Royal Society B 273:1545–1550.

216. Phillips, B. L., G. P. Brown, J. K. Webb, and R. Shine 2006. Invasion and the evolution of speed in toads. Nature 439:803.

217. Piaggio, A. J., R. M. Engeman, M. W. Hopken, J. S. Humphrey, K. L. Keacher, W. E. Bruce, and M. L. Avery. 2014. Detecting an elusive invasive species: a diagnostic PCR to detect Burmese python in Florida waters and an assessment of persistence of environmental DNA. Molecular Ecology Resources 14:374–380.

218. Pontius, J. U., J. C. Mullikin, D. R. Smith, K. Lindblad-Toh, S. Gnerre, M. Clamp, J. Chang et al. 2007. Initial sequence and comparative analysis of the cat genome. Genome Research 17:1675–1689.

219. Porco, D., A. Bedos, P. Greenslade, C. Janion, D. Skarzynski, M. I. Stevens, B. Jansen van Vuuren, and L. Deharveng. 2012. Challenging species delimitation in Collembola: cryptic diversity among common springtails unveiled by DNA barcoding. Invertebrate Systematics 26:470477.

220. Porco, D., T. Decaëns, L. Deharveng, S. W. James, D. Skarzynski, C. Erséus, K. R. Butt et al. 2013. Biological invasions in soil: DNA barcoding as a monitoring tool in a multiple taxa survey targeting European earthworms and springtails in North America. Biological Invasions 15:899–910.

221. Præbel, K., K. O. Gjelland, E. Salonen, and P. A. Amundsen 2013. Invasion genetics of vendace (Coregonus albula (L.)) in the Inari-Pasvik watercourse: revealing the origin and expansion pattern of a rapid colonization event. Ecology and Evolution 3:1400–1412.

222. Prentis, P. J., and A. Pavasovic 2013. Understanding the genetic basis of invasiveness. Molecular Ecology 22:2366–2368.

223. Prentis, P. J., J. R. U. Wilson, E. E. Dormontt, D. M. Richardson, and A. J. Lowe 2008. Adaptive evolution in invasive species. Trends in Plant Science 13:288–294.

224. Pritchard, J. K., J. K. Pickrell, and G. Coop 2010. The genetics of human adaptation: hard sweeps, soft sweeps, and polygenic adaptation. Current Biology 20:R208–R215.

225. Puth, L. M., and D. M. Post 2005. Studying invasion: have we missed the boat? Ecology Letters 8:715–721.

226. Pyšek, P., D. M. Richardson, M. Rejmánek, G. L. Webster, M. Williamson, and J. Kirschner 2004. Alien plants in checklists and floras: towards better communication between taxonomists and ecologists. Taxon 53:131–143.

227. Pyšek, P., V. Jarošík, P. E. Hulme, J. Pergl, M. Hejda, U. Schaffner, and M. Vilà 2012. A global assessment of invasive plant impacts on resident species, communities and ecosystems: the interaction of impact measures, invading species' traits and environment. Global Change Biology 18:1725–1737.

228. Pyšek, P., P. E. Hulme, L. A. Meyerson, G. F. Smith, J. S. Boatwright, N. R. Crouch, E. Figueiredo et al. 2013. Hitting the right target: taxonomic challenges for, and of, plant invasions. AOB Plants 5:plt042.

229. Ramirez, K. S., J. W. Leff, A. Barberán, S. T. Bates, J. Betley, T. W. Crowther, E. F. Kelly et al. 2014. Biogeographic patterns in below-ground diversity in New York City's Central Park are similar to those observed globally. Proceedings of the Royal Society B 281:20141988.

230. Randall, J. M., L. E. Morse, N. Benton, R. Hiebert, S. Lu, and T. Killeffer 2008. The invasive species assessment protocol: a tool for creating regional and national lists of invasive nonnative plants that negatively impact biodiversity. Invasive Plant Science and Management 1:36–49.

231. Rasic, G., I. Filipovic, A. R. Weeks, and A. A. Hoffmann 2014. Genome-wide SNPs lead to strong signals of geographic structure and relatedness patterns in the major arbovirus vector, Aedes aegypti. BMC Genomics 15:12.

232. Rees, H. C., B. C. Maddison, D. J. Middleditch, J. R. M. Patmore, and K. C. Gough 2014. The detection of aquatic animal species using environmental DNA – a review of eDNA as a survey tool in ecology. Journal of Applied Ecology 51:1450–1459.

233. Ricciardi, A., M. F. Hoopes, M. P. Marchetti, and J. L. Lockwood 2013. Progress toward understanding the ecological impacts of nonnative species. Ecological Monographs 83:263–282.
234. Richards, C. L., A. W. Schrey, and M. Pigliucci 2012. Invasion of diverse habitats by few Japanese knotweed genotypes is correlated with epigenetic differentiation. Ecology Letters 15:1016–1025.
235. Richardson, D. M. 2011. Fifty Years of Invasion Ecology. The Legacy of Charles Elton. Wiley-Blackwell, Oxford.
236. Richardson, D. M., and P. Pyšek 2006. Plant invasions: merging the concepts of species invasiveness and community invasibility. Progress in Physical Geography 30:409–431.
237. Richardson, D. M., and P. Pyšek 2012. Naturalization of introduced plants: ecological drivers of biogeographical patterns. New Phytologist 196:383–396.
238. Richardson, D. M., and A. Ricciardi 2013. Misleading criticisms of invasion science: a field guide. Diversity and Distributions 19:1461–1467.
239. Richardson, D. M., P. Pyšek, and J. T. Carlton 2011. A compendium of essential concepts and terminology in invasion ecology. In D. M. Richardson, ed. Fifty years of invasion ecology. The legacy of Charles Elton, pp. 409–420. Oxford: Wiley-Blackwell.
240. Rieseberg, L. H., O. Raymond, D. M. Rosenthal, Z. Lai, K. Livingstone, T. Nakazato, J. L. Durphy et al. 2003. Major ecological transitions in wild sunflowers facilitated by hybridization. Science 301:1211–1216.
241. Rieseberg, L. H., S.-C. Kim, R. A. Randell, K. D. Whitney, B. L. Gross, C. Lexer, and K. Clay 2007. Hybridization and the colonization of novel habitats by annual sunflowers. Genetica 129:149–165.
242. Rignot, E., I. Velicogna, M. R. van den Broeke, A. Monaghan, and J. T. M. Lenaerts 2011. Acceleration of the contribution of the Greenland and Antarctic ice sheets to sea level rise. Geophysical Research Letters 38:L05503.
243. Rius, M., and J. A. Darling 2014. How important is intraspecific genetic admixture to the success of colonising populations? Trends in Ecology & Evolution 29:233–242.
244. Robertson, K., E. E. Goldberg, and B. Igic 2011. Comparative evidence for the correlated evolution of polyploidy and self-compatibility in Solanaceae. Evolution 65:139–155.
245. Rocha, L. A., M. A. Bernal, M. R. Gaither, and M. E. Alfaro 2013. Massively parallel DNA sequencing: the new frontier in biogeography. Frontiers of Biogeography 5:67–77.
246. Rohfritsch, A., N. Bierne, P. Boudry, S. Heurtebise, F. Cornette, and S. Lapegue 2013. Population genomics shed light on the demographic and adaptive histories of European invasion in the Pacific oyster, Crassostrea gigas. Evolutionary Applications 6:1064–1078.
247. Roman, J., and J. A. Darling 2007. Paradox lost: genetic diversity and the success of aquatic invasions. Trends in Ecology and Evolution 22:454–464.
248. Rowe, K. C., S. Singhal, M. D. Macmanes, J. F. Ayroles, T. L. Morelli, E. M. Rubidge, K. Bi, and C. C. Moritz 2011. Museum genomics: low-cost and high-accuracy genetic data from historical specimens. Molecular Ecology Resources 11:1082–1092.

249. Roy, H. E., D. B. Roy, and A. Roques 2011. Inventory of terrestrial alien arthropod predators and parasites established in Europe. BioControl 56:477–504.
250. Salmon, A., M. L. Ainouche, and J. F. Wendel 2005. Genetic and epigenetic consequences of recent hybridization and polyploidy in Spartina (Poaceae). Molecular Ecology 14:1163–1175.
251. Sampson, J. F., and M. Byrne 2008. Outcrossing between an agroforestry plantation and remnant native populations of Eucalyptus loxophleba. Molecular Ecology 17:2769–2781.
252. Sax, D. F., J. J. Stachowicz, and S. D. Gaines. 2005. Species invasions. Insights into ecology, evolution, and biogeography. Sinauer Associates Inc. Publishers, Massachusetts.
253. Scheffer, S. J., M. L. Lewis, and R. C. Joshi 2006. DNA barcoding applied to invasive leafminers (Diptera:Agromyzidae) in the Philippines. Annals of the Entomological Society of America 99:204–210.
254. Schierenbeck, K. A., and N. C. Ellstrand 2009. Hybridization and the evolution of invasiveness in plants and other organisms. Biological Invasions 11:1093–1105.
255. Schilthuizen, M., and V. Kellermann 2014. Contemporary climate change and terrestrial invertebrates: evolutionary versus plastic changes. Evolutionary Applications 7:56–67.
256. Schmitz, R. J., M. D. Schultz, M. A. Urich, J. R. Nery, M. Pelizzola, O. Libiger, A. Alix et al. 2013. Patterns of population epigenomic diversity. Nature 495:193–198.
257. Scholes, R. J., J. Settele, R. Betts, S. Bunn, P. Leadley, D. Nepstad, J. Overpeck et al. 2014. IPCC Working Group II Assessment Report 5 Chapter 4 Terrestrial and Inland Water Systems. Intergovernmental Panel on Climate Change, Geneva.
258. Schwartz, M. A., J. J. Hellmann, J. M. McLachlan, D. F. Sax, J. O. Borevitz, J. Brennan, A. E. Camacho et al. 2012. Managed relocation: integrating the scientific, regulatory, and ethical challenges. BioScience 62:732–743.
259. Scoville, A. G., and M. E. Pfrender. 2010. Phenotypic plasticity facilitates recurrent rapid adaptation to introduced predators. Proceedings of the National Academy of Sciences of the USA 107:4260–4263.
260. Scriber, J. M. 2011. Impacts of climate warming on hybrid zone movement: geographically diffuse and biologically porous "species borders". Insect Science 18:121–159.
261. Shade, A., H. Peter, S. D. Allison, D. L. Baho, M. Berga, H. Bürgmann, D. H. Huber et al. 2012. Fundamentals of microbial community resistance and resilience. Frontiers in Microbiology 3:47.
262. Shirk, R. Y., J. L. Hamrick, C. Zhang, and S. Qiang 2014. Patterns of genetic diversity reveal multiple introductions and recurrent founder effects during range expansion in invasive populations of Geranium carolinianum (Geraniaceae). Heredity 112:497–507.
263. Sikora, M., M. L. Carpenter, A. Moreno-Estrada, B. M. Henn, P. A. Underhill, F. Sanchez-Quinto, I. Zara et al. 2014. Population genomic analysis of ancient and modern genomes yields new insights into the genetic ancestry of the Tyrolean iceman and the genetic structure of Europe. PloS Genetics 10:12.
264. Simberloff, D. 2009. The role of propagule pressure in biological invasions. Annual Review of Ecology Evolution and Systematics 40:81–102.

265. Simberloff, D., J. L. Martin, P. Genovesi, V. Maris, D. A. Wardle, J. Aronson, F. Courchamp et al. 2013. Impacts of biological invasions: what's what and the way forward. Trends in Ecology and Evolution 28:58–66.

266. Smith, M. A., and B. L. Fisher 2009. Invasions, DNA barcodes, and rapid biodiversity assessment using ants of Mauritius. Frontiers in Zoology 6:31.

267. Smith, J. M., and J. Haigh 2007. The hitch-hiking effect of a favourable gene. Genetics Research 89:391–403.

268. Smith, C. D., A. Zimin, C. Holt, E. Abouheif, R. Benton, E. Cash, V. Croset et al. 2011. Draft genome of the globally widespread and invasive Argentine ant (Linepithema humile). Proceedings of the National Academy of Sciences of the USA 108:5673–5678.

269. Smith, S., L. Bernatchez, and L. B. Beheregaray 2013. RNA-seq analysis reveals extensive transcriptional plasticity to temperature stress in a freshwater fish species. BMC Genomics 14:12.

270. Sork, V. L., S. N. Aitken, R. J. Dyer, A. J. Eckert, P. Legendre, and D. B. Neale 2013. Putting the landscape into the genomics of trees: approaches for understanding local adaptation and population responses to changing climate. Tree Genetics and Genomes 9:901–911.

271. Spear, D., L. C. Foxcroft, H. Bezuidenhout, and M. A. McGeoch 2013. Human population density explains alien species richness in protected areas. Biological Conservation 159:137–147.

272. Stachowicz, J. J., J. R. Terwin, R. B. Whitlatch, and R. W. Osman. 2002. Linking climate change and biological invasions: ocean warming facilitates nonindigenous species invasions. Proceedings of the National Academy of Sciences of the USA 99:15497–15500.

273. Steane, D. A., B. M. Potts, E. McLean, S. M. Prober, W. D. Stock, R. E. Vaillancourt, and M. Byrne 2014. Genome-wide scans detect adaptation to aridity in a widespread forest tree species. Molecular Ecology 23:2500–2513.

274. Stewart, C. N. Jr, P. J. Tranel, D. P. Horvath, J. V. Anderson, L. H. Rieseberg, J. H. Westwood, C. A. Mallory-Smith et al. 2009. Evolution of weediness and invasiveness: charting the course for weed genomics. Weed Science 57:451–462.

275. Strauss, S. Y., J. A. Lau, and S. P. Carroll 2006. Evolutionary responses of natives to introduced species: what do introductions tell us about natural communities? Ecology Letters 9:354–371.

276. Stukely, M. J. C., and C. E. Crane 1994. Genetically based resistance of Eucalyptus marginata to Phytophthora cinnamomi. Phytopathology 84:650–656.

277. Sutherland, W. J., M. Clout, I. M. Côté, P. Daszak, M. H. Depledge, L. Fellman, E. Fleishman et al. 2010. A horizon scan of global conservation issues for 2010. Trends in Ecology and Evolution 25:1–7.

278. Sutherland, W. J., R. Aveling, L. Bennun, E. Chapman, M. Clout, I. M. Côté, M. H. Depledge et al. 2012. A horizon scan of global conservation issues for 2012. Trends in Ecology and Evolution 27:12–18.

279. Taberlet, P., E. Coissac, M. Hajibabaei, and L. H. Rieseberg 2012a. Environmental DNA. Molecular Ecology 21:1789–1793.

280. Taberlet, P., E. Coissac, F. Pompanon, C. Brochmann, and E. Willerslev 2012b. Towards next-generation biodiversity assessment using DNA-metabarcoding. Molecular Ecology 21:2045–2050.
281. Takahara, T., T. Minamoto, and H. Doi 2013. Using environmental DNA to estimate distribution of an invasive fish species in ponds. PLoS ONE 8:e56584.
282. Tatem, A. J. 2009. The worldwide airline network and the dispersal of exotic species: 2007-2010. Ecography 32:94–102.
283. Tatem, A. J., and S. I. Hay 2007. Climatic similarity and biological exchange in the worldwide airline transportation network. Proceedings of the Royal Society of London B 274:1489–1496.
284. Taylor, H. R., and W. E. Harris 2012. An emergent science on the brink of irrelevance: a review of the past 8 years of DNA barcoding. Molecular Ecology Resources 12:377–388.
285. Taylor, S. A., T. A. White, W. M. Hochachka, V. Ferretti, R. L. Curry, and I. Lovette 2014. Climate-mediated movement of an avian hybrid zone. Current Biology 24:671–676.
286. Telonis-Scott, M., B. van Heerwaarden, T. K. Johnson, A. A. Hoffmann, and C. M. Sgrò 2013. New levels of transcriptome complexity at upper thermal limits in wild Drosophila revealed by exon expression analysis. Genetics 195:809–830.
287. Teshler, M. P., S. A. Dernovici, A. Ditommaso, D. Coderre, and A. K. Watson 2004. A novel device for the collection, storage, transport, and delivery of beneficial insects, and its application to Ophraella communa (Coleoptera: Chrysomelidae). Biocontrol Science and Technology 14:347–357.
288. Thompson, J. D. 1991. The biology of an invasive plant - what makes Spartina anglica so successful. BioScience 41:393–401.
289. Thompson, G. D., M. P. Robertson, B. L. Webber, D. M. Richardson, J. J. Le Roux, and J. R. U. Wilson 2011. Predicting the subspecific identity of invasive species using distribution models: Acacia saligna as an example. Diversity and Distributions 17:1001–1014.
290. Thompson, G. D., D. U. Bellstedt, M. Byrne, M. A. Millar, D. M. Richardson, J. R. U. Wilson, and J. J. Le Roux 2012. Cultivation shapes genetic novelty in a globally important invader. Molecular Ecology 21:3187–3199.
291. Thornton, K. R., J. D. Jensen, C. Becquet, and P. Andolfatto 2007. Progress and prospects in mapping recent selection in the genome. Heredity 98:340–348.
292. Tian, C., P. K. Gregersen, and M. F. Seldin 2008. Accounting for ancestry: population substructure and genome-wide association studies. Human Molecular Genetics 17:R143–R150.
293. Treasure, A. M., and S. L. Chown 2014. Antagonistic effects of biological invasion and temperature change on body size of island ectotherms. Diversity and Distributions 20:202–213.
294. Treier, U. A., O. Broennimann, S. Normand, A. Guisan, U. Schaffner, T. Steinger, and H. Mueller-Schaerer 2009. Shift in cytotype frequency and niche space in the invasive plant Centaurea maculosa. Ecology 90:1366–1377.
295. Turchin, M. C., C. W. K. Chiang, C. D. Palmer, S. Sankararaman, D. Reich, and J. N. Hirschborn; GIANT Consortium 2012. Evidence of widespread selection on standing variation in Europe at height-associated SNPs. Nature Genetics 44:1015–1019.

296. Valdez-Moreno, M., C. Quintal-Lazama, R. Gómez-Lazano, and M. d. C. García-Rivas. 2012. Monitoring of an alien invasion: DNA barcoding and the identification of Lionfish and their prey on coral reefs of the Mexican Caribbean. PLoS ONE 7:e36636.

297. Valéry, L., H. Fritz, J.-C. Lefeuvre, and D. Simberloff 2008. In search of a real definition of the biological invasion phenomenon itself. Biological Invasions 10:1345–1351.

298. Van der Putten, W. H., J. N. Klirononos, and D. A. Wardle. 2007. Microbial ecology of biological invasions. The ISME Journal 1:28–37.

299. Van Kleunen, M., W. Dawson, D. Schlaepfer, J. M. Jeschke, and M. Fischer 2010. Are invaders different? A conceptual framework of comparative approaches for assessing determinants of invasiveness. Ecology Letters 13:947–958.

300. Van Wilgen, B. W. 2009. The evolution of fire and invasive alien plant management practices in fynbos. South African Journal of Science 105:335–342.

301. Vandepitte, K., O. Honnay, J. Mergeay, P. Breyne, I. Roldán-Ruiz, and T. De Meyer 2012. SNP discovery using Paired-End RAD-tag sequencing on pooled genomic DNA of Sisymbrium austriacum (Brassicaceae). Molecular Ecology Resources 13:269–275.

302. Vandepitte, K., T. De Meyer, K. Helsen, K. Van Acker, I. Roldan-Ruiz, J. Mergeay, and O. Honnay 2014. Rapid genetic adaptation precedes the spread of an exotic plant species. Molecular Ecology 23:2157–2164.

303. Veldtman, R., T. F. Lado, A. Botes, S. Proches, A. E. Timm, H. Geertsema, and S. L. Chown 2011. Creating novel food webs on introduced Australian acacias: indirect effects of galling biological control agents. Diversity and Distributions 17:958–967.

304. Vellinga, E. C., B. E. Wolfe, and A. Pringle 2009. Global patterns of ectomycorrhizal introductions. New Phytologist 181:960–973.

305. Vorsino, A. E., A. M. Wieczorek, M. G. Wright, and R. H. Messing 2012. Using evolutionary tools to facilitate the prediction and prevention of host-based differentiation in biological control: a review and perspective. Annals of Applied Biology 160:204–216.

306. Walther, G.-R., A. Roques, P. E. Hulme, M. T. Sykes, P. Pyšek, I. Kühn, M. Zobel et al. 2009. Alien species in a warmer world: risks and opportunities. Trends in Ecology and Evolution 24:686–693.

307. Wang, Z., M. Gerstein, and M. Snyder 2009. RNA-Seq: a revolutionary tool for transcriptomics. Nature Reviews Genetics 10:57–63.

308. Wang, X., D. Wheeler, A. Avery, A. Rago, J.-H. Choi, J. K. Colbourne, A. G. Clark, and J. H. Werren. 2013. Function and evolution of DNA methylation in Nasonia vitripennis. PloS Genetics 9:e1003872.

309. Waterston, R. H., K. Lindblad-Toh, E. Birney, J. Rogers et al. 2002. Initial sequencing and comparative analysis of the mouse genome. Nature 420:520–562.

310. Webber, B. L., and J. K. Scott 2012. Rapid global change: implications for defining natives and aliens. Global Ecology and Biogeography 21:305–311.

311. Weber, E., and B. Schmid 1998. Latitudinal population differentiation in two species of Solidago (Asteraceae) introduced into Europe. American Journal of Botany 85:1110–1121.

312. Weeks, A. R., Y. J. Fripp, and A. A. Hoffmann 1995. Genetic structure of Halotydeus destructor and Penthaleus major populations in Victoria (Acari: Penthaleidae). Experimental and Applied Acarology 19:633–646.

313. White, T. A., S. E. Perkins, G. Heckel, and J. B. Searle 2013. Adaptive evolution during an ongoing range expansion: the invasive bank vole (Myodes glareolus) in Ireland. Molecular Ecology 22:2971–2985.

314. Whitney, K. D., and C. A. Gabler 2008. Rapid evolution in introduced species, 'invasive traits' and recipient communities: challenges for predicting invasive potential. Diversity and Distributions 14:569–580.

315. Willi, Y., J. Van Buskirk, and A. A. Hoffmann 2006. Limits to the adaptive potential of small populations. Annual Review of Ecology, Evolution and Systematics 37:433–458.

316. Williams, C. K., and R. J. Moore 1989. Phenotypic adaptation and natural selection in the wild rabbit, Oryctolagus cuniculus, in Australia. Journal of Animal Ecology 58:495–507.

317. Wood, T. E., N. Takebayashi, M. S. Barker, I. Mayrose, P. B. Greenspoon, and L. H. Rieseberg 2009. The frequency of polyploid speciation in vascular plants. Proceedings of the National Academy of Sciences of the USA 106:13875–13879.

318. Wright, S. I., and B. S. Gaut 2005. Molecular population genetics and the search for adaptive evolution in plants. Molecular Biology and Evolution 22:506–519.

319. Wurm, Y., J. Wang, O. Riba-Grognuz, M. Corona, S. Nygaard, B. G. Hunt, K. K. Ingram et al. 2011. The genome of the fire ant Solenopsis invicta. Proceedings of the National Academy of Sciences of the USA 108:5679–5684.

320. Yamazaki, Y., N. Shimada, and Y. Tago 2005. Detection of hybrids between masu salmon Oncorhynchus masou masou and amago salmon O. m. ishikawae occurred in the Jinzu River using a random amplified polymorphic DNA technique. Fisheries Science 71:320–326.

321. Yeaman, S., and M. C. Whitlock 2011. The genetic architecture of adaptation under migration-selection balance. Evolution 65:1897–1911.

322. Yeaman, S., K. A. Hodgins, H. Suren, K. A. Nurkowski, L. H. Rieseberg, J. A. Holliday, and S. N. Aitken 2014. Conservation and divergence of gene expression plasticity following c. 140 million years of evolution in lodgepole pine (Pinus contorta) and interior spruce (Picea glauca x Picea engelmannii). New Phytologist 203:578–591.

323. Yoccoz, N. 2012. The future of environmental DNA in ecology. Molecular Ecology 21:2031–2038.

324. Yoccoz, N. G., K. A. Brathen, L. Gielly, J. Haile, M. E. Edwards, T. Goslar, H. Von Stedingk et al. 2012. DNA from soil mirrors plant taxonomic and growth form diversity. Molecular Ecology 21:3647–3655.

325. You, M., Z. Yue, W. He, X. Yang, G. Yang, M. Xie, D. Zhan et al. 2013. A heterozygous moth genome provides insights into herbivory and detoxification. Nature Genetics 45:220–225.

326. Yu, D. W., Y. Ji, B. C. Emerson, X. Wang, C. Ye, C. Yang, and Z. Ding 2012. Biodiversity soup: metabarcoding of arthropods for rapid biodiversity assessment and biomonitoring. Methods in Ecology and Evolution 3:613–623.

327. Yun, L., S. R. Larson, I. W. Mott, K. B. Jensen, and J. E. Staub 2014. Genetic control of rhizomes and genomic localization of a major-effect growth habit QTL in perennial wildrye. Molecular Genetics and Genomics 289:383–397.
328. Zhang, W., X. Fan, S. Zhu, H. Zhao, and L. Fu 2013a. Species-specific identification from incomplete sampling: applying DNA barcodes to monitoring invasive Solanum plants. PLoS ONE 8:e55927.
329. Zhang, J., P. Kapli, P. Pavlidis, and A. Stamatakis 2013b. A general species delimitation method with applications to phylogenetic placements. Bioinformatics 29:2869–2876.
330. Zhang, Y.-Y., M. Fischer, V. Colot, and O. Bossdorf 2013c. Epigenetic variation creates potential for evolution of plant phenotypic plasticity. New Phytologist 197:314–322.
331. Zhang, B., O. Edwards, L. Kang, and S. Fuller 2014. A multi-genome analysis approach enables tracking of the invasion of a single Russian wheat aphid (Diuraphis noxia) clone throughout the New World. Molecular Ecology 23:1940–1951.
332. Zhu, Y. Y., H. R. Chen, J. H. Fan, Y. Y. Wang, Y. Li, J. B. Chen, J. X. Fan et al. 2000. Genetic diversity and disease control in rice. Nature 406:718–722.
333. Ziska, L. H., D. M. Blumenthal, G. B. Runion, E. R. Hunt, and H. Diaz-Soltero 2011. Invasive species and climate change: an agronomic perspective. Climatic Change 105:13–42.

CHAPTER 4

Physiological and Genomic Basis of Mechanical-Functional Trade-Off in Plant Vasculature

SONALI SENGUPTA AND ARUN LAHIRI MAJUMDER

4.1 INTRODUCTION

A green plant is unique in its hydraulic architecture. Hydraulic conductivity of the xylem is closely linked to the minimum leaf area, which it must supply with water and nutrients for survival. Hydraulic conductivity, as quantified by Zimmermann (1974), is generally measured as leaf specific conductivity (flow rate per unit pressure gradient) divided by the leaf area supplied by the xylem pipeline segment. This measure is a key for quick evaluation of pressure gradients within a plant. Modeling the functional and natural architecture of plant water flow pipeline takes more traits in consideration than merely the physical attributes of a mechanical pump. The contribution of living cells and more specifically, genes and proteins, for maintenance of the "green pump" remains largely unaddressed.

Physiological and Genomic Basis of Mechanical-Functional Trade-Off in Plant Vasculature. © Sengupta S and Majumder AL. Frontiers in Plant Science 5,224 (2014), doi: 10.3389/fpls.2014.00224. *Licensed under a Creative Commons Attribution 3.0 Unported License, http://creativecommons.org/licenses/by/3.0/.*

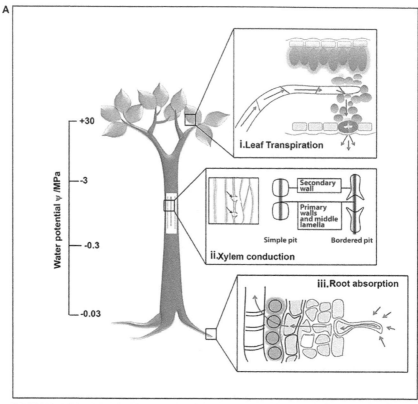

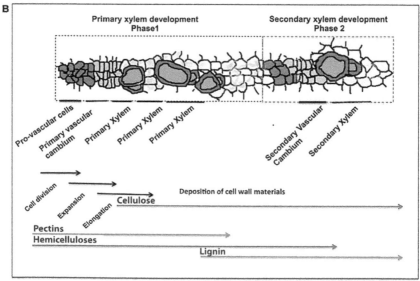

FIGURE 1: (A) The soil-plant-air continuum functioning in maintenance of water transport column. The plant root takes up water from soil, and the water column is maintained continuous along the xylem. The continuity across the xylem vessel is maintained by several intrinsic physical properties of water, input from the adjoining living cells and transpirational pool. The rough estimate of pressure along the vascular cylinder is presented in the scale bar (image not to actual scale). (B) A schematic of xylogenesis, adapted and modified from Hertzberg et al., 2001. The two phases of xylem development (primary and secondary); and the tissues involved in the process are shown within respective dotted boxes. The biological processes (cell division, expansion, elongation, deposition of cell wall) involved are shown by black arrows, under corresponding tissue types. The cell wall materials that are deposited are also shown under corresponding tissue types during xylogenesis. The order of such differentiation may be traced from left to right in the figure, though their actual time frame may differ from species to species.

Several theories have been proposed to explain ascent of sap. The operation of the green pump is simple yet elegant and is best described by the Cohesion-Tension Theory (CTT) (Dixon, 1914) but also synthesized from the work of many scientists over the last few decades. Besides physical explanations, the living parenchyma cells around xylem were originally proposed to be of importance by Bose (1923) in his pulsation theory. Later, the living xylem parenchyma cells indeed proved of high importance for the continuous ascent of sap.

The major governing factors are the physical properties of aqueous solution, means of transport and xylem anatomy, consideration of all of which makes the "sap conducting system" comparable to basic hydraulic systems such as pumps and irrigations in household or human blood vasculature. Components of such system are mainly (i) a driving force, (ii) a pipeline system, (iii) a reservoir and other regulating factors. To establish a soil-water-atmosphere continuum, an uninterrupted "water network" is necessary, which is built in the plant where transpirational evaporation is the driving force (Figure 1A). The evaporation of water from the porous green tissue surface creates a capillary pull in the water menisci (Figure 1Ai) and a curvature is induced in them, which is sufficient to support a huge water column against gravity in the stem and root vascular cylinder (Figure 1Aii). The water reservoir is the soil, wherefrom the root draws its supply (Figure 1Aiii). The empirical Jurin law says that a menisci radius of 0.12 μm can support a column of 120 m (Sperry, 2013; Zimmermann, 1983). The pull creates sub-atmospheric pressure in the xylem vessels. As the height of a plant increases, the water potential drops, and it is expected that leaves, twigs and upper extremities will display a 10–1000 times drop of pressure (Figure 1A, Tyree and Sperry, 1989). Sixty five percentage of the water potential drop occurs in tree trunk xylem, with a 20% contribution from root and 14% from leaves (Tyree and Sperry, 1989). This explains why big tree trunks can survive severe localized damages near the base.

4.2 PLANT ARCHITECTURE AND THE GREEN PUMP

Architecture of a plant is defined by its height, girth, woodiness, root system design and shoot disposition. Such architecture varies across the plant kingdom, along which varies the plants' hydraulic nature. Secondary thickening is a major player that governs the green pump. It has been shown that root pressure plays little or no part in maintenance of this column in woody plants. Severing the root may not hamper upward movement of water, if there is a direct supply to the vessels; however leaves are necessary. Even the best vacuum pump is able to pull water to not more than 10.4 m, considering that a Sequoia tree may have to pull water up to 100 m. However, in the monocots, root pressure is considered to be a major player of sap pull.

Considering the physical properties of green-pump, cavitation and embolism are major threats to the water column in xylem and subsequently, to survival, across the kingdom. To successfully transport water and minerals from soil to leaf, existing pressure in xylem conduits needs to remain sub-atmospheric (negative), in contrast to animal system where long distance transport is actively under positive pressure. The molecular property of cohesion gives a high strength to water. Ultrapure water confined to tubes of very small bore will need a tension comparable to the strength needed to break steel columns of the same diameter. Cohesion imparts strength comparable to solid wires in a water column. The vice is: once air is introduced in such system, the column will snap apart. To prevent such snapping, xylem properties play an important role.

4.3 PHYSIOLOGY OF XYLOGENESIS: THE BIPHASIC DEVELOPMENT IN XYLEM

The biphasic development of xylem in plants is critical to understand the hydraulic architecture as well as the air-water-soil continuum (Figure 1B). Procambium develops into xylem precursor cells that eventually differentiate into xylem fiber cells, xylem parenchyma, and tracheary elements, consisting of vessels and tracheids in the first phase. The second phase deposits secondary xylem walls onto the primary xylem walls (Fukuda, 1997; De Boer and Volkov, 2003), derived from vascular cambium and made of cellulose microfibrils impregnated with lignin, structural proteins, hemicellulose and pectin (Figure 1B, Ye, 2002; Fukuda, 2004; Yokoyama and Nishitani, 2006). Prior to secondary development, the tracheary components elongate and with the advent of secondary wall deposition, the cellular components in the living tracheid undergo programmed cell death (Fukuda, 2004) living only the hollow pipeline (Fukuda, 1997; Zhang et al., 2011) composed of vessels interconnected by pits (De Boer and Volkov, 2003; Choat and Pittermann, 2009). The paired pits are often bordered (Figure 1A); from secondary deposition forming two overarched secondary walls, in between which a fine pit membrane with small pores persist. Pit membranes are made up of meshes of polysaccharide (Tyree and Zimmermann, 2002; Pérez-Donoso et al., 2010) and allow axial passage of water and small molecules. Besides, they act as safety protection against spread of air seeds (Tyree and Zimmermann, 2002; De Boer and Volkov, 2003; Choat et al., 2008; Pérez-Donoso et al., 2010).

4.4 PHYSIOLOGY OF CAVITATION

The negative pressure in the xylem may descend low enough to make the water metastable. To achieve non-disrupted flow in such system, water must remain liquid below its vapor pressure. This metastable state induces nucleation of vaporization, or cavitation. Cavitation is the introduction of air spaces into the continuous water column and under physical metastable state water is prone to form air bubbles easily. Introduced in a xylem lumen, air cavities rupture the water column and in its worst, block the transport of water and minerals to the leaf. This blockage is known as "embolism" and may lead the plant to a lethal fate.

Cavitation is known to occur in plants frequently. Paradoxically, occurrence of cavitation is the strongest support for CTT. It is only natural to observe cavitation if water is under such negative pressure. The root vessels of field grown, well watered maize plants have been known to embolize daily and then refill. Vessels that were filled by dawn may embolize at mid-afternoon and by sunset they are again refilled (McCully et al., 1998). When transpiration rate is high and water scarcity is at bay, trees display cavitation, which means that embolism can well be induced by water stress. Large metaxylem vessels show a higher rate of embolism, and evidence suggest that water stress-induced embolism is of the frequent most sort (Tyree and Sperry, 1989). It is a prerequisite for cavitation that some vessels are embolized to start with; which is met by bubbles introduced in some of the vessels by mechanical damage, harbivory and insect attack.

4.5 STRESS-INDUCED EMBOLISM IN PLANTS

Both abiotic and biotic stresses can induce embolism in a plant. Drought and frost—induced embolisms are most prevalent, while mechanical stress and pathogen-induced damage are often the primary inducers.

Desert plants and dry-season crops are most threatened by drought-induced embolism. Air-seeding increases during drought as the sap pressure becomes increasingly negative due to high suction. The evaporation from leaf surface increases and the porous conduit wall may release air inside the functional conduits. They behave as nucleation centers and cause the sap pressure to increase to atmospheric level. The bubble is then likely to start an embolism that fills up the diameter of conduit, as the surrounding water is pulled up by transpiration.

Interconduit pit membranes with nano-scale pores normally restrict passage of air bubble from affected to functional conduits but at a high pressure difference they fail to stop the propagation. The rate of this propagation is important to measure the cavitation resistance in a plant.

Freezing is another cause of embolism, specially in woody temperate species. Freeze-thaw cycles may lead to 100% loss of water transport due to embolism in some species (Scholander et al., 1961). The primary governing factor in damage intensity seems to be the mean diameter of the conduits. Smaller vessel diameters are more vulnerable to damage.

Frost-induced air seeding is caused by segregation of gas by ice. There is a certain amount of salting out from the sap during freezing of sap, and if the salts are not able to move through the walls, they raise the osmotic pressure of remaining solution (Sevanto et al., 2012). This embolism can be more severe if there is functional drought prevailing. Freezing-induced embolism is a primary stress in forests where seasonal freeze-thaw is observed. Herbaceous plants, on the other hand, hardly survive freezing and are mostly at threat from drought-induced embolism.

Vascular wilt pathogens can wipe out entire crop. It is known that vascular pathogens induce water stress in their hosts; but can embolism be a cause of such stress? All vascular wilt pathogens break into rigid secondary xylem walls to enter the vessels as well as the pit membranes. Generally vascular wilt pathogens or their spores and conidia are too large to pass through pit membrane pores (Mollenhauer and Hopkins, 1974; Choat et al., 2003, 2004; Qin et al., 2008). Even when they manage to break into the vessel the milieu is not friendly. The microenvironment of xylem pipeline is nutritionally very poor and the pathogens surviving in xylem niche are not too many in number. It is speculated that they prefer this environment to minimize competition. Nevertheless, fungal and bacterial pathogens can extract the little amount of ions and nutrients available in the xylem stream and are able to break through and digest secondary wood to leech nutrition from living cells. Doing so, they weaken the pressurized cell wall and their infestation within the dead pipeline makes the water stream reactive and prone to cavitation. They may as well block the vessels and pit membranes, occluding parts of functional conduit network.

There is also an internal mechanical stress associated with ascent of sap. The high negative tension within the xylem pipeline causes an inward pool. Depending on the sapwood elasticity, there is a daily diameter change of tree trunk correlated to transpiration and daylight. In Scots pine, Perämäki et al. (2001) described daily changes in the sapwood diameter. The pull causes pressure on a stem surface element directed toward the center of the stem and the tracheal structure resists the movement of the surface element. The mechanical strength of the tracheary wall and its composition is, hence, an important factor in maintaining normal xylem activity as is the plasticity of pit membrane structure and composition.

4.6 VULNERABILITY OF XYLEM TO CAVITATION

Xylem seems to be vulnerable to cavitation in many different ways. This vulnerability can vary depending on the species, season, and availability, state and temperature of water. Broadly, the vulnerability of plants to cavitation is often plotted on xylem vulnerability curves, which is a function of decline in xylem hydraulic conductivity due to increasingly negative xylem pressure. Such declines are typically expressed relative to the maximum decline possible as the Percentage Loss of Conductivity (PLC). Comparisons of the vulnerability to cavitation among species are made using the xylem pressure at 50% loss of conductivity (P_{50}) with the traditional plotting of vulnerability curve (Meinzer and McCulloh, 2013). There remain controversies related to the techniques used for measurement of vulnerability described elsewhere in details (McElrone et al., 2012; Cochard et al., 2013; Wheeler et al., 2013).

The vulnerability curve for a number of tree species, as put forward by Tyree et al. (1999) shows a typical exponential shape, indicating that sub-zero pressure is a direct inducer of cavitation. This makes cavitation a regular process and necessitates a resistance mechanism in plants. It has also been claimed that cavitation is rapidly repaired by a miraculous mechanism (Holbrook and Zwieniecki, 1999) known as "refilling." We can thus categorize cavitation resistance under two proposed mechanisms; one, by refilling the air bubbles efficiently; and two, by modulating pit membrane properties. The possible genetic controls of both are worthy of discussion.

4.7 CAVITATION RESITANCE BY REFILLING: A QUESTIONABLE TRAIT

The removal of air seeds from lumen to turn a non-functional vessel to functional is known as refilling. The idea, though widely observed, recently was confronted with a serious doubt voiced by the plant hydraulic scientists. The long-established experimental procedure that has been followed to measure cavitation has been pronounced faulty (Sperry, 2013). It has been claimed that the standard procedure of xylem hydraulic conductivity measurement, by excising the stem under water to avoid air aspiration in the open conduits, is not a valid observation procedure. It has been suggested that in many species, significant amount of cavitation is introduced even when the stem is cut under water. The consequences of this artifact on previous datasets were significant, as it may be reflected in all vulnerability to cavitation curves obtained in other species for a long period of time; and perturb our analysis of refilled vessels.

However debatable the issue may be, recent high resolution and real-time imaging studies (Holbrook et al., 2001; Windt et al., 2006; Scheenen et al., 2007; Brodersen et al., 2010) also satisfy the requirements of the hypothesis that plant has some kind of resistance strategies to protect itself from embolism. It has been proposed that plants have an osmotically driven embolism repair mechanism and existing rehydration pathways through the xylem. The mechanisms were predicted to be largely of two types: (i) "novel" refilling, a refilling mechanism without "positive root pressures, even when xylem pressures are still substantially negative"; (ii) root pressure aiding the refilling of vessels raising the pressure inside vessels near atmospheric (Salleo et al., 1996; Holbrook and Zwieniecki, 1999; Tyree et al., 1999; Hacke and Sperry, 2003; Stiller et al., 2005). The first type is common among woody dicots whereas evidence of the second type is common among annual herbaceous species.

A

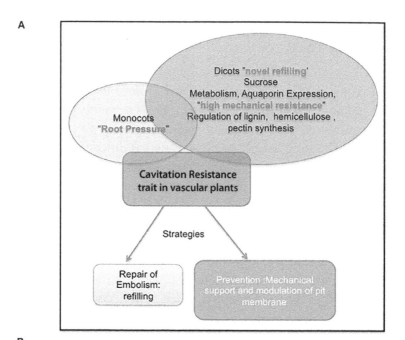

B

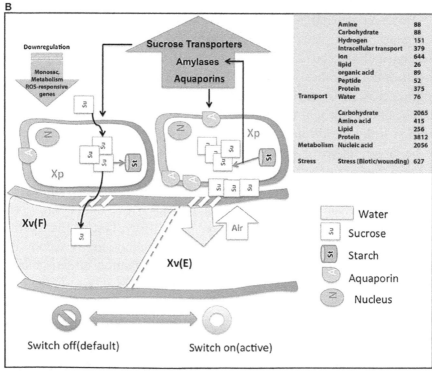

Figure 2. (A) The strategies of vascular plant in a battle against embolism. Monocots often employ root pressure, while dicots employ novel refilling mechanism, and mechanical resistance to resist cavitation. There is no clear demarcation between the strategies employed by the two groups, and the strategies may overlap. (B) The sugar sensing model of embolism refilling process, modified from Secchi et al. (2011). For detail explanations of the model, refer text and Secchi et al. (2011). Briefly, when vessels are filled and functional, a default "switch off" mode is active. Sucrose is continuously transported from accompanying xylem parenchyma cells into the vessels. Cavitation induces a "switch on" mode of sensing. When a vessel is filled with air, free passage of sucrose to the vessel lumen is hindered, and the sucrose molecules are deposited on vessel wall. This, with a positive feedback loop generates a cascade of high starch to sucrose conversion (Bucci et al., 2003; Salleo et al., 2004; Regier et al., 2009). The increased sucrose pool would be maintained by upregulation of amylases and sugar transporters. The genes up/downregulated during the sensing process are mentioned in the figure. Abbreviations used: Xv(F), Xylem Vessel Filled; Xv(E), Xylem Vessel Embolized; Xp, Xylem Parenchyma. Other abbreviations are explained in the figure.

4.8 GENETIC CONTROL OF REFILLING MECHANISM

Bay leaf tree, *Laurus nobilis* is an aromatic shrub in which mechanism of refilling is proposed to be linked to starch to sugar conversion. Reserve carbohydrate depletion from xylem parenchyma induces phloem unloading in a radial manner via ray parenchyma (Salleo et al., 2009; Nardini et al., 2011). Xylem-phloem solute exchange has been found to occur along both symplastic and apoplastic paths (Van Bel, 1990). It has been hypothesized that solutes might move radially along the ray cell walls, enter the embolized xylem conduits and increase the solute concentration of the residual water within them, thus promoting xylem refilling by altering osmoticum. The role of xylem parenchyma in refilling is significant. Lianas, shrubs and vine fibers are often observed to have living protoplasts and starch granules (Fahn and Leshem, 1963; Brodersen et al., 2010). Repeated cycles of embolism and repair are correlated to cyclic depletion of starch in xylem during drought (Salleo et al., 2009; Secchi et al., 2011). Debatably, repeated cycles of embolism formation and repair may disable the refilling mechanism and ultimately lead to carbon starvation (Sala et al., 2010, 2012; McDowell, 2011). The hydrolyzed starch movement from xylem is yet unresolved.

Water stressed *Populus trichocarpa* plants revealed an upregulation of ion transporters, aquaporins, and carbon metabolism related genes (Secchi et al., 2011; Secchi and Zwieniecki, 2012). A putative sucrose-cation co-transporter may aid the refilling process as suggested by the chemical profiling of vessel lumen. Grapevine refilling petioles show strong upregulation of carbon metabolism and aquaporin expression (Perrone et al., 2012).

A basic assumption is made that in dicots, to enhance refilling ability trait, one might target carbohydrate metabolizing genes in a localized manner to improve sucrose release. Sucrose may be used as an osmoticum inside non-functional lumens or may be used as energy currency. Localization of increased aquaporins (PIPs and TIPs) within axial parenchyma surrounding conduits may prove important. It is now proved by imaging studies (Brodersen et al., 2010) that living cells play a central role in embolism refilling and restoring transport, and by further prevention of air seed and pathogen by sealing off conduits with tyloses. Further detailed work is needed to identify the stress signals that mediate talk between xylem vessels and parenchyma.

In monocots, root pressure is the most important mechanism for refilling reported till date. Grasses exhibit root pressure more often, and with the increase of plant height the basal root pressure increases (Cao et al., 2012). Monocots do not exhibit secondary thickening and ray cells thus the osmoticum and sucrose transport theory do not apply to monocots (Andre, 1998). Selection for root pressure in these species solves the embolism repair problem and negates the need for carbohydrate transport along the pathway common in woody angiosperms (Brodersen et al., 2013). However, Stiller et al. (2005) showed the presence of "novel" refilling in rice in presence of high negative pressure and suggested that in upland or low-rainfed rice this mechanism can serve side by side of a positive root pressure. Root pressure may involve a stronger mechanical tissue, and whether or not any trade-off between safety and efficiency is involved is unclear. Study of more vascular function mutants in monocot crops may resolve the genes involved in this process.

4.9 GENOMIC PERSPECTIVE: GENES, PROTEINS AND MODELS IMPLICATED IN REFILLING

The battle with cavitation is fought either with efficient refilling or fine structural modulation of pit membrane and strength of vascular cylinder wall. The genomic, transcriptomic and proteomic studies may thus come under two broad sections: genomic basis of refilling and genomic basic of mechanical strength (Figure 2A).

4.10 GENOMIC BASIS OF REFILLING

The process of refilling or repair of embolism requires pumping water in an air-filled cavity. Physically this will require an empty or air-filled vessel, functional neighbor vessels, a source of energy to drive the refilling and a source of water to refill. In the previous sections, the physical and physiological components of embolism repair have been discussed in detail. However, a reductionist biologist looks further beyond for the possible identities of molecular candidates that repair the non-functional vessel. It is hypothesized that refilling is a result of an intricate interaction of xylem parenchyma, (even possibly phloem), vessel wall chemistry, and the composition and flexibility of pit membranes (Holbrook and Zwieniecki, 1999). The signals that are sensed when embolism occurs and the cascades that follow the primary signal transduction event, involve interconnected molecular regulators; that has been subject of several studies. The most recent model of refilling puts forward a role of sugar signaling in embolism sensing and refilling mechanism, the involved gene families being Aquaporins, Sucrose transporters and enzymes related to starch breakdown, Alpha and Beta Amylase (Secchi and Zwieniecki, 2010).

4.11 AQUAPORINS

Aquaporins are conservedly implicated in the refilling process of angiosperms and gymnosperms from the very beginning. The refilling of vessels in *Populus trichocarpa* is accompanied by selective upregulation of PIPs (Plasma Membrane Intrinsic Proteins). Secchi et al. (2011) proposed that the sensing of embolism and accomplishment of refilling is mediated by sugar signals, specifically sucrose. According to their proposed model, when a vessel is filled with air, free passage of sucrose to the vessel lumen is hindered, and the sucrose molecules are deposited on vessel wall. This, with a positive feedback loop generate a cascade of high starch to sucrose conversion (Bucci et al., 2003; Salleo et al., 2004; Regier et al., 2009). The increased sucrose pool would be maintained by upregulation of amylases and sugar transporters. Secchi et al. (2011) showed a distinct upregulation in aquaporins and sucrose transporter (*PtSuc 2.1*) in air injected or artificially high osmotica-treated vessels. *Ptsuc2.1* shows a high homology to walnut sucrose transporter, which, on upregulation is able to relieve freeze-thaw induced embolism (Decourteix et al., 2006). The increased sucrose and the upregulation of aquaporins are correlated spatially and temporally, but connections are difficult to establish. The model hence proposed is schematically represented in Figure 2B. Almeida-Rodriguez et al. (2011) showed a gene expression profile of 33 Aquaporins in fine roots of hybrid poplar saplings and compared light and high transpiration induced vascular hydraulics physiology with respect to Aquaporin expression. Dynamic changes were observed in expression pattern of at least 11 aquaporins from poplar; and some of them were localized in the root tissue. In *Arabidopsis*, Postaire et al. (2010) showed that, hydraulic conductivity of excised rosettes and roots are correlated wih expression of aquaporins. AtPIP1; 2, AtPIP2;1, and AtPIP2;6 are the most highly expressed PIP genes in the *Arabidopsis* rosette (Alexandersson et al., 2005) and under long night, AtPIP1;2 knockout plants loose 21% hydraulic conductivity in the rosette(Postaire et al., 2010). The disturbed hydraulics phenotype is a genetic dissection of the direct relation between aquaporin expression and plant water transport; although there may be components other than Aquaporin that may serve an important role (Sack and Holbrook, 2006; Heinen et al., 2009). It has been shown in hybrid poplar *Populus trichocarpa* × *deltoides*, increasing evaporation from leaf surface and perturbed hydraulics is correlated with high aquaporin expression (Plavcová et al., 2013). In common grapevine, *Vitis vinifera* L. (cv Chardonnay) inhibitors of aquaporin-mediated transport greatly affects both leaf hydraulic conductance and stomatal conductance (Pou et al., 2013). Of 23–28 Aquaporin isoforms in grapevine, a subset including VvPIP2;2, VvTIP1;1 plays important role during early water stress, while VvPIP2;1, VvPIP2;3, VvTIP2;1 are highly expressed during recovery(Pou et al., 2013). In Maize roots, radial water transport are diurnally regulated by proteins from the PIP2 group (Lopez et al., 2003). It is evident, though, that not all aquaporins participate in the refilling process. The sugar signal initiation is one important component; as originally described by Secchi et al. (2011) and must induce embolism-related aquaporin isoforms. The transcriptomic studies show that a very high number of Carbohydrate Metabolism related genes were upregulated during embolism (Secchi et al., 2011). Upregulation of the disaccharide metabolism gene group was observed, along with downregulation of monosaccharide metabolism; that suggests an accumulation of sucrose pool on the vessel wall (Secchi et al., 2011). Further upregulation of ion transporters and downregulation of carbohydrate transporters build up an osmoticum inside the cell to facilitate efflux of water. Figure 2B (inset) shows a summary of the number of gene categories showing differential expression during embolism (Secchi et al., 2011). The energy required for the pumping in comes from starch hydrolysis and one can presume, xylem specific isoforms of aquaporin, Starch synthetase and sucrose transporters will be highly expressed during refilling in plants. For critical evaluation of the model parameters, and its feasibility across the plant kingdom we extracted all aquaporin gene sequences from *Arabidopsis* and the *Arabidopsis* homologs of *Populus trichocarpa* sucrose transporters and amylases implicated in embolism Secchi et al., 2009, 2011; Secchi and Zwieniecki, 2010, 2012, 2013, 2014. The accession numbers of the fetched *Arabidopsis* genes are presented in Tables 1A,B. We subjected the gene sequences to protein-protein interaction network interaction analysis in String software in Expasy, without suggested functional neighbors (Szklarczyk et al., 2010). Generated interaction network for *Arabidopsis* gene subsets (mentioned in Table 1) clearly shows three interaction network clusters, connected to each other (Figure 3), the middle cluster (termed 'a' in Figure 3) shows evidenced network of PIPs as well as a RD28, dehydration stress related protein. Two other clusters (b and c in Figure 3) exhibit sucrose transporters and NIPs. Amylases form an un-joined node (d in Figure 3). We further localized the genes in *Arabidopsis* publicly available transcriptome analysis database in different tissues and observed shared enrichment in root endodermis, cortex and stele using e-northern (Figure 4A, Toufighi et al., 2005). A co-expression profile (Figure 4B) was obtained using string software, and the common n-mers present in the genes to induce a co-expression in certain tissues has been analyzed using promomer tool (Figure 4C; Table 2, Supplementary Table 1, Toufighi et al., 2005). Many of the enriched cis-elements contribute to dehydration and sugar stress. Overall, the genomic and transcriptomic data and candidate-gene based data emphasizes the high probability of sugar sensing of embolism. Secchi and Zwieniecki (2014) also showed that in hybrid poplar, downregulation of PIP1 delimits the recovery of the plant from water-stress-induced embolism, and thus is probably manages the vulnerability of xylem in negative pressure under control condition. The sugar content in the plant tissue strengthens the view further (Secchi and Zwieniecki, 2014).

TABLE 1A: Genes, families and members important in refilling experimentally reported in Populus trichocarpa.

Gene families	Specific genes				
	Family	Subfamily	Gene name	JGlv2.0 annotation name	*Arabidopsis* homologs
Aquaporins	PIP (Plasma Intrinsic Protein)	PoptrPIP1	PoptrPIP1.1	POPTR_008s06580	For analysis, the entire aquaporin family of *Arabidopsis* has been used instead of only specific homologs, refer to Table 1B
			PoptrPIP1.2	POPTR_003s12870	
			PoptrPIP1.3	POPTR_0010s19930	
			PoptrPIP1.4	POPTR_0006s09920	
			PoptrPIP1.5	POPTR_0016s12070	
		PoptrPIP2	PoptrPIP2.1	POPTR_0006s09910	
			PoptrPIP2.2	POPTR_0009s13890	
			PoptrPIP2.3	POPTR_0004s18240	
			PoptrPIP2.4	POPTR_0016s09090	
			PoptrPIP2.5	POPTR_0010s22950	
			PoptrPIP2.6	POPTR_0006s12980	
			PoptrPIP2.7	POPTR_0008s03950	
			PoptrPIP2.8	POPTR_0009s01940	

TABLE 1A: *Cont.*

Gene families	Specific genes				
	Family	Subfamily	Gene name	JGIv2.0 annotation name	*Arabidopsis* homologs
Alpha-beta amylases	Alpha-amylase	PoptrA-MY	PtAMY1	POPTR_0515s00220	AT4G25000
			PtAMY2	POPTR_0002s01570	AT1G76130
			PtAMY3	POPTR0010s10300	AT1G69830
	Beta amylase	Pop-trBMY	PtBMY1a	POPTR_0008s17420	AT3G23920
			PtBMY1b	POPTR_0001s11000	AT3G23920
			PtBMY2	POPTR_0003s10570	AT5G45300
			PtBMY3	POPTR_0008s20870	AT5G18670
			PtBMY4	POPTR_0003s08360	AT2G02860
			PtBMY5	POPTR_0017s06840	AT1G09960
Sucrose transporters	Sucrose transporter		PtSUC2.1	POPTR_0019s11560	AT5G55700
			PtSUT1.2	POPTR_0013s11950	AT4G15210
			PtSUT2.a	POPTR_0008s14750	AT1G22710

Gene id data compiled from Secchi et al. (2011); TAIR and phyrozome public database,

TABLE 1B: The entire aquaporin family in *Arabidopsis* extracted from TAIR.

Gene family name	Accession	TIGR protein type
Delta tonoplast integral protein family	At1g31880	Major intrinsic protein, putative
	At1g80760	Nodulin-like protein
	At1g73190	Tonoplast intrinsic protein, alpha (alpha-TIP)
	At2g45960	Aquaporin (plasma membrane intrinsic protein 1B)
	AT3g06100	Putative major intrinsic protein
	AT5g47450	Membrane channel protein-like; aquaporin (tonoplast intrinsic protein)-like
	AT3g53420	Plasma membrane instrinsic protein 2a
	At2g36830	Putative aquaporin (tonoplast intrinsic protein gamma)
	At2g37170	Aquaporin (plasma membrane intrinsic protein 2B)

TABLE 1B: *Cont.*

Gene family name	Accession	TIGR protein type
	At2g37180	Aquaporin (plasma membrane intrinsic protein 2C)
	AT4g35100	Plasma membrane intrinsic protein (SIMIP)
	At2g29870	Putative aquaporin (plasma membrane intrinsic protein)
	At1g01620	Plasma membrane intrinsic protein 1c, putative
	AT3g61430	Plasma membrane intrinsic protein 1a
	AT3g54820	Aquaporin/MIP-like protein
	At1g17810	Tonoplast intrinsic protein, putative
	AT3g47440	Aquaporin-like protein
	At2g16850	Putative aquaporin (plasma membrane intrinsic protein)
	At2g39010	Putative aquaporin (water channel protein)
	AT3g16240	Delta tonoplast integral protein (delta-TIP)
	At1g53180	Aquaporin, putative
	AT4g23400	Water channel-like protein
	At2g25810	Putative aquaporin (tonoplast intrinsic protein)
	AT4g00430	Probable plasma membrane intrinsic protein 1c
	AT5g37810	Membrane integral protein (MIP)-like
	AT5g37820	Membrane integral protein (MIP)-like
	AT4g17340	Membrane channel like protein
	AT4g10380	Major intrinsic protein (MIP)-like

4.12 TRANSCRIPTION FACTORS

The corregulation of sugar metabolism and water transport pathways require a complex transcriptional switch. Indeed, a large number of transcription factors control the refilling process, and they may regulate the diurnal pattern, the temporal accuracy and spatial distribution of the pathways involved. The role of TFs is shared; However, a look at the cis elements of pathway components may elucidate the nature of such sharing. The transcription factors important for xylogenesis and probably embolism are: AP2/EREBP, bZIP, C3HHD-ZIPIII, NAC, MYB, bHLH, WRKY, AP2/ERF, WRKY, HD, AUX/IAA, ARF, ZF, AP2, MYC, (*Arabidopsis*);

HD-ZIPIII, MYB, MADS, and LIM in *Populus*, MYB and Hap5a in Pine and HRT in *Hordeum* (Dharmawardhana et al., 2010). With the onset of genomic approaches, much more intensive analysis have been made possible. In a comprehensive genome-wide transcriptome analysis of *P. trichocarpa*, with snapshots from each elongating internode from a sapling stage (Internode1 through Internode11) a large number of differential representation of transcription factors have been obtained (Dharmawardhana et al., 2010). No less than 1800 transcription factors were readily detectable in at least one growth phase, of which, 439 are differentially regulated during xylogenesis (Dharmawardhana et al., 2010); some of which are represented in Table 3. Another study identified 588 differentially changed transcripts during shoot organogenesis in *Populus* (Bao et al., 2009, 2013). While the refilling process is majorly governed by sugar and dehydration signaling, NAC and Myb TF families remain singularly important in both xylem maturation and lignin biosynthesis. Aspects of xylogenesis that may be linked with mechanical-functional trade-off of vascular bundle revolve around lignin. There have been studies on genomics and transcriptomics of xylogenesis and secondary wood formation; however the genes responsible to maintain integrity of the vascular cylinder are not clearly known. In Supplementary Table 2, a comparative snapshot of some selected transcripts and emanating studies revealing the xylogenesis transcriptome in gymnosperms and angiosperms is provided. Several recent studies address the genomics of xylogenesis excellently; some of which are summarized in Table 4.

4.13 CAVITATION RESISTANCE INTRODUCED BY PIT MEMBRANE

The major key of cavitation resistance is pit membrane adaptation. To survive, ultrastructure of pit membrane needs to balance between minimizing vascular resistance and limiting invasion by pathogen and microbes. While the first is favored by thin and highly porous membrane, the later needs thick membrane and narrower pores. This calls for a trade-off between water transport function and biotic invasion resistance.

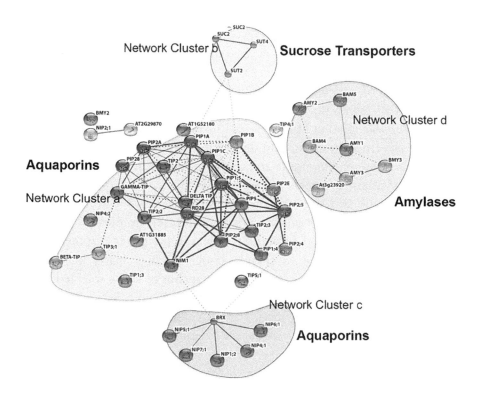

FIGURE 3: The protein-protein interaction network of Arabidopsis sucrose transporters, amylases and aquaporins, generated using String database. Thicker lines indicate stronger reaction (Szklarczyk et al., 2010).

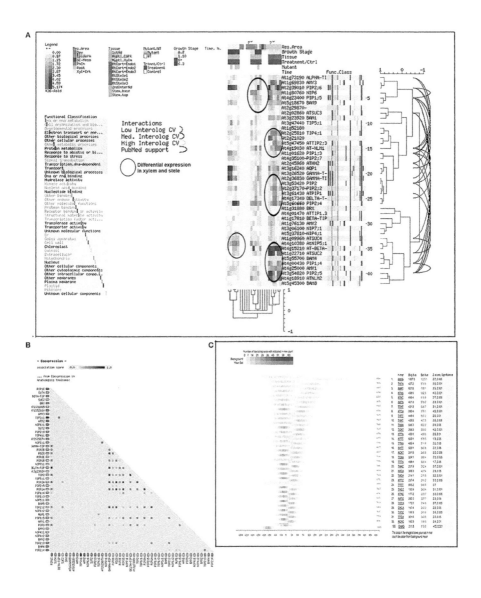

FIGURE 4: (A) Localization of the genes from Tables 1, 2 in various Arabidopsis tissue, from public microarray databases, and e-northern tool at Botany Array Resource (Toufighi et al., 2005). (B) Co-expression profile of the genes in Arabidopsis (Szklarczyk et al., 2010). (C) Distribution of relevant n-mers in the promoters of the above genes. That may induce shared expression. The results are generated using String and Promomer tools in Botany Array Resource (Toufighi et al., 2005). A tabulated form of the results are presented in Supplementary Table 5.

TABLE 2: Representative common n-mer details over represented in the embolism with respective transcription factors and their probable roles.

n-mers	Z-score	Regulation mode	Probable role	Consensus matches to n-mer in the PLACE 25.0.1 database
AAAT**	3.5	Positive	Dehydration responsive	Matched AAAT at offset 4 in CACTAAATTGTCAC 14BPATERD1: "14bp region" (from −599 to −566) necessary for expression erd1 (early responsive to dehydration) in dehydrated *Arabidopsis*
ATAA**	4.0	Positive	Sugar responsive	Matched ATAA at offset 2 in ACATAAAATAAAAAAGGCA −314MOTIFZMSBE1: located between −314 and −295 region of maize (Z.m.) Sbe1 gene promoterl critical positive cis element; important for the high-level, sugar-responsive expression of the Sbe1 gene in maize endosperm cells; recognized by nuclear proteins.
ATAT**	2.7	Positive/negative	MADS domain	Matched [AT][AT][AT][AT] at offset 5 in TTDCCWWWWWWWGGHAAAGAMOUSAT-CONSENSUS: binding consensus sequence of *Arabidopsis* (A.t.) AGAMOUS MADS domain
AATA	3.3	Positive	Sugar responsive	Matched AATA at offset 6 in ACATAAAATAAAAAAGGCA −314MOTIFZMSBE1: Located between −314 and −295 region of maize (Z.m.) Sbe1 gene promoter; critical positive cis element; important for the high-level, sugar responsive expression of the Sbe1 gene in maize endosperm cells; regognized by nuclear proteins
TTAT	3.1	Positive	Sugar responsive, binding activity to Myb core	Matched AATA at offset 6 in ACATAAAATAAAAAAGGCA −314MOTIFZMSBE1: Located between −314 and −295 region of maize (Z.m.) Sbe1 gene promoter; critical positive cis element; important for the high-level, sugar responsive expression of the Sbe1 gene in maize endosperm cells; regognized by nuclear proteins; matched TATT at offset 2 in TTTATTTACCAAACGGTAACATC23BPUASNSCYCB1: "23 bp UAS (Upstream activating sequence)" found in the promoter of *Nicotiana sylvestris* (N.s.) CycB1 gene; located between −386 and −409; contains a 5 bp element identical to the MYB binding core (ACGT); required for M-phase-specific expression; binds protein complexes in a cell cycle-regulate manner

TABLE 2: *Cont.*

n-mers	Z-score	Regulation mode	Probable role	Consensus matches to n-mer in the PLACE 25.0.1 database
ATCA**	4.5	Positive/negative	MADS domain, homeobox binding domain	Matches [AT][AT][ACGT][ACGT] at offset 8 in NTTDCCWWWWNNGGWAAN (AGAMOUS-like 1); AGL1 contains consensus sequences of *Arabidopsis* (A.t.) AGL1 domain; see S000339; AGL20 is a MADS domain gene from *Arabidopsis* that is activated in shoot apical meristem during the transition to flowering; AGL20 is also regulated by the Gibberellin pathway; complex regulatory networks involving several MADS-genes underlie development of vegetative structures
GAAG**	4.0	Positive	ABA-responsive, MADS	Matched GAAG at offset 6 in ATGTACGAAGC ABAREG2: motif related to ABA regulation; gene: sunflower helianthinin; transacting factor: bZIP? Matched [AT][ACGT] at offset 0 in NNWNCCAWWWWTRGWWAN AGL2ATCONSENSUS: binding consensus sequence of *Arabidopsis* (A.t) AGL2 (AGAMOUS-like 2); AGL2 contains MADS domain; AGL2 binds DNA as a dimer. Matched to CGAA at offset 5 in ATGTACGAAGC ABAREG2: motif related to ABA regulation; gene: sunflower helian-thin; transacting factor: bZIP?
CGAA	2.4	Positive	ABA-responsive	

*An html table for all n-mers is presented in Supplementary Table 1. **denotes overrepresentation.*

TABLE 3: Some representative transcription factors in Populus trichocarapa Xylem Maturation (Dharmawardhana et al., 2010).

WRKY family transcription factor
DRE binding protein (DREB1A)
Ethylene responsive element binding factor
Putative AP2 domain transcription factor
Ethylene responsive element binding factor 4 (aterf4,9)
Homeodomain-like protein 1
Auxin response transcription facotr (ARF1,9)
WRKY family transcription factor
ATPAO4 (POLYAMINE OXIDASE 4); amine oxidase
Ethylene-responsive transcriptional coactivator
Lateral root primordia (LRP1)
Transcription factor TINY, putative
MADS-box protein
Putative CCCH-type zinc finger protein
bHLH protein/contains helix-loop-helix DNA binding motif
Zinc finger protein Zat12
WRKY family transcription factor
BEL1-like homeobox 4 protein (BLH4)
TINY-like protein
Myb family transcription factor
Putative squamosa-promoter binding protein
Putative transcription factor/similar to transcription factor SF3
ES43 like protein/ES43 protein
AP2 domain protein RAP2.1
Abscisic acid responsive elements-binding factor (ABF3)
bHLH protein/contains helix-loop-helix DNA binding motif
Myb family transcription factor
CCAAT-binding transcription factor subunity A (CBF-A)

The thickness range of the pit membranes in the angiosperms is very broad, almost 70–1900 nm and so are the diameter of the pores (10–225 nm). Species with thicker pit membrane and smaller pores prevent seeding and embolism more successfully and thus may represent the group of species which has higher drought resistance.

Pit membrane porosity is not the only determinant of air bubble propagation among conduits. The other factor which serve equally important role is the contact angle between pit membrane and air water interface. This particular property is a direct function of pit membrane composition. The more hydrophobic the membranes are the more the contact angle and subsequently lower the pressure needed for air-seeding. Additionally, high lignin content, though required for mechanical strength, interrupt with the hydrogeling of pectins. Pectic substances can swell or shrink in presence or absence of water and thus they control the porosity of membranes. Polygalacturonase mutants in *Arabidopsis* showed a higher P50 value (−2.25MPa), suggesting a role for pectins in vulnerability to cavitation (Tixier et al., 2013). Mechanically stronger pit membranes thus may resist stretching and expansion of pore membranes indicating a compromise in function. Water stress has been reported to exhibit a direct relation to low lignin synthesis (Donaldson, 2002; Alvarez et al., 2008) although it is not known whether this low lignin help the water transport better.

TABLE 4: Representative transcriptome studies in literature

Xylogenesis	Embolism	Lignin biosynthesis
Li et al., 2013	Secchi et al., 2011	Hertzberg et al., 2001
Carvalho et al., 2013		Zhong et al., 2011
Pesquet et al., 2005		Lu et al., 2005
Li et al., 2012		Schrader et al., 2004
Dharmawardhana et al., 2010		
Karpinska et al., 2004		
Bao et al., 2009		
Rengel et al., 2009		
Mishima et al., 2014		
Plavcova et al., 2013		
Zhong et al., 2011		

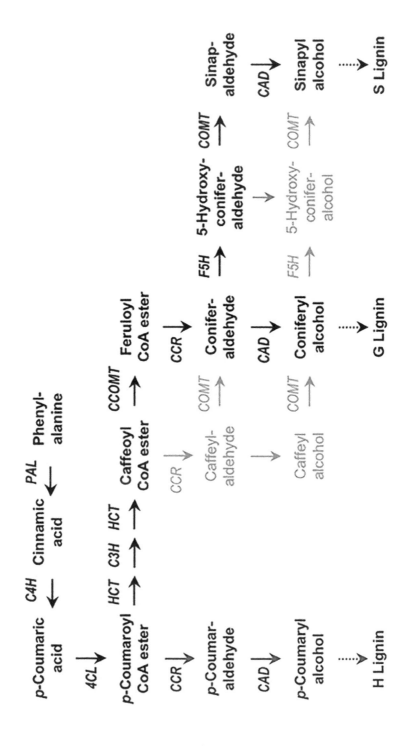

FIGURE 5: Simplified scheme for monolignol synthesis. The main pathway in dicotyledonous plants is highlighted in black, involving phenylalanine ammonia-lyase (PAL), cinnamate 4-hydroxylase (C_4H), 4-coumarate CoA ligase (4CL), p-hydroxycinnamoyl-CoA: quinate shikimate p-hydroxycinnamoyltransferase (HCT), p-coumarate 3-hydroxylase (C_3H), caffeoyl-CoA O-methyltransferase (CCOMT), hydroxycinnamyl-CoA reductase (CCR), ferulate 5-hydroxylase (F_5H), caffeate O-methyltransferase (COMT), and cinnamyl alcohol dehydrogenase (CAD). Alternate pathways are in light gray. H subunits are only minor lignin components in dicots. Adapted from Quentin et al. (2009).

4.14 SUGGESTED GENETIC BASIS OF CAVITATION RESISTANCE BY PIT MEMBRANE MODULATION AND MECHANICAL SUPPORT

Identification of genes and proteins behind the structural and mechanical controls of pit membrane formation has not progressed so far as repair mechanism of embolism is concerned. Genetic aspects of plant hydraulics are little studied, since most of the xylem studies are done in woody trees and study of herbaceous crops is rather scant. It is hard to obtain mutants in trees as the generation time is high, and the study process is long and laborious. Also, hydraulics in plants is not a simple structural or functional trait but is a complex physiological phenomenon. Figuring out the multi-trait control switch of this function is thus difficult.

4.15 CAN LIGNIN BIOSYNTHESIS BE CONSIDERED AS A CONTROL SWITCH?

Among the living cell processes that may take active part in controlling hydraulics, lignin biosynthesis is a major candidate and highly deciphered. In chemical nature, it is a polymer of phenylpropanoid compounds synthesized through a complex biosynthetic route (Figure 5; Hertzberg et al., 2001; Vanholme et al., 2010). Luckily enough, the genes on the metabolic grid are sequenced in plants like *Arabidopsis* and *Populus*, which is helpful to understand their modulation under stress. Till date, both biotic and abiotic stressors have been implicated in modulation of lignin biosynthesis, as well as seasonal, developmental and varietal changes (Anterola and Lewis, 2002; Zhong and Ye, 2009). Representing a large share of non-fossil organic carbon in biosphere, lignification provides mechanical support and defends the plant against pests and pathogens. The mechanical support, further, is mostly linked to xylem vessels and hydraulics.

Lignin is made from monolignols (hydroxy-cinnamyl alcohol), sinapyl alcohol, coniferyl alchol, and p-coumaryl alcohol in a smaller quantity. The complex metabolic grid and the transcriptional switches are described in details elsewhere (Hertzberg et al., 2001). The major metabolic pathway channeling into this grid is phenylpropanoid pathways through phenylalanine (Phe). Phe, synthesized in plastid through shikimic acid biosynthesis pathway, eventually generates p-coumaric acid by the activity Phenylalanine Ammonia-Lyase (PAL) and Cinnamate 4-Hydroxylase (C_4H). p-coumaric acid empties itself into the lignin biosynthesis grid to result into three kinds of lignin units; guaiacyl (G), syringyl (S), and p-hydroxyphenyl (H) units. Gymnosperm lignin polymer is majorly composed of G and H units, angiosperms show G and S units and H is elevated in compressed softwood and grasses (Boerjan et al., 2003).

There are stresses in nature that change plant lignin content. For example, lignin amount in *Picea abies* is predicted to correlate positively with annual average temperature (Gindl et al., 2000). Temperate monocots as well show an increase of lignin in response to increasing temperature (Ford et al., 1979). In *Triticum aestivum*, 2°C chilling stress decreases leaf lignin but increases in root is observed (Olenichenko and Zagoskina, 2005). Curiously, some studies have shown that although no changes in the levels of lignin or its precursors were observed in plants maintained at low temperatures, there was an increase in related enzyme activities as well as an increase in gene expression. Cold acclimatization in Rhododendron shows upregulation of C_3H, a cytochrome P450-dependent monooxygenase without further functional characterization (El Kayal et al., 2006). It has been argued that expression of C_3H could result in changes in the composition of lignin, altering the stiffness of the cell wall albeit without a definitive proof. The basal part of the maize roots show a growth reduction and low plasticity of cell wall associated with upregulation of two genes in lignin grid (Fan et al., 2006) in response to drought. The increase of free lignin precursors in the xylem sap and reduced anionic peroxidase activity in maize has been associated with low lignin synthesis in drought (Alvarez et al., 2008). It is possible that reducing lignin may directly affect the vascular tissue, encouraging water transport, lowering air seeding and increasing cavitation resistance; however it is not known what share of reduced lignin actually amount to stem vasculature, water column support and pit membrane plasticity.

4.16 BIOTECHNOLOGICAL MODIFICATION OF LIGNIN METABOLISM

With the advancement of genomic data, it is now possible to map the genetic changes which may influence hydraulic architecture. However, the model systems are questionable. Among the woody plant species, the genome of poplar has been sequenced; and the lignin biosynthesis network is fully characterized in *Arabidopsis* and rice. It is expected that change in lignin content may result differently in herbaceous and woody plants. There are controversial results obtained so far. In free-standing transgenic poplar trees, a 20–40% reduction in lignin content was associated with increased xylem vulnerability to embolism, shoot dieback and mortality (Voelker et al., 2011). Similarly the severe inhibition of cell wall lignification produced trees with a collapsed xylem phenotype, resulting in compromised vascular integrity, and displayed reduced hydraulic conductivity and a greater susceptibility to wall failure and cavitation (Coleman et al., 2008). A study on the xylem traits of 316 angiosperm trees in Yunnan, and their correlations with climatic factors claimed that wood density and stem hydraulic traits are independent variables (Zhang et al., 2013).

A weak pipeline and less lignification compromises vascular integrity as observed from the above results. On the other hand, low lignin helps to increase the plasticity of the pit membrane pectin. Thus compromising lignin quantity may have serious impact on strength of the vascular cylinder; on the other hand, it may increase the pit membrane hydrophilic property and may offer resistance toward cavitation.

Lately, *Arabidopsis* has been taken in as a model for secondary tissue development, although it lacks formation of secondary wood. Tixier et al. (2013) argued that *Arabidopsis* might be as well considered to be a model of xylem hydraulics. They regarded the inflorescence stem of *A. thaliana* as a model for xylem hydraulics despite its herbaceous habit, as it has been shown previously that the inflorescence stem achieves secondary growth (Altamura et al., 2001; Ko et al., 2004), allows long-distance water transport from the roots to the aerial parts of plant, and experience gravity and other mechanical perturbations (Telewski, 2006). There are distinct similarities between woody dicots and *Arabidopsis* inflorescence stems with respect to vessel length and diameter as well as presence of simple perforation plates and border (Sperry et al., 2005; Hacke et al., 2006; Schweingruber, 2006; Wheeler et al., 2007; Christman and Sperry, 2010). It has a genetic potential to develop ray cells and rayless wood is observed in juvenile trees (Carlquist, 2009; Dulin and Kirchoff, 2010). Having *Arabidopsis* as a full proof model for woodiness may open numerous possibilities. The best among them are study of environmental stresses on hydraulic characters. A number of mutants can be generated and screened in *Arabidopsis* with deviant safety vs. efficiency phenotype with little effort. The *Arabidopsis thaliana* irregular xylem 4 phenotype (irx4) a mutant for cinnamoyl-CoA reductase 1 (CCR1) gene, has provided us with valuable insight in the role of lignin reduction and associated phenotypic changes in vasculature. As reported by Jones (2001), near-half decrease of lignin component with no associated change in cellulose or hemicellulose content gives the plant an aberrant vascular phenotype. Most of the cell interior is filled up with expanded cell wall and the xylem vessels collapse. Abnormal lignin gives the cell wall a weak ultrastructure and less structural integrity (Jones et al., 2001; Patten et al., 2005). Later it has been claimed that by modulating the CCR gene, irx4 mutant has obtained a delayed albeit normal pattern of lignification program (Laskar et al., 2006). It thus has to be borne in mind that not only the content but the spatiotemporal pattern of lignin deposition may change the xylem ultrastructure and change the safety-efficiency trade-off limit.

There are a few transcriptional control switches in lignin production which can be used in modification of vascular conductance. Modulation of co-ordinate expression of cellulose and lignin in rice is an important study regarding such transgene opportunities. Expression of the *Arabidopsis* SHN2 gene (Aharoni et al., 2004) under a constitutive promoter in rice alters its lignocellulosic properties along with introduction of drought resistance and enhanced water use efficiency (Karaba, 2007). The *Arabidopsis* SHINE/WAX INDUCER (SHN/WIN) transcription factor belongs to the AP2/ERF TF family, and besides wax regulation, control drought tolerance in *Arabidopsis* (Aharoni et al., 2004; Broun et al., 2004; Kannangara et al., 2007). Expression analysis of cell wall biosynthetic genes and their putative transcriptional regulators shows that moderated lignocellulose coordinated regulation of the cellulose and lignin pathways which decreases lignin but compensates mechanical strength by increasing cellulose. All the processes ascribed to master control switch SHN may be directed toward evolution of land plants; waxy cover to lignin synthesis for erect disposition and water transport. However, no xylem irregularities are seen in this mutant (Aharoni et al., 2004).

As the best studied pathway related to secondary cell wall formation, lignin biosynthesis should offer the best metabolic grid that can be tweaked in plants to genetically understand mechanical functional trade-off and resistance to cavitation. General reduction of PAL (Phenylalanine ammonia lyase, E.C. 4.3.1.5) activities in developing plants may be one possible point of interest. PAL is a "metabolic branch- point" where Phe is directed toward either lignins or proteins (Rubery and Fosket, 1969). However, according to Anterola et al. (1999, 2002) and other such studies there are other pathways originating from pentose phosphate or glycolysis that may directly end into lignin biosynthesis and PAL may not serve as rate limiting step at all. Cinnamate 4-hydroxylase (C_4H) is another candidate that has been downregulated with decrease in overall lignin content, however, with no effect on vascular integrity or function (Fahrendorf and Dixon, 1993; Nedelkina et al., 1999). p-Coumarate-3-hydroxylase (C_3H) in *Arabidopsis* (CYP98A3) may be necessary and rate-limiting step in the monolignol pathway (Schoch et al., 2001). Its expression is correlated with the onset of lignification and a mutant line results in dwarfed phenotype with reduced lignin (Schoch et al., 2001). Cinnamoyl CoA O-methyltransferase (CCOMT), 4-coumarate:CoA ligase (4CL), cinnamoyl CoA reductase (CCR), and cinnamyl alcohol dehydrogenase (CAD) isoforms are downstream pathways in monolignol formation, and their relation to vascular integrity are yet to establish, though phenotypes associated with their mutations are tall/dwarf stature, altered lignin composition, and reduced mechanical support. Conclusive data are yet to be obtained from these studies.

4.17 CONCLUSION

Hydraulic safety margin in a plant is clearly driven by its phylogenetic origin. Conifers have developed minimal hydraulic resistance which is a necessity for water transport through short unicellular tracheids. The unique torus-margo anatomy of the conifer pit membrane let them adaptively overpower multicellular vessels in angiosperms in certain cases. Conifer stems are proposed to have larger hydraulic safety margins when compared with most angiosperm stems (Meinzer et al., 2009; Choat et al., 2012; Johnson et al., 2012) although it is also suggested that they recover poorly from drought-induced embolism (Brodribb et al., 2010). The refilling mechanisms vary greatly between monocots and dicots and herbaceous and woody plants. Resistance to cavitation is thus closely related to many factors: such as nature of the mechanical tissue, the vasculature, the height of the plant, the systematic position of the plant, developmental stage and stresses the plant must face. It can be further emphasized that though, in certain dicots a trade-off within the water transport ability and mechanical strength (efficiency vs. safety) has been observed, the genomic factors which may control the trade-off are not identified till date completely; and the observation is far from universal. The two major physiological phenomena which seem to be linked to embolism resistance are lignification and solute transport between xylem parenchyma, vessel and phloem. The genes and proteins behind these physiological traits are many, and even the obtained transgenic plants and mutants have only been scantily characterized. The effects of assembly of the components are poorly understood and the models proposed do not address all plant families universally. Overall, although a phylogenetic trend is observed among the plants for the evolutionary establishment of hydraulic safety margins, the mechanisms behind have not been understood enough till date to predict the molecular basis and evolution in genomic scale. However, the best metabolic pathway to offer advantageous biotechnological outputs appears to be the lignin synthesis network, which should be assessed by mutant screening as well as by tissue specific overexpression studies in the plant. In case of monocots, drought-induced root- specific overexpression may be of advantage in generating better crops, as root pressure seems to be the major regulator. Crop biotechnology is largely benefitted when the gene pool and their interaction behind a biological process is better known. Overexpressing aquaporins along with the sugar sensing network under a dehydration-responsive promoter could be a formidable strategy to prevent embolism-induced wilting. An approach toward modulation of lignin biosynthesis grid regulation may yield better woody, or even herbaceous crops. The overwhelming knowledge emanating from transcriptomic and genomic studies build the platform where biologists can attempt crop modification for such complex traits as vascular integrity and water transport, without or marginally limiting other beneficial traits, in near future.

REFERENCES

1. Aharoni, A., Dixit, S., Jetter, R., Thoenes, E., van Arkel, G., and Pereira, A. (2004). The SHINE clade of AP2 domain transcription factors activates wax biosynthesis, alters cuticle properties, and confers drought tolerance when overexpressed in *Arabidopsis*. Plant Cell 16, 2463–2480. doi: 10.1105/tpc.104.022897

2. Alexandersson, E., Fraysse, L., Sjovall-Larsen, S., Gustavsson, S., Fellert, M., Karlsson, M., et al. (2005). Whole gene family expression and drought stress regulation of aquaporins. Plant Mol. Biol. 59, 469–484. doi: 10.1007/s11103-005-0352-1

3. Almeida-Rodriguez, A. M., Hacke, U. G., and Laur, J. (2011). Influence of evaporative demand on aquaporin expression and root hydraulics of hybrid poplar. Plant Cell Environ. 34, 1318–1331. doi: 10.1111/j.1365-3040.2011.02331.x

4. Altamura, M. M., Possenti, M., Matteucci, A., Baima, S., Ruberti, I., and Morelli, G. (2001). Development of the vascular system in the inflorescence stem of *Arabidopsis*. New Phytol. 151, 381–389. doi: 10.1046/j.0028-646x.2001.00188.x

5. Alvarez, S., Marsh, E. L., Schroeder, S. G., and Schachtman, D. P. (2008). Metabolomic and proteomic changes in the xylem sap of maize under drought. Plant Cell Environ. 31, 325–340. doi: 10.1111/j.1365-3040.2007.01770.x

6. Andre, J. P. (1998). A study of the vascular organization of bamboos (Poaceae-Bambuseae) using a microcasting method. IAWA J. 19, 265–278. doi: 10.1163/22941932-90001529

7. Anterola, A. M., Jeon, J. H., Davin, L. B., and Lewis, N. G. (2002). Transcriptional control of monolignol biosynthesis in Pinus taeda: factors affecting monolignol ratios and carbon allocation in phenylpropanoid metabolism. J. Biol. Chem. 277, 18272–18280. doi: 10.1074/jbc.M112051200

8. Anterola, A. M., and Lewis, N. G. (2002). Trends in lignin modification: a comprehensive analysis of the effects of genetic manipulations/mutations on lignification and vascular integrity. Phytochemistry 61, 221–294. doi: 10.1016/S0031-9422(02)00211-X

9. Anterola, A. M., van Rensburg, H., van Heerden, P. S., Davin, L. B., and Lewis, N. G. (1999). Multi-sitemodulationoffluxduringmonolignol formation in loblolly pine (Pinus taeda). Biochem. Biophys. Res. Commun. 261, 652–657. doi: 10.1006/bbrc.1999.1097

10. Bao, H., Li, E., Mansfield, S. D., Cronk, Q. C. B., El-Kassaby, Y. A., and Douglas, C. J. (2013). The developing xylem transcriptome and genome-wide analysis of alternative splicing in Populus trichocarpa (black cottonwood) populations. BMC Genomics 14:359. doi: 10.1186/1471-2164-14-359

11. Bao, Y., Dharmawardhana, P., Mockler, T., and Strauss, S. H. (2009). Genome scale transcriptome analysis of shoot organogenesis in Populus. BMC Plant Biol. 9:132. doi: 10.1186/1471-2229-9-132

12. Boerjan, W., Ralph, J., and Baucher, M. (2003). Lignin biosynthesis. Annu. Rev. Plant Biol. 54, 519–546. doi: 10.1146/annurev.arplant.54.031902.134938

13. Bose, J. C. (1923). The Physiology of the Ascent of Sap. London: Longmans, Green and Co.

14. Brodersen, C., Mcelrone, A., Choat, B., Lee, E., Shackel, K., and Matthews, M. (2013). In vivo visualizations of drought-induced embolism spread in Vitis vinifera. Plant Physiol. 161, 1820–1829. doi: 10.1104/pp.112.212712

15. Brodersen, C. R., McElrone, A. J., Choat, B., Matthews, M. A., and Shackel, K. A. (2010). The dynamics of embolism repair in xylem: in vivo visualizations using high-resolution computed tomography. Plant Physiol. 154, 1088–1095. doi: 10.1104/pp.110.162396

16. Brodribb, T. J., Bowman, D., Nichols, S., Delzon, S., and Burlett, R. (2010). Xylem function and growth rate interact to determine recovery rates after exposure to extreme water deficit. New Phytol. 188, 533–542. doi: 10.1111/j.1469-8137.2010.03393.x

17. Broun, P., Poindexter, P., Osborne, E., Jiang, C. Z., and Riechmann, J. L. (2004). WIN1, a transcriptional activator of epidermal wax accumulation in *Arabidopsis*. Proc. Natl. Acad. Sci. U.S.A. 101, 4706–4711. doi: 10.1073/pnas.0305574101

18. Bucci, S. J., Scholz, F. G., Goldstein, G., Meinzer, F. C., Da, L., and Sternberg, S. L. (2003). Dynamic changes in hydraulic conductivity in petioles of two savanna tree species: factors and mechanisms contributing to the refilling of embolized vessels. Plant Cell Environ. 26, 1633–1645. doi: 10.1046/j.0140-7791.2003.01082.x

19. Cao, K. F., Yang, S. J., Zhang, Y. J., and Brodribb, T. J. (2012). The maximum height of grasses is determined by roots. Ecol. Lett. 15, 666–672. doi: 10.1111/j.1461-0248.2012.01783.x

20. Carlquist, S. (2009). Xylem heterochrony: an unappreciated key to angiosperm origin and diversifications. Bot. J. Linn. Soc. 161, 26–65. doi: 10.1111/j.1095-8339.2009.00991.x

21. Carvalho, A., Paiva, J., Louzada, J., and Lima-Brito, J. (2013). The transcriptomics of secondary growth and wood formation in conifers. Mol. Biol. Int. 2013:974324. doi: 10.1155/2013/974324

22. Choat, B., Ball, M., Luly, J., and Holtum, J. (2003). Pit membrane porosity and water stress-induced cavitation in four co-existing dry rainforest tree species. Plant Physiol. 131, 41–48. doi: 10.1104/pp.014100

23. Choat, B., Cobb, A. R., and Jansen, S. (2008). Structure and function of bordered pits: new discoveries and impacts on whole-plant hydraulic function. New Phytol. 177, 608–626. doi: 10.1111/j.1469-8137.2007.02317.x

24. Choat, B., Jansen, S., Brodribb, T. J., Cochard, H., Delzon, S., Bhaskar, R., et al. (2012). Global convergence in the vulnerability of forests to drought. Nature 491, 752–756. doi: 10.1038/nature11688

25. Choat, B., Jansen, S., Zwieniecki, M. A., Smets, E., and Holbrook, N. M. (2004). Changes in pit membrane porosity due to deflection and stretching: the role of vestured pits. J. Exp. Bot. 55, 1569–1575. doi: 10.1093/jxb/erh173

26. Choat, B., and Pittermann, J. (2009). New insights into bordered pit structure and cavitation resistance in angiosperms and conifers. New Phytol. 182, 557–560. doi: 10.1111/j.1469-8137.2009.02847.x

27. Christman, M. A., and Sperry, J. S. (2010). Single-vessel flow measurements indicate scalariform perforation plates confer higher flow resistance than previously estimated. Plant Cell Environ. 33, 431–443. doi: 10.1111/j.1365-3040.2009.02094.x

28. Cochard, H., Badel, E., Herbette, S., Delzon, S., Choat, B., and Jansen, S. (2013). Methods for measuring plant vulnerability to cavitation: a critical review. J. Exp. Bot. 64, 4779–4791. doi: 10.1093/jxb/ert193

29. Coleman, H. D., Samuels, A. L., Guy, R. D., and Mansfield, S. D. (2008). Perturbed lignification impacts tree growth in hybrid poplar - a function of sink strength, vascular integrity and photosynthetic assimilation. Plant Physiol. 148, 1229–1237. doi: 10.1104/pp.108.125500

30. De Boer, A. H., and Volkov, V. (2003). Logistics of water and salt transport through the plant: structure and functioning of the xylem. Plant Cell Environ. 26, 87–101. doi: 10.1046/j.1365-3040.2003.00930.x

31. Decourteix, M., Alves, G., Brunel, N., Ameglio, T., Guilliot, A., Lemoine, R., et al. (2006). JrSUT, a putative xylem sucrose transporter, could mediate sucrose influx into xylem parenchyma cells and be upregulated by freeze–thaw cycles over the autumn–winter period in walnut tree (Juglans regia L.). Plant Cell Environ. 29, 36–47. doi: 10.1111/j.1365-3040.2005.01398.x

32. Dharmawardhana, P., Brunner, A. M., and Strauss, S. H. (2010). Genome-wide transcriptome analysis of the transition from primary to secondary stem development in Populus trichocarpa. BMC Genomics 11:150. doi: 10.1186/1471-2164-11-150

33. Dixon, H. (1914). Transpiration and the Ascent of Sap in Plants. New York, NY: Macmillian.

34. Donaldson, L. A. (2002). Abnormal lignin distribution in wood from severely drought stressed Pinus radiata trees. IAWA J. 23, 161–178. doi: 10.1163/22941932-90000295

35. Dulin, M. W., and Kirchoff, B. K. (2010). Paedomorphosis, secondary woodiness, and insular woodiness in plants. Bot. Rev. 76, 405–490. doi: 10.1007/s12229-010-9057-5

36. El Kayal, W., Keller, G., Debayles, C., Kumar, R., Weier, D., Teulieres, C., et al. (2006). Regulation of tocopherol biosynthesis through transcriptional control of

tocopherol cyclase during cold hardening in Eucalyptus gunnii. Physiol. Plantarum 126, 212–223. doi: 10.1111/j.1399-3054.2006.00614.x

37. Fahn, A., and Leshem, B. (1963). Wood fibres with living protoplasts. New Phytol. 62, 91–98. doi: 10.1111/j.1469-8137.1963.tb06317.x

38. Fahrendorf, T., and Dixon, R. A. (1993). Stress responses in alfalfa (Medicago sativa L.) XVIII: molecular cloning and expression of the elicitor-inducible cinnamic acid 4-hydroxylase cytochrome P450. Arch. Biochem. Biophys. 305, 509–515. doi: 10.1006/abbi.1993.1454

39. Fan, L., Linker, R., Gepstein, S., Tanimoto, E., Yamamoto, R., and Neumann, P. M. (2006). Progressive inhibition by water deficit of cell wall extensibility and growth along the elongation zone of maize roots is related to increased lignin metabolism and progressive stelar accumulation of wall phenolics. Plant Physiol. 140, 603–612. doi: 10.1104/pp.105.073130

40. Ford, C. W., Morrison, I. M., and Wilson, J. R. (1979). Temperature effects on lignin, hemicellulose and cellulose in tropical and temperate grasses. Aust. J. Agr. Res. 30, 621–633. doi: 10.1071/AR9790621

41. Fukuda, H. (1997). Programmed cell death during vascular system formation. Cell Death Differ. 4, 684–688. doi: 10.1038/sj.cdd.4400310

42. Fukuda, H. (2004). Signals that control plant vascular cell differentiation. Nat. Rev. Mol. Cell Biol. 5, 379–391. doi: 10.1038/nrm1364

43. Gindl, W., Grabner, M., and Wimmer, R. (2000). The influence of temperature on latewood lignin content in treeline Norway spruce compared with maximum density and ring width. Trees 14, 409–414. doi: 10.1007/s004680000057

44. Hacke, U. G., and Sperry, J. S. (2003). Limits of xylem refilling under negative pressure in Laurus nobilis and Acer negundo. Plant Cell Environ. 26, 303–311. doi: 10.1046/j.1365-3040.2003.00962.x

45. Hacke, U. G., Sperry, J. S., and Wheeler, J. K., Castro, L. (2006). Scaling of angiosperm xylem structure with safety and efficiency. Tree Physiol. 26, 689–701. doi: 10.1093/treephys/26.6.689

46. Heinen, R. B., Ye, Q., and Chaumont, F. (2009). Role of aquaporins in leaf physiology. J. Exp. Bot. 60, 2971–2985. doi: 10.1093/jxb/erp171

47. Hertzberg, M., Aspeborg, H., Schrader, J., Andersson, A., Erlandsson, R., Blomqvist, K., et al. (2001). A transcriptional roadmap to wood formation. Proc. Natl. Acad. Sci. U.S.A. 98, 14732–14737. doi: 10.1073/pnas.261293398

48. Holbrook, N. M., Ahrens, E. T., Burns, M. J., and Zwieniecki, M. A. (2001). In vivo observation of cavitation and embolism repair using magnetic resonance imaging. Plant Physiol. 126, 27–31. doi: 10.1104/pp.126.1.27

49. Holbrook, N. M., and Zwieniecki, M. A. (1999). Embolism repair and xylem tension: do we need a miracle? Plant Physiol. 120, 7–10. doi: 10.1104/pp.120.1.7

50. Johnson, D. M., McCulloh, K. A., Woodruff, D. R., and Meinzer, F. C. (2012). Hydraulic safety margins an embolism reversal in stems and leaves: why are conifers and angiosperms so different? Plant Sci. 195, 48–53. doi: 10.1016/j.plantsci.2012.06.010

51. Jones, A. M. (2001). Programmed cell death in development and defense. Plant Physiol. 125, 94–97. doi: 10.1104/pp.125.1.94

52. Jones, L., Ennos, A. R., and Turner, S. R. (2001). Cloning and characterization of irregular xylem4 (irx4): a severely lignin-deficient mutant of *Arabidopsis*. Plant J. 26, 205–216. doi: 10.1046/j.1365-313x.2001.01021.x

53. Kannangara, R., Branigan, C., Liu, Y., Penfield, T., Rao, V., Mouille, G., et al. (2007). The transcription factor WIN1/SHN1 regulates cutin biosynthesis in *Arabidopsis* thaliana. Plant Cell 19, 1278–1294. doi: 10.1105/tpc.106.047076

54. Karaba, A. (2007). Improvement of Water use Efficiency in Rice and Tomato Using *Arabidopsis* Wax Biosynthetic Genes and Transcription Factors. Ph.D. thesis, Wageningen University, Wageningen.

55. Karpinska, B., Karlsson, M., Srivastava, M., Stenberg, A., Schrader, J., Sterky, F., et al. (2004). MYB transcription factors are differentially expressed and regulated during secondary vascular tissue development in hybrid aspen. Plant Mol. Biol. 56, 255–270. doi: 10.1007/s11103-004-3354-5

56. Ko, J. H., Han, K. H., Park, S., and Yang, J. (2004). Plant body weight-induced secondary growth in *Arabidopsis* and its transcription phenotype revealed by whole-transcriptome profiling. Plant Physiol. 135, 1069–1083. doi: 10.1104/pp.104.038844

57. Laskar, D. D., Jourdes, M., Patten, A. M., Helms, G. L., Davin, L. B., and Lewis, N. G. (2006). The *Arabidopsis* cinnamoyl CoA reductase irx4 mutant has a delayed but coherent (normal) program of lignification. Plant J. 48, 674–686. doi: 10.1111/j.1365-313X.2006.02918.x

58. Li, X., Wu, H. X., and Southerton, S. G. (2012). Identification of putative candidate genes for juvenile wood density in Pinus radiate. Tree Physiol. 32, 1046–1057. doi: 10.1093/treephys/tps060

59. Li, X., Yang, X., and Wu, H. X. (2013). Transcriptome profiling of radiata pine branches reveals new insights into reaction wood formation with implications in plant gravitropism. BMC Genomics 14:768. doi: 10.1186/1471-2164-14-768

60. Lopez, F., Bousser, A., Sissoëff, I., Gaspar, M., Lachaise, B., Hoarau, J., et al. (2003). Diurnal regulation of water transport and aquaporin gene expression in maize roots: contribution of PIP2 proteins. Plant Cell Physiol. 44, 1384–1395. doi: 10.1093/pcp/pcg168

61. Lu, S., Sun, Y.-H., Shi, R., Clark, C., Li, L., and Chiang, V. L. (2005). Novel and mechanical stress-responsive MicroRNAs in Populus trichocarpa that are absent from *Arabidopsis*. Plant Cell 17, 2186–2203. doi: 10.1105/tpc.105.033456

62. McCully, M. E., Huang, C. X., and Ling, L. E. (1998). Daily embolism and refilling of xylem vessels in the roots of field-grown maize. New Phytol. 138, 327–342. doi: 10.1046/j.1469-8137.1998.00101.x

63. McDowell, N. G. (2011). Mechanisms linking drought, hydraulics, carbon metabolism, and vegetation mortality. Plant Physiol. 155, 1051–1059. doi: 10.1104/pp.110.170704

64. McElrone, A. J., Brodersen, C. R., Alsina, M. M., Drayton, W. M., Matthews, M. A., Shackel, K. A., et al. (2012). Centrifuge technique consistently overestimates vulnerability to water stress-induced cavitation in grapevines as confirmed with high-resolution computed tomography. New Phytol. 196, 661–665. doi: 10.1111/j.1469-8137.2012.04244.x

65. Meinzer, F. C., Johnson, D. M., Lachenbruch, B., McCulloh, K. A., and Woodruff, D. R. (2009). Xylem hydraulic safety margins in woody plants: coordination of

stomatal control of xylem tension with hydraulic capacitance. Funct. Ecol. 23, 922–930. doi: 10.1111/j.1365-2435.2009.01577.x

66. Meinzer, F. C., and McCulloh, K. A. (2013). Xylem recovery from drought-induced embolism: where is the hydraulic point of no return? Tree Physiol. 33, 331–334. doi: 10.1093/treephys/tpt022

67. Mishima, K., Fujiwara, T., Iki, T., Kuroda, K., Yamashita, K., Tamura, M., et al. (2014). Transcriptome sequencing and profiling of expressed genes in cambial zone and differentiating xylem of Japanese cedar (Cryptomeria japonica). BMC Genomics 15:219. doi: 10.1186/1471-2164-15-219

68. Mollenhauer, H. H., and Hopkins, D. L. (1974). Ultrastructural study of Pierce's disease bacterium in grape xylem tissue. J. Bacteriol. 119, 612–618.

69. Nardini, A., Salleo, S., and Jansen, S. (2011). More than just a vulnerable pipeline: xylem physiology in the light of ion-mediated regulation of plant water transport. J. Exp. Bot. 62, 4701–4718. doi: 10.1093/jxb/err208

70. Nedelkina, S., Jupe, S. C., Blee, K. A., Schalk, M., Werck-Reichhart, D., and Bolwell, G. P. (1999). Novel characteristics and regulation of a divergent cinnamate 4-hydroxylase (CYP73A15) from French bean: engineering expression in yeast. Plant Mol. Biol. 39, 1079–1090. doi: 10.1023/A:1006156216654

71. Olenichenko, N., and Zagoskina, N. (2005). Response of winter wheat to cold: production ofphenolic compounds and l-phenylalanine ammonia lyase activity. Appl. Biochem. Microbiol. 41, 600–603. doi: 10.1007/s10438-005-0109-2

72. Patten, A. M., Cardenas, C. L., Cochrane, F. C., Laskar, D. D., Bedgar, D. L., Davin, L. B., et al. (2005). Reassessment of effects on lignification and vascular development in the irx4 Arabidopsis mutant. Phytochemistry 66, 2092–2107. doi: 10.1016/j.phytochem.2004.12.016

73. Perämäki, M., Nikinmaa, E., Sevanto, S., Ilvesniemi, H., Siivola, E., Hari, P., et al. (2001). Tree stem diameter variations and transpiration in Scot pine: analysis using a dynamic sap flow model. Tree Physiol. 21, 889–897. doi: 10.1093/treephys/21.12-13.889

74. Pérez-Donoso, A. G., Sun, Q., Roper, M. C., Greve, L. C., Kirkpatrick, B., and Labavitch, J. M. (2010). Cell wall-degrading enzymes enlarge the pore size of intervessel pit membranes in healthy and Xylella fastidiosa-infected grapevines. Plant Physiol. 152, 1748–1759. doi: 10.1104/pp.109.148791

75. Perrone, I., Pagliarani, C., Lovisolo, C., Chitarra, W., Roman, F., and Schubert, A. (2012). Recovery from water stress affects grape leaf petiole transcriptome. Planta 235, 1383–1396. doi: 10.1007/s00425-011-1581-y

76. Pesquet, E., Ranocha, P., Legay, S., Digonnet, C., Barbier, O., Pichon, M., et al. (2005). Novel markers of xylogenesis in zinnia are differentially regulated by auxin and cytokinin. Plant Physiol. 139, 1821–1839. doi: 10.1104/pp.105.064337

77. Plavcová, L., Hacke, U. G., Almeida-Rodriguez, A. M., Li, E., and Douglas, C. J. (2013). Gene expression patterns underlying changes in xylem structure and function in response to increased nitrogen availability in hybrid poplar. Plant Cell Environ. 36, 186–199. doi: 10.1111/j.1365-3040.2012.02566.x

78. Postaire, O., Tournaire-Roux, C., Grondin, A., Boursiac, Y., Morillon, R., Schäffner, A. R., et al. (2010). A PIP1 aquaporin contributes to hydrostatic pressure-induced

water transport in both the root and rosette of *Arabidopsis*. Plant Physiol. 152, 1418–1430. doi: 10.1104/pp.109.145326

79. Pou, A., Medrano, H., Flexas, J., and Tyerman, S. D. (2013). A putative role for TIP and PIP aquaporins in dynamics of leaf hydraulic and stomatal conductances in grapevine under water stress and re- watering. Plant Cell Environ. 36, 828–843. doi: 10.1111/pce.12019

80. Qin, G. M., Vallad, G. E., and Subbarao, K. V. (2008). Characterization of Verticillium dahliae and V. tricorpus isolates from lettuce and artichoke. Plant Dis. 92, 69–77. doi: 10.1094/PDIS-92-1-0069

81. Quentin, M., Allasia, V., Pegard, A., Allais, F., Ducrot, P.-H., and Favery, B. (2009). Imbalanced lignin biosynthesis promotes the sexual reproduction of homothallic oomycete pathogens. PLoS Pathog. 5:e1000264. doi: 10.1371/journal.ppat.1000264

82. Regier, N., Streb, S., Cocozza, C., Schaub, M., Cherubini, P., Zeeman, S. et al. (2009). Drought tolerance of two black poplar (Populus nigra L.) clones: contribution of carbohydrates and oxidative stress defence. Plant Cell Environ. 32, 1724–1736. doi: 10.1111/j.1365-3040.2009.02030.x

83. Rengel, D., Clemente, H. S., Servant, F., Ladouce, N., Paux, E., Wincker, P., et al. (2009). A new genomic resource dedicated to wood formation in Eucalyptus. BMC Plant Biol. 9:36. doi: 10.1186/1471-2229-9-36

84. Rubery, P. H., and Fosket, D. E. (1969). Changes in phenylalanine ammonia lyase activity during xylem differentiation in Coleus and soybean. Planta 87, 54–62. doi: 10.1007/BF00386964

85. Sack, L., and Holbrook, N. M. (2006). Leaf hydraulics. Annu. Rev. Plant Biol. 57, 361–381. doi: 10.1146/annurev.arplant.56.032604.144141

86. Sala, A., Piper, F., and Hoch, G. (2010). Physiological mechanisms of drought-induced tree mortality are far from being resolved. New Phytol. 186, 274–281. doi: 10.1111/j.1469-8137.2009.03167.x

87. Sala, A., Woodruff, D. R., and Meinzer, F. C. (2012). Carbon dynamics in trees: feast or famine? Tree Physiol. 32, 764–775. doi: 10.1093/treephys/tpr143

88. Salleo, S., Lo Gullo, M. A., De Paoli, D., and Zippo, M. (1996). Xylem recovery from cavitation-induced embolism in young plants of Laurus nobilis: a possible mechanism. New Phytol. 132, 47–56. doi: 10.1111/j.1469-8137.1996.tb04507.x

89. Salleo, S., Lo Gullo, M. A., Trifilo, P., and Nardini, A. (2004). New evi- dence for a role of vessel-associated cells and phloem in the rapid xylem refilling of cavitated stems of Laurus nobilis L. Plant Cell Environ. 27, 1065–1076. doi: 10.1111/j.1365-3040.2004.01211.x

90. Salleo, S., Trifilò, P., Esposito, S., Nardini, A., and Lo Gullo, M. A. (2009). Starch-to-sugar conversion in wood parenchyma of field-growing Laurus nobilis plants: a component of the signal pathway for embolism repair? Funct. Plant Biol. 36, 815–825. doi: 10.1071/FP09103

91. Scheenen, T. W., Vergeldt, F. J., Heemskerk, A. M., and Van As, H. (2007). Intact plant magnetic resonance imaging to study dynamics in long-distance sap flow and flow-conducting surface area. Plant Physiol. 144, 1157–1165. doi: 10.1104/pp.106.089250

92. Schoch, G., Goepfert, S., Morant, M., Hehn, A., Meyer, D., Ullmann, P., et al. (2001). CYP98A3 from *Arabidopsis thaliana* is a 30-hydroxylase of phenolic esters,

a missing link in the phenyl- propanoid pathway. J. Biol. Chem. 276, 36566–36574. doi: 10.1074/jbc.M104047200

93. Scholander, P. F., Hemmingsen, E. A., and Garey, W. (1961). Cohesive lift of sap in the Rattan vine. Science 134, 1835–1838. doi: 10.1126/science.134.3493.1835

94. Schrader, J., Nilsson, J., Mellerowicz, E. E., Berglund, A., Nilsson, P. Hertzberg, M., et al. (2004). A high-resolution transcript profile across the wood-forming meristem of poplar identifies potential regulators of cambial stem cell identity. Plant Cell 16, 2278–2292. doi: 10.1105/tpc.104.024190

95. Schweingruber, F. H. (2006). Anatomical characteristics and ecological trends in the xylem and phloem of Brassicaceae and Resedaceae. IAWA J. 27, 419–442. doi: 10.1163/22941932-90000164

96. Secchi, F., Gilbert, M. E., and Zwieniecki, M. A. (2011). Transcriptome response to embolism formation in stems of Populus trichocarpa provides insight into signaling and the biology of refilling. Plant Physiol. 157, 1419–1429. doi: 10.1104/pp.111.185124

97. Secchi, F., MacIver, B., Zeidel, M. L., and Zwieniecki, M. A. (2009). Functional analysis of putative genes encoding the PIP2 water channel subfamily in Populus trichocarpa. Tree Physiol. 29, 1467–1477. doi: 10.1093/treephys/tpp060

98. Secchi, F., and Zwieniecki, M. A. (2010). Patterns of PIP gene expression in Populus trichocarpa during recovery from xylem embolism suggest a major role for the PIP1 aquaporin subfamily as moderators of refilling process. Plant Cell Environ. 33, 1285–1297. doi: 10.1111/j.1365-3040.2010.02147.x

99. Secchi, F., and Zwieniecki, M. A. (2012). Analysis of xylem sap from functional (non-embolized) and non- functional (embolized) vessels of Populus nigra -chemistry of refilling. Plant Physiol. 160, 955–964. doi: 10.1104/pp.112.200824

100. Secchi, F., and Zwieniecki, M. A. (2013). The physiological response of Populus tremula x alba leaves to the down-regulation of PIP1 aquaporin gene expression under no water stress. Front. Plant Sci. 4:507. doi: 10.3389/fpls.2013.00507

101. Secchi, F., and Zwieniecki, M. A. (2014). Down-regulation of plasma intrinsic protein1 aquaporin in poplar trees is detrimental to recovery from embolism. Plant Physiol. 164, 1789–1799. doi: 10.1104/pp.114.237511

102. Sevanto, S., Holbrook, N. M., and Ball, M. C. (2012). Freeze/Thaw-induced embolism: probability of critical bubble formation depends on speed of ice formation. Front. Plant Sci. 6:107. doi: 10.3389/fpls.2012.00107

103. Sperry, J. S. (2013). Cutting-edge research or cutting-edge artefact? An overdue control experiment complicates the xylem refilling story. Plant Cell Environ. 36, 1916–1918. doi: 10.1111/pce.12148

104. Sperry, J. S., Hacke, U. G., and Wheeler, J. K. (2005). Comparative analysis of end wall resistivity in xylem conduits. Plant Cell Environ. 28, 456–465. doi: 10.1111/j.1365-3040.2005.01287.x

105. Stiller, V., Sperry, J. S., and Lafitte, R. (2005). Embolized conduits of rice (Oryza sativa, Poaceae) refill despite negative xylem pressure. Am. J. Bot. 92, 1970–1974. doi: 10.3732/ajb.92.12.1970

106. Szklarczyk, D., Franceschini, A., Kuhn, M., Simonovic, M., Roth, A., Minguez, P., et al. (2010). The STRING database in 2011: functional interaction networks of pro-

teins, globally integrated and scored. Nucleic Acids Res. 39(Database issue), D561–D568. doi: 10.1093/nar/gkq973

107. Telewski, F. W. (2006). A unified hypothesis of mechanoperception in plants. Am. J. Bot. 93, 1466–1476. doi: 10.3732/ajb.93.10.1466

108. Tixier, A., Cochard, H., Badel, E., Dusotoit-Coucaud, A., Jansen, S., and Herbette, S. (2013). *Arabidopsis thaliana* as a model species for xylem hydraulics: does size matter? J. Exp. Bot. 64, 2295–2305. doi: 10.1093/jxb/ert087

109. Toufighi, K., Brady, S. M., Austin, R., Ly, E., and Provart, N. J. (2005). The Botany Array Resource: e-Northerns, Expression Angling, and promoter analyses. Plant J. 43, 153–163. doi: 10.1111/j.1365-313X.2005.02437.x

110. Tyree, M. T., Sallo, S., Nardini, A., Gullo, M. A. L., and Mosca., R. (1999). Refilling of embolized vessels in young stems of laurel. Do we need a new paradigm? Plant Physiol. 120, 11–21. doi: 10.1104/pp.120.1.11

111. Tyree, M. T., and Sperry, J. S. (1989). Vulnerability of xylem to cavitation and embolism. Annu. Rev. Plant Physiol. Mol. Biol. 40, 19–38. doi: 10.1146/annurev. pp.40.060189.000315

112. Tyree, M. T., and Zimmermann, M. H. (2002). Xylem Structure and the Ascent of Sap. Berlin: Springer. doi: 10.1007/978-3-662-04931-0

113. Van Bel, A. J. E. (1990). Xylem-phloem exchange via the rays: the undervalued route of transport J. Exp. Bot. 41, 631–644. doi: 10.1093/jxb/41.6.631

114. Vanholme, R., Demedts, B., Morreel, K., Ralph, J., and Boerjan, W. (2010). Lignin Biosynthesis and Structure. Plant Physiol. 153, 895–905. doi: 10.1104/pp.110.155119

115. Voelker, S. L., Lachenbruch, B., Meinzer, F. C., Kitin, P., and Strauss, S. H. (2011). Transgenic poplars with reduced lignin show impaired xylem conductivity, growth efficiency and survival. Plant Cell Environ. 34, 655–668. doi: 10.1111/j.1365-3040.2010.02270.x

116. Wheeler, E. A., Baas, P., and Rodgers, S. (2007). Variations in dicot wood anatomy: a global analysis based on the insidewood database. IAWA J. 28, 229–258.

117. Wheeler, J. K., Huggett, B. A., Tofte, A. N., Rockwell, F. E., and Holbrook, N. M. (2013). Cutting xylem under tension or supersaturated with gas can generate PLC and the appearance of rapid recovery from embolism. Plant Cell Environ. 36, 1938–1949. doi: 10.1111/pce.12139

118. Windt, C. W., Vergeldt, F. J., De Jager, P. A., and Van AS, H. (2006), MRI of long-distance water transport: a comparison of the phloem and xylem flow characteristics and dynamics in poplar, castor bean, tomato and tobacco. Plant Cell Environ. 29, 1715–1729. doi: 10.1111/j.1365-3040.2006.01544.x

119. Ye, Z. H. (2002). Vascular tissue differentiation and pattern formation in plants. Annu. Rev. Plant Biol. 53, 183–202. doi: 10.1146/annurev.arplant.53.100301.135245

120. Yokoyama, R., and Nishitani, K. (2006). Identification and characterization of *Arabidopsis thaliana* genes involved in xylem secondary cell walls. J. Plant Res. 119, 189–194. doi: 10.1007/s10265-006-0261-7

121. Zhang, J., Elo, A., and Helariutta, Y. (2011). *Arabidopsis* as a model for wood formation. Curr. Opin. Biotechnol. 22, 293–299. doi: 10.1016/j.copbio.2010.11.008

122. Zhang, Z., Fradin, E., Jonge, R., van Esse, H. P., Smit, P., Liu, C.-M., et al. (2013). Optimized agroinfiltration and virus-induced gene silencing to study Ve1-mediated

Verticillium resistance in tobacco. Mol. Plant Microbe Interact. 26, 182–190. doi: 10.1094/MPMI-06-12-0161-R

123. Zhong, R., McCarthy, R. L., Lee, C., and Ye, Z.-H. (2011). Dissection of the transcriptional program regulating secondary wall biosynthesis during wood formation in poplar. Plant Physiol. 157, 1452–1468. doi: 10.1104/pp.111.181354

124. Zhong, R., and Ye, Z. H. (2009). Transcriptional regulation of lignin biosynthesis. Plant Signal. Behav. 4, 1028–1034. doi: 10.4161/psb.4.11.9875

125. Zimmermann, M. H. (1974). Long distance transport. Plant Physiol. 54, 472–479. doi: 10.1104/pp.54.4.472

126. Zimmermann, M. H. (1983). Xylem Structure and the Ascent of Sap. New York, NY: Springer-Verlag, 143. doi: 10.1007/978-3-662-22627-8

There are several supplemental files that are not available in this version of the article. To view this additional information, please use the citation on the first page of this chapter.

PART II

GENOMIC TOOL KIT
FOR CROP GENOMICS

CHAPTER 5

Silicon Era of Carbon-Based Life: Application of Genomics and Bioinformatics in Crop Stress Research

MAN-WAH LI, XINPENG QI, MENG NI, AND HON-MING LAM

5.1 INTRODUCTION

According to the Food and Agricultural Organization of the United Nations (FAO), food production must be increased by 70% in the next 40 years to meet the increasing global demand [1]. Abiotic and biotic stresses are major limiting factors hampering crop productivity. Therefore, understanding the stress responses of crops using genomic information is important in bringing forth more effective crop improvement strategies.

The publishing of the *Arabidopsis thaliana* genome in 2000 is a cornerstone of the plant genomics era [2]. Taking advantage of the high-throughput data acquisition platforms of the next generation sequencing technology, additional crop genomes have been subsequently decoded. So

Silicon Era of Carbon-Based Life: Application of Genomics and Bioinformatics in Crop Stress Research. © *Li M-W, Qi X, Ni M, and Lam H-M*; licensee MDPI, Basel, Switzerland. International Journal of Molecular Sciences *14,6 (2014). doi:10.3390/ijms140611444. Licensed under a Creative Commons Attribution 3.0 Unported License, http://creativecommons.org/licenses/by/3.0/.*

far, the draft genomes of more than 40 plants have been completed, including those processed in the 1000 Plant and Animal Project [3]. Other "-omics" technologies such as transcriptomics, proteomics, metabolomics, and phenomics (Figure 1) have also undergone rapid development in recent years. Together, there is a large volume of accumulated data, and hence data management and data mining have become a bottleneck for "-omics" researches.

To convert the great amount of data into manageable information, it is essential to establish standard formats and methods for storing, retrieving, and sharing data. Algorithms based on mathematical and statistical models are needed to handle biological data. This review aims to provide a systematic summary of the currently available databases and bioinformatics resources and highlight some challenges and advancements in the study of genomics and other "-omics", with emphasis on their implications on crop stress research.

5.2 GENERAL BIOINFORMATICS RESOURCES

5.2.1 DATABASES

Various databases have been developed to accommodate the comprehensive -omics data and some of them also provide onsite analytical tools (Table 1). The three commonly used sequence databases are GenBank in USA, European Nucleotide Archive (ENA) in Europe, and DNA Data Bank of Japan (DDBJ). They are collaboratively accommodated by the International Nucleotide Sequence Databases (INSD), and the deposited data are frequently synchronized. There are also repositories designated specifically for plants, such as Phytozome that holds the genomic information of more than 40 plant species, including all the sequenced crops. Besides basic genomic information, databases such as Legume Information System (LIS) facilitate synteny analyses and comparative genomic studies between closely related crop plants.

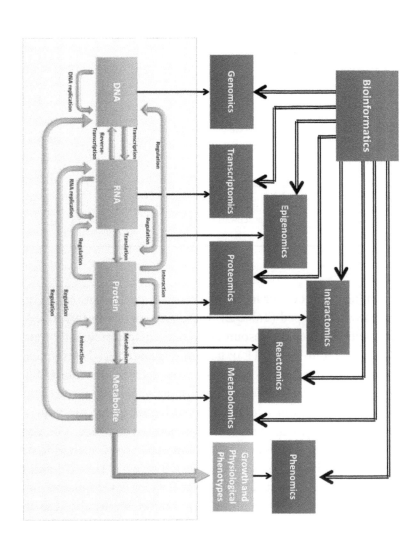

FIGURE 1: Infusion of biological "-omics" with bioinformatics.

TABLE 1: Example of some commonly used databases.

Database name	URL	Reference
GenBank	http://www.ncbi.nlm.nih.gov/genbank/	[4]
ENA	http://www.ebi.ac.uk/ena/	[5]
DDBJ	http://www.ddbj.nig.ac.jp/	[6]
Phytozome	http://www.phytozome.net/	[7]
Gramene	http://www.gramene.org/	[8]
KEGG	http://www.genome.jp/kegg/	[9]
PlantGDB	http://www.plantgdb.org/	[10]
EnsemblPlants	http://plants.ensembl.org/index.html	[11]
VISTA	http://genome.lbl.gov/vista/index.shtml	[12]
PLAZA	http://bioinformatics.psb.ugent.be/plaza/	[13]
GigaDB	http://gigadb.org/	[14]
SGN	http://solgenomics.net/	[15]
GrainGenes	http://wheat.pw.usda.gov	[16]
LIS	http://www.comparative-legumes.org/	[17]

Online resources for individual crops, together with massive datasets, have been developed (Table 2) where systematically integrated information including: genetic resources (genetic maps, molecular markers, and quantitative trait loci (QTL)); genomic resources (DNA sequences, gene models, and regulatory elements); gene expression data (ESTs, cDNA sequences, and transcriptomes); and functional units (proteomic and metabolomic data), is provided. Crops of higher economic values are usually accompanied with a more comprehensive database. The genomic sequences of some economically less important crops, such as foxtail millet, sorghum, and barley, have been released recently [18–20] and their corresponding integrated databases are still under development.

Some data repositories also provide information related to abiotic and biotic stress responses. For example, in MaizeGDB, there are well documented records for tropical maize exhibiting tolerance to drought stress [21]. In SoyBase, genetic markers associated with salt tolerance, drought tolerance, and cyst nematode resistance are incorporated with genomic and expression information. Databases for individual crops could also facilitate the unveiling of the genetic basis of specific traits. For example, the tomato genome sequence helped identify the R-genes which were then incorporated in the Plant Resistance Genes database [22].

TABLE 2: Data repositories for crop plants.

Crop	Database name	URL of related database	Ref
Rice	RAP-DB	http://rapdb.dna.affrc.go.jp/	[23]
Maize	MaizeGDB	http://www.maizegdb.org/	[24]
Medicago	*Medicago truncatula*		
	SEQUENCING RESOURCE	http://www.medicago.org/genome/index_old.php	-
Wheat	GrainGenes	http://wheat.pw.usda.gov/	[16]
Potato	Solanaceae Genomics Resource	http://solanaceae.plantbiology.msu.edu/index.shtml	-
Soybean	SoyBase	http://soybase.org	[25]
Tomato	TOMATO FUNCTIONAL GENOMICS DATABASE	http://ted.bti.cornell.edu/	[26]

5.2.2 BIOLOGICAL ONTOLOGIES RELATED TO CROP STRESS RESEARCH

The standardization of ontology is important for the structuring of huge datasets, interconnection between databases, merging resources, and curation of information. Each ontology term has its own name, identifier/ID/accession number and definition. The identifier/ID/accession number is usually made up of a prefix and a number. For example, the Gene Ontology term "lipid binding" has the accession number GO:0008289. The definition of "lipid binding" is a gene product that can interact selectively and non-covalently with a lipid.

The Gene Ontology (GO) project provides a well-established and controlled vocabulary database for describing the function of a gene and its gene product. The ontology covers three aspects, including cellular component, molecular function, and biological process. GO is used in genome annotation to provide information on gene products. An evidence code (by Evidence Code ontology) is used to describe the evidence that links the GO annotation with the gene product. The Evidence Ontology (EO) suggests whether an annotation has been made manually by a curator or by automated electronic annotation. For example, EXP refers to: "inferred from experiment"; IBA refers to: "inferred from biological aspect of ancestor"; and IEA refers to: "inferred from electronic annotation". All this information can be found in the Gene Ontology website [27].

TABLE 3: List of ontologies containing information related to crop stress responses.

Domain	Prefix	Description	Reference/website
Plant Environmental Conditions	EO	Controlled vocabulary for the representation of plant environmental conditions	http://www.gramene. org/db/ontology/ search?id=EO:0007359
Gene Ontology	GO	Controlled vocabulary for genes and gene products	[28]
Taxonomy Ontology	GR_tax	Representation of the taxonomic tree of plants in the ontology format	http://www.gramene. org/db/ontol-ogy/search?id=GR_ tax:090165
The Plant-Associated Microbe Gene Ontology	PAMGO	Controlled vocabulary for the interaction of microbes with their hosts	[29]
Plant Ontology	PO	Controlled vocabulary for anatomy, morphology and stages of development for all plants	[30]
Sequence Ontology	SO	Controlled vocabulary for sequence annotations, for the exchange of annotation data and for the description of sequence objects in databases	[31]
Plant Trait Ontology	TO	Controlled vocabulary for phenotypic traits in plants	http://www.gramene. org/db/ontology/ search?id=TO:0000387

Plant Trait Ontology (TO) is a controlled vocabulary for describing the plant trait and phenotype. In addition to anatomical and morphological traits, TO also includes a subset of controlled vocabularies for abiotic and biotic stress traits. For example, the yellow dwarf disease resistance

(TO:0000292) is the child term of resistance to disease by mycophasma-like organism (TO:0000013) under the lineage of stress trait (TO:0000164).

There are many other biological ontology projects for different research fields. Ontologies listed in Table 3 contain the information related to crop stress responses.

5.3 RECENT ADVANCES AND CHALLENGES IN CROP GENOMICS

5.3.1 POLYPLOIDY AS A MAJOR CHALLENGE IN CROP GENOME ASSEMBLY

Polyploidy is a major hindrance in crop genome assembly. One of the ways to tackle the highly polyploid genomes is to make references to the closely related, putative progenitor diploid genomes if they are available. The Catalogue of Life [32] and the Integrated Taxonomic Information System [33] may help to identify such related species. For example, the fiber-producing cotton (*Gossypium hirsutum*) is tetraploid, comprising an A-genome and a D-genome. To assist the assembly of the tetraploid genome, the diploid D-genome of *G. raimondii* was first sequenced and assembled [34]. A second example is strawberry (*Fragaria* × *ananassa*), with an estimated genome size of about 600 Mb. Although this is much smaller than other crop genomes, it is an octaploid (AAA′A′BBB′B′) [35]. Therefore, the genome sequence of the woodland strawberry (*Fragaria vesca*), a potential progenitor of *Fragaria* × *ananassa*, was completed in 2012 to provide the first diploid model for the genomes of F. spp. [35,36]. Wheat is another example of polyploid crop genomes. The hexaploid bread wheat (*Triticum aestivum*) contains the A, B and D genomes, which probably originated from *Triticum urartu* (A genome), *Aegilops tauschii* (D genome), and an unknown species related to *Aegilops speltoides* (B genome). The genomic sequence information of *T. aestivum, T. monococcum* (a community standard line related to the A-genome donor), and *Ae. Tauschii*, as well as the cDNA sequence information of *T. aestivum* and *Ae. Speltodies*, were obtained [37]. With reference to the respective diploid genome information, over 90% of the wheat genes were successfully assembled into the A, B, or D genome with over 70% precision [37]. The drafted de novo genomes of *T. urartu* and *Ae. tauschii* were recently published, representing 94.3% and 97.0% of the predicted genome sizes respectively [38,39]. Although the lack of a good reference for the B genome is still an obstacle in building the *T. aestivum* genome, these pieces of work have built a good framework for the further whole genome assembly of bread wheat, and established a model for the study of other polyploid genomes.

5.3.2 REDUCED GENETIC DIVERSITY OF MODERN CROPS

Modern crops originated from a small number of plants. Bottleneck effects during domestication and prolonged human selection together have significantly reduced the genetic diversity of modern crops. Such a reduction in genetic diversity has been confirmed by several genomic studies (Supplementary Table S1). For example, whole-genome resequencing of 14 cultivated and 17 wild soybean genomes revealed that the wild soybeans have higher numbers of SNPs and genetic diversity compared to those of the cultivated ones [40]. The domesticated rice cultivars (*Oryza sativa indica* and *Oryza sativa japonica*) also show a lower genetic diversity than their wild relatives (*O. rufipogon* and *O. nivara*) in a study on 50 accessions of cultivated and wild rice [41]. More interestingly, even though both indica rice and japonica rice are cultivated, the japonica rice shows significantly lower genetic diversity than the indica rice, suggesting that the japonica rice has suffered from a stronger bottleneck effect under domestication [41]. On the other hand, although maize landraces and improved lines have retained a higher nucleotide diversity from their wild progenitor, as compared to other self-fertilizing crop species, a weak bottleneck effect can still be observed [42]. Reduced genomic diversity of major staple crops limits their adaptability to the changing environment and reduces the room for crop improvement. Therefore, crop improvement programs should turn their focus to the genetically compatible wild species, which have higher biodiversity and can serve as natural genetic reservoirs.

5.3.3 SEQUENCE AND STRUCTURAL VARIATIONS IN GENOMES PROVIDING CLUES FOR STRESS STUDIES

Sequence differences and structural variations in genomes are usually identified by comparing the genomes of wild species to their related land-races and modern cultivars, and also to other model plants. These differences can, on the one hand, provide information about genome evolution, and, on the other hand, serve as molecular markers for genetic mapping. Sequence differences and structural variations that affect gene structure, gene expression, and gene copy number are major determinants shaping the diversity among different varieties of the same species. For instance, wild soybeans and some rice accessions possess some present/absent variations or unmapped contigs that contain bona fide genes annotated to be involved in abiotic and biotic stresses [40,41,43,44]. One specific example relating to biotic stresses is the enrichment and over-representation of LRR (leucine-rich repeat) and NB-ARC (nucleotide-binding adaptor shared by APAF-1, certain *R* gene products and CED-4) domain-containing genes in some crop genomes [19,45]. In plant genomes, disease resistance (*R*) genes are responsible for defense responses [46]. LRR and NB-ARC are two important domains found on the R proteins [46]. The LRR domain-containing proteins play important roles in pathogen-host interactions and the activation of defense responses [47,48]. On the other hand, the NB-ARC domain is responsible for the mulitmerization and autoactivation of the R proteins upon stimulus [49]. The LRR and NB-ARC-containing genes exhibit higher ratios of nonsynonymous-to-synonymous SNPs than the genome average in crops such as soybean [40], rice [41,50], and sorghum [51]. In maize, 101 out of 3490 large-effect SNPs detected are located on 49 LRR domain-containing genes [44]. LRR and NB-ARC domain-containing genes are important components in the plant defense response system [46,49,52] while the high nonsynonymous-to-synonymous SNP ratio of LRR or NB-ARC domain-containing genes suggests a dynamic evolution of these genes to combat pathogens.

In addition to disease resistance genes, some transcription factors are found to be over-retained after the whole-genome duplication in *Musa*

α/β (banana) [53]. Some of these transcription factors such as Myb, AP2/ERF, and WRKY are known to be important regulators in abiotic stress responses [54]. On the other hand, compared to rice, sorghum, and maize, there are more genes encoding for cytochrome P450, CCAT-binding factor transcription factors, late-embryogenesis-abundant proteins, and osmoprotectant biosynthesis proteins in the *Ae. tauschii* genome (progenitor B genome of wheat) [38]. These genes are important for the adaptation to cold and physiological drought. Moreover, a significantly higher number of transmembrane ATPase subunits, which are probably involved in Na^+ exclusion and mineral uptake, have been detected in *Ae. tauschii* than in wheat [37,38]. The extra genes in *Ae. tauschii* may be good candidates for wheat improvement.

5.3.4 ADVANCES IN ULTRA-HIGH-DENSITY GENETIC MAPPING USING SNPS

Genetic mapping using genetic populations is one classical strategy to identify genes related to stress responses. Members in the mapping population can either be related (e.g., QTL mapping using bi-parental populations) or unrelated (e.g., genome-wide association study (GWAS) using germplasm collections) (for population structure, data characteristics and methods, see reviews [55,56]). There are some successful cases in identifying stress tolerance causal genes through mapping [57–59]. For example, a salt tolerance-conferring sodium transporter from rice was identified through QTL mapping [58]. The SKC1 locus corresponding to shoot K^+ content was mapped with a BC_2F_2 population generated from a cross between a salt-tolerant indica variety and a susceptible japonica variety [58]. The *SKC1* locus was further confined to a 7.4-kb stretch by the BC_3F_4 progeny testing of fixed recombinant plants. The locus contains only a single open reading frame, which encodes for a HKT-type transporter. *SKC1* near-isogenic lines accumulated less Na^+ under salt treatment compared to the susceptible parent. Voltage-clamp also supports the notion that the SKC1 protein functions as a Na^+-selective transporter that probably regulates K^+/Na^+ homeostasis under salt stress [58].

Classical molecular markers for mapping such as AFLP, RFLP, and SSR markers are sparsely distributed in the genome, and hence limit the mapping resolution and pose difficulties in pinpointing the phenotype-causal genes. With the availability of genomic sequence data, SNP markers become more accessible for use in mapping, to help achieve a much better resolution. However, conventional PCR-based methods are laborious and time-consuming while the resolution of array-based methods is limited by the number of probes on the array.

High-resolution genotyping by whole-genome resequencing has been established [60,61], making the ultra-high-density genetic mapping more attainable. In principle, this method can achieve the highest resolution, provided that there are enough resources to capture all the SNPs in a population. In reality, polymorphic SNPs are usually captured by low-coverage sequencing (~1X for unrelated populations [58] and <0.1X for recombinant populations [60,62,63]).

In a QTL study of recombinant inbred populations originating from indica and japonica rice, SNPs between the parental reference genomes were first identified using DiffSeq in the EMBOSS package and cleaned by SSAHASNP in the ssaha2 package. Low-depth sequencing reads of recombinant inbred lines (RILs) were mapped to the parents' pseudomolecules by using the SSAHA2 software [64] to determine the genotype of each RIL. SNPs were analyzed by a sliding window approach to determine the recombinant break points within the genome of every single line in the population to form a bin map [60]. This sliding window strategy can accommodate the high error rate of next generation sequencing and allow missing data resulting from low-coverage sequencing [60]. Each "bin" will serve as a "marker" in the subsequent linkage map construction using MAPMAKER/EXP and in QTL mapping using QTL Cartographer. In this study, using 150 rice RILs, the sequencing-based method increased the resolution by 35-fold and greatly reduced the time needed for genotyping, compared to the map generated from 287 PCR-based markers [57]. The power of this method was further illustrated in a study using 210 rice RILs to map the *GS3* and *GW5/qSW5* loci related to the grain length and grain width, respectively [62].

Since missing genotypes in low-depth sequencing would reduce the effectiveness of GWAS, after SNPs have been identified by mapping the sequencing reads, the k-nearest neighbor method (KNN) that uses in-house algorithms for data-imputation can be adopted in addition to increasing the sequencing depth, in order to reduce the missing genotypes [61]. GWAS has been conducted in mapping 14 agronomic traits, including drought tolerance, using 373 indica rice lines. One to seven loci have been mapped for each trait, and some of them overlap with the previously known loci/genes identified through bi-parental QTL mapping or mutant studies [61]. With the great reduction of sequencing cost (<US$0.1 per raw megabase in 2012) [65], we anticipate that mapping by sequencing will become a popular method to obtain high resolution maps for stress-related loci/genes.

5.3.5 GENOMIC SELECTIONS

Genomic selection (GS) is introduced to evaluate the overall effects of all contributing loci genome-wide [66]. During the process of GS, a training population will be used for computational model training to obtain the genomic estimated breeding values (GEBVs) [67]. Complex traits such as drought tolerance are usually determined by multiple small-effect QTLs. GEBV associates markers and QTLs by regarding all the markers as variables contributing to the trait and the effect of each marker allele towards the complex QTLs is quantified (it can be zero). GEBV determines the sum of the marker effects and thus indicates the breeding value of an individual; favorable individuals with high GEBVs from breeding populations will be selected for field application. Genotypic and phenotypic information of the breeding population can be used to further improve the computational model to form a training-breeding cycle [66]. Unlike GWAS and QTL studies, which are designed to reduce the breeding time by selecting plants with desired molecular markers at early growth stages instead of evaluating the actual phenotypes at a later stage, GEBVs serve only as selection criteria but do not lead to target markers or causal genes.

As high-throughput genotyping and phenotyping have accelerated GS studies by increasing marker density and selection capacity, one of the major challenges of GS is selection accuracy. Evaluations of GS accuracy have been performed in maize [68], wheat [69,70], barley [71], and cassava [72]. Several statistical models for GEBV calculations, including best linear unbiased prediction (BLUP) [67], Bayesian shrinkage regression (BayesA, BayesB, etc.) [73], and mixed models have been employed. There is no agreement on which model is the most efficient, because many factors such as population size and genetic background may affect statistical power [71]. It is believed that GS is a valuable approach for plant breeding [74], however, it will take some time for this concept to develop into a practical tool [75]. A GS-based breeding scheme has already been proposed and is considered to be an important tool for developing durable stem rust-resistant wheat [76].

5.3.6 IDENTIFICATION OF STRESS-RELATED GENE FAMILIES

When properly annotated genomes are available, the genome-wide identification of all members of a gene family will become feasible. Since genome duplication (polyploidy or paleopolyploidy) and single gene duplication are common in crops [77], genes usually exist in multiple copies and/or in gene families. Identifying all members of a gene family may give a more comprehensive view on the possible functions of a group of evolutionarily related genes. Bioinformatics tools such as Fgenesh [78], GAZE [79], and JIGSAW [80] have been adopted for searching gene families in crops.

Two typical ways to identify members of gene families from within a genome are keyword search and pattern/homology search. Keyword search usually requires precise keywords including gene names and controlled vocabularies. The most commonly used controlled vocabularies are Gene Ontology, as mentioned in section 2.2, and the functional classification by Pfam, InterPro and KEGG [81–83].

A genome-wide pattern search usually begins with searching sequence databases using programs like BLASTP or TBLASTN [84]. Databases can either be online resources (Table 1) or in-house databases. The occurrence of the desired functional domains in the potential sequences can then be verified using the Pfam protein families database [81], SMART database [85], or HMMER [86]. When the BLAST results are associated with unannotated sequences, these will require further analyses to determine the putative gene structures. One example of applying the above strategy to identify stress-related genes is the analysis of AP2/EREBPs in the rice genome [87]. "AP2/EREBP" was used as the keyword in searching databases, including DRTF, MSU NCBI, and KOMBE. Any non-redundant sequences obtained were then used as query terms in the TBLAST and BLASTP searches of the MSU and NCBI databases. Four genes with an incomplete AP2 domain were excluded after Pfam and SMART analyses because of their very small AP2/ERF domain. A total of 163 genes were identified using this method, in contrast to the 139 genes as suggested previously [87]. Expression studies revealed that a number of the members are responsive to abiotic or biotic stresses. A few of them can even be induced by multiple stresses, suggesting their possible involvement in stress responses [87].

Supplementary Table S2 summarizes the strategies and tools used in recent literature on genome-wide analyses of gene families related to stress responses in major crops.

5.4 FUNCTIONAL GENOMICS

5.4.1 TRANSCRIPTOME

There are two major technologies for obtaining the overall transcription map of specific plant tissues: hybridization-based microarray technology [88] and next generation RNA sequencing technology (RNA-seq) [89]. RNA-seq technology, in conjunction with efficient bioinformatics tools, is now more widely used to support predicted gene models, extract differentially expressed genes, and find novel transcripts in de novo assemblies. Public repositories such as ArrayExpress [90] are designed for the storage of expression data. Standard data formats including Minimum Information about Microarray Experiments (MIAME) or Minimum Information about Sequencing Experiments (MINSEQE) are unified to facilitate transcriptome data submission/downloading. Bioinformatics tools dealing with transcriptome alignment, splicing event prediction, and de novo assembly are also available (Table 4).

TABLE 4: Widely used bioinformatics tools for the analysis of transcriptome data.

Software	Description	Download URL	Reference
ABMapper	RNA-seq data alignment	http://hkbic.cuhk.edu.hk/software/abmapper	[91]
Bowtie	RNA-seq data alignment	http://bowtie-io.sourceforge.net/bowtie2/index.shtml	[92]
Cufflinks	Transcript assembly	http://cufflinks.cbcb.umd.edu/	[93]
DEGseq	Differential gene expression detection	http://www.bioconductor.org/packages/2.11/bioc/html/DEGseq.html	[94]
Infernal	RNA-seq data alignment	http://infernal.janelia.org/	[95]
Oases	De novo assembly	www.ebi.ac.uk/~zerbino/oases/	[96]
Tophat	RNA-seq data alignment & Alternative splicing detection	http://tophat.cbcb.umd.edu/	[97]
Trans-AByss	De novo assembly	http://www.bcgsc.ca/platform/bioinfo/software/	[98]
Trinity	De novo assembly	http://trinityrnaseq.sourceforge.net/	[99]

Crops such as maize [100] and soybean [101] have their own transcriptome atlases, compiled from sub-transcriptomes from multiple tissues and different developmental stages. For the transcriptome atlas of soybean, plant ontology (PO) was used to describe the developmental stage of each experimental tissue, providing a common ground for readers and users to discuss and perform further analyses. The cDNA short reads generated by Illumina Genome Analyzer were aligned to the soybean reference genome sequence assembly using GSNAP, released in 2005. The digital expression counts were determined using the R programming language and normalized using a variation of RPKM methods [101]. The global inventory of expressed transcripts of crops under stress is dynamic, both temporally and spatially. Time series sampling is a typical experimental design to trace the trajectory of such differentially expressed transcripts of crops under stress conditions. A typical example was the study of the soybean transcriptome under alkaline stress. Soybean plants were treated with $NaHCO_3$ and transcriptomes were analyzed using microarray [102]. GO terms were successfully assigned to the 1380 significantly changed probe sets that are related to metabolism, signal transduction, energy, transcription, secondary metabolism, transporter, as well as disease and defense. A time series study revealed the interplay of signal transduction and metabolism during the progression of the treatment. MapMan tools were used to visualize these changes [102]. Other time series studies include the studies of rice root under low potassium [103], cassava under cold stress [104], and soybean subjected to *Pseudomonas syringae* infection [105]. The other widely reported experimental design is the comparative transcriptome study performed among crop accessions with different degrees of stress tolerance, such as the study of soybean accessions exhibiting differential tolerance toward low potassium [106], rice cultivars with contrasting abilities to withstand drought [107] and chilling [108], wheat with differential drought tolerance [109], and Medicago [110] and foxtail millet [111] cultivars with differential salt tolerance.

Another strategy to associate transcript abundance to genomic variations is the expression QTL (eQTL), which use differentially expressed transcripts as the quantitative traits [112]. The eQTL maps of maize root [113] and rice shoots [114] have identified thousands of cis and trans regulation factors by population transcriptome screening. The eQTLs colocalizing with traditional QTL regions could give supportive evidence explaining the genetic basis of the targeted phenotypic characters. One successful example is the eQTL study of the partial resistance toward *Puccinia hordei* in barley [115], in which some eQTLs were reported to colocalize with previously known rust resistance QTL regions.

5.4.2 PROTEOME

Due to the alternative splicing of RNA transcripts and post-translational modifications of the proteins themselves, the proteome within a cell can be much more complicated than the corresponding genome. The gel-based proteomics technology will soon be obsolete due to its limited sensitivity and semi-quantitative nature [116]. The rise of the next generation proteomics systems such as Orbitrap and QStar, together with the application of isotopic tag-based quantitative proteomics (ICATs [117], SILAC [118], isobaric tag-based quantitative proteomics (ITRAQ [119]), and label-free quantitative proteomics (MaxQuant [120], Serac [121], SIEVE (Thermo Scientific, San Jose CA, USA)) have expedited the development of high-throughput proteomic studies. Nevertheless, the pace of adopting these platforms in plant stress studies is far behind studies in humans.

TABLE 5: Bioinformatics resources commonly used in crop proteomic studies.

Program	URL
Mascot	http://www.matrixscience.com
SEQUEST	http://fields.scripps.edu/sequest/index.html
X!Tandem	http://www.thegpm.org/TANDEM/index.html
Database	URL
Expasy	http://www.uniprot.org/help/uniprotkb
UniprotKB/SwissProt and UniProtKB/ TrEMBL database	http://www.uniprot.org/help/uniprotkb
Protein Information Resource (PIR)	http://pir.georgetown.edu/
RCSB Protein Data Bank (RCSB PDB)	http://www.rcsb.org/pdb/download/download.do
EMBL-EBI's Protein Data Bank in Europe (PDBe)	http://www.ebi.ac.uk/pdbe/
SWISS-2DPAGE	http://world-2dpage.expasy.org/swiss-2dpage/
The Plant Proteome Database (PPDB)	http://ppdb.tc.cornell.edu/
Plant Protein Phosphorylation DataBase (P³DB)	http://www.p3db.org/
RIKEN Plant Phosphoproteome Database (RIPP)	https://database.riken.jp/sw/links/en/ria102i/
Secretom—The Tomato Fruit Glyco-proteome	http://solgenomics.net/secretom/detail/glycoproteome
Plant Secretome KnowledgeBase (PlantSecKB)	http://proteomics.ysu.edu/secretomes/plant.php

Despite the advancement in the proteomics platforms, the application of de novo peptide sequencing is still limited. Protein identifications still largely rely on database searches in which experimental peptide mass spectra are compared with theoretical peptide mass spectra generated from existing sequence databases. Some commonly used databases and useful algorithms are summarized in Table 5. Since the genomes of many crops have not been completely sequenced, and some others are still unknown, proteins of species without a genome database are frequently identified by referring to cross-species databases. In these cases, it is not uncommon that molecular weights and isoelectric points (pI) of the identified proteins may deviate from the actual spot position on the 2D gel, despite the high protein scores.

Comprehensive reviews summarizing plant proteomic studies from 2006 to 2008 are available [122,123]. We have also summarized the plant proteomic studies in 2012 (Supplementary Table S3). Recently, plant proteomic investigations have been subdivided into several areas, including subcellular proteomics and proteomics-related post-translational modifications. For example, 21 differentially expressed proteins were identified from salt-treated wheat chloroplasts [124], and 13 and 11 differentially expressed microsomal proteins, respectively, were identified from two distinct cadmium-accumulating soybean cultivars [125].

Stress-induced posttranslational modifications of proteins are common. They are either the results of deleterious damage from the stress, or beneficial modifications to regulate the functions of the proteins in order to cope with the stress. To study posttranslational modifications of proteins, special techniques within proteomics are used. Redox proteomics requires special labeling methods, including the reduction and subsequent labeling of the oxidized thiol groups with 5-iodoacetamidofluorescein (IAF) [126]. Twenty-two highly oxidized proteins involved in a wide range of biological processes were identified in ozone-treated rice using this method [127]. Phosphoproteome [128], glycoproteome, and secretome [129–132] are sub-categories of proteomics that require special staining and enrichment techniques. Post-translational modifications involved in gene expression regulations will be discussed in the Epigenomics section below.

5.4.3 INTERACTOME

Protein-protein interactions determine the contextual functions of a protein and hence play a crucial role in regulation and signal transduction [133]. There are several commonly used experimental systems to identify protein-protein interactions, including: (1) yeast two hybrid (Y2H) (reviewed in [134]); (2) biomolecular fluorescence complementation (BiFC) (reviewed in [134]); (3) affinity pull-down coupled with mass spectrometry (AP-MS) (reviewed in [134]); (4) blue native PAGE [135]; and (5) structural analysis of protein crystals [136,137]. In addition, literature curation involving tedious literature searches can be used to supplement the experimental efforts [138] and in silico prediction can be done by searching for orthologous pairs which interact in other systems, to identify possible interologues [134,139]. Multiple systems are generally adopted to authenticate the interactions.

The concept of the plant interactome was initiated years ago, and was based mainly on the information collected through literature curation [140]. Subsequently, an experimentally constructed interactome map of *A. thaliana* was established via intensive screening, recording a total of 6200 high-confidence interactions among 2700 proteins through the screening of proteins encoded by 8000 open reading frames in the Arabidopsis genome [141]. It is estimated that this screening only captured around 2% of the binary protein-protein interactome in *A. thaliana* [141]. Using the in silico interolog prediction method, more than 37,000 interactions among 4567 rice proteins were predicted, 168 of which have been experimentally confirmed [139]. In this piece of work, the INPARANOID 3.0 program was used to predict high-confidence protein orthologues in 12 species including rice. With the assumption that protein-protein interactions are retained in evolutionarily conserved orthologous proteins, rice protein-protein interactions were compiled using the predicted orthologous proteins and the known interactions in interactome databases [139]. Only a few studies directly related to crop stress interactomes have been published (Table 6). In the search for rice stress-related interactomes, 4 stress proteins related to disease (XA21 and NH1) and flooding (SUB1A and SUB1C) were used as baits for the initial interactome screens by Y2H [142]. Preys identified from the initial screens were then used as baits for subsequent screens. Together with the information from literature curation, an interactome network consisting of 100 proteins were constructed. The interactomes of the two kinds of stresses were linked by proteins such as SNRK1A, which has been shown to be related to ABA, a positive regulator of abiotic stress responses and a negative regulator of biotic stress responses [142].

TABLE 6: Recent large-scale stress interactome studies in crop plants.

Species	Stress	Strategy	Reference
Rice	Abiotic and biotic	Using stress components as bait in Y2H	[142,148]
Wheat	Cold and dehydration	Using stress components as bait in Y2H	[149]
Soybean	SCN infection	In silico prediction	[150]

Online resources such as PRIN [143] can help to predict rice interactomes, while BioGRID [144], DIP [145], PlaPID [146], and InAct [147] can be queried for some pre-determined interactomes in certain plant species. Recent large-scale stress interactome studies in crop plants are shown in Table 6.

5.4.4 EPIGENOME

In addition to the genetic information encoded by DNA, epigenetic modifications of DNA and histones provide another dimension of regulation to influence gene expressions. Chromatin-associated proteins, including DNA methylase, histones, and histone-modifying enzymes, are cataloged in the ChromDB [151]. Technological platforms for epigenomic research can be considered as an extension of genomic and proteomic studies with modifications in analysis protocols.

For example, cytosine DNA methylation, one of the epigenetic modifications, plays an important role in gene silencing and genomic imprinting [152,153]. The transcriptional levels of endogenous genes are highly correlated with the methylation status within their promoter or transcribed regions [154,155]. One way to detect DNA methylation is to capture and enrich the methylated DNA fragments by immunoprecipitation [156]. Bisulfite treatment is another way to distinguish between methylated and unmethylated DNA. The bisulfite treatment converts unmethylated (but not methylated) cytosines to uracils [157]. Both immunoprecipitation-enriched and bisulfite-treated DNA can be analyzed by microarray- or sequencing-based methods to the single-base level of resolution [157–159]. A number of bioinformatics tools are designed to handle the bisulfite sequencing data (Table 7).

TABLE 7: Bioinformatics tools for the analysis of bisulfite sequencing data.

Tools	Descriptions	Reference
BiQ Analyzer HT	Quantitative study of locus-specific DNA methylation patterns from bisulfite sequencing data.	[166]
Bismark	Mapping of bisulfite-sequencing reads and methylation calling.	[167]
BRAT-BW	Genome-wide single base-resolution methylation data analysis	[168]
BSmooth	Providing estimate of methylation profiles with low-coverage whole-genome bisulfite-sequencing data.	[169]
BS seeker	Mapping of bisulfite-sequencing reads.	[170]
BSMAP	Bisulfite reads mapping algorithm.	[171]
CpG_MPs	Analysis of biosulfite-sequencing read and identification of genome-wide methylation pattern	[172]
GBSA	Both gene-centric or gene-independent analyses of whole-genome bisulfite sequencing data	[173]
Kismeth	Analysis of plant bisulfite sequencing results, with a tool for designing bisulfite sequencing primers.	[174]
QUMA	Quantification tool for methylation analysis	[175]
RRBSMAP	Derivative of BSMAP—a specific tool for reduced-representation bisulfite sequencing	[176]

Both biotic and abiotic stresses will lead to massive changes in the DNA methylation status [160–162]. Some stress-induced DNA methylations can be inherited by the next generation. The mechanism for trans-generation DNA methylation may be partially mediated by small RNAs [163]. This trans-generation DNA methylation has been observed in some crops in response to stress [164,165], as a way of pre-acquiring immunity toward the upcoming stresses via designed parental priming [164].

Histone proteins are responsible for the packing of DNA. The epigenetic modifications of core histones affecting the tightness of DNA packing are called histone codes that can relay important information to affect gene expressions [177]. The Histone Sequence Database provides a comprehensive collection of histone sequences and structural information [178].

The addition of acetyl groups to histones neutralizes the positive charges and hence loosens the condensed DNA, leading to transcriptional activation [179], while the methylation of histones results in gene deactivation or repression [180]. The phosphorylation of histones causes the relaxation of chromatin and modulates histone acetylation and methylation [181].

Individual types of histone modifications on specific amino acid residues can be detected using specific antibodies or various mass spectrometries while genome-wide histone-DNA associations can be captured by chromatin immunoprecipitation (ChIP) and subsequently analyzed using either microarray (ChIP-Chip) [182] or sequencing (ChIP-seq) [183].

Some histone-modifying enzymes are induced in crops under stress. For example, a trithorax-like H3K4 methyltransferase was found to be induced by drought in drought-tolerant barley cultivars [184] while a histone deacetylase was found to be induced by compatible infections and repressed by incompatible infections [185]. The methylation statuses of four transcription factors were affected by salt stress. The expression of three of these transcription factors were also found to be correlated with their H3 methylation and acetylation statuses [186]. A genome-wide study in rice identified 4837 genes that harbor differential H3K4me3 modification under drought stress, in which the expression of 609 genes were significantly correlated with the H3K4me3 modification [187].

5.4.5 PHENOME

Every observable biological characteristic beyond the genotype can be regarded as the phenotype. Phenotypes can be observed at the molecular, cellular, organismal, and population levels. Phenotypes also vary throughout the organism's lifecycle, spanning different growth stages, and during different periods of stress. The environment can also exert significant influences on the phenotype. The total sum of phenotypes of an organism or a population constitutes the phenome.

As mentioned in section 2.2, to make phenotypic data in public databases more searchable and accessible to users of bioinformatics tools, ontologies are used to describe the setup of the experiment and the phenotypic data. For example, one may study salt tolerance (TO:0006001) at the whole-plant flowering stage (PO:0007016) and the days to flower (TO:0000344) of *Oryza sativa* (GR_tax:013681), in a greenhouse study (EO:0007248) under a sodium chloride regimen (EO:0007048). These ontologies provide a common language to describe an experiment and render it understandable by both researchers and computational algorithms. For instance, some people may record certain phenotypes during the flowering stage. However, what does it mean by "flowering stage"? Some people refer to "flowering stage" as the time when the first flower opens. Others may refer to it as having half of the individual plants with flowers opened. In this case, the flowering time is well defined in plant ontology. PO:0007026, PO:0007034, PO:0007053 and PO:0007052 refer to the stage at which the first flower, 1/4 of the flowers, 1/2 of the flowers, and 3/4 of the flowers, open, respectively. PO:0007024 marks the end of the flowering stage. The application of these ontologies can thus reduce the discrepancies in annotating the phenotypes and treatment conditions.

High-quality phenotypic information is essential for mapping, association studies, gene identifications, gene functional studies and genomic selections. To design experiments to collect phenotypic information, some critical parameters have to be considered, such as the sample/population sizes, experimental conditions, phenotypes to be assessed, and the data acquisition methods.

The size of a population can vary from a few plants for functional studies, several hundred lines for mapping and GS, to as many as a thousand germplasms for GWAS. Some public collections of germplasms or populations are available for public requests. The United States Department of Agriculture National Plant Germplasm System has a collection of over 500,000 germplasm accessions from 10,000 plant species including rice, soybean, tomato and many other staple crops. Table 8 summarized some publicly available mutant and germplasm collections, some of which also provide phenotypic descriptions and photos of the mutant.

Since the phenome is the overall outcome of the interactions between the genotype and the environment, whether the phenotypic data are collected in a controlled environment or not can greatly affect the final interpretation of results. Field experiments can better mimic the actual conditions of crop production, but the consistency of the phenotype greatly depends on the location of the field, the soil composition, weather conditions, season, and so on. The interpretation of results can thus be complicated. For example, a change in the transpiration rate in some of the plants may not solely be the result of the stress treatment, but also the result of localized changes in light intensity and/or temperature in the field [188]. In general, a larger number of replications are required to compensate for the effects due to environmental variations. A controlled environment such as that in a greenhouse or a growth chamber can minimize the effects of environmental fluctuations and hence will emphasize the contribution of the genotype. However, data from controlled experiments are usually limited in scale and may overlook the fast-changing environment in the real production field.

TABLE 8: Databases for mutant and germplasm resources.

Species	Databases	URL
Barley	Barley DB	http://www.shigen.nig.ac.jp/barley/
Barley	NordGen Plant Collection	http://www.nordgen.org
Maize	RescueMu Maize Mutant Phe-notype Database	http://maizegdb.org/rescuemu-phenotype.php
Rice	Oryza Tag Line	http://oryzatagline.cirad.fr/
Rice	Rice Tos17 Insertion Mutant Database	http://pfg101.nias.affrc.go.jp/
Soybean	SoyBase—Fast Neutron Mutants	http://www.soybase.org/mutants/index.php
Tomato	Genes that Make Tomatoes	http://zamir.sgn.cornell.edu/mutants/index.html
Tomato	LycoTILL	http://www.agrobios.it/tilling/index.html
Tomato	TOMATOMA	http://tomatoma.nbrp.jp/index.jsp
Tomato/Pea	URGV TILLING Database	http://urgv.evry.inra.fr/UTILLdb
Tomato/Potato	SOL genomics network	http://solgenomics.net/
Wheat	The Scottish Wheat Variety Database	http://wheat.agricrops.org/menu.php

Choosing the appropriate phenotypes to be assessed is also important. For example, stomatal conductance and pathogen titer are good indicators of osmotic stress tolerance and disease resistance, respectively. However, they are not quite applicable in large-scale experiments due to the limitation of the machine and the laborious procedures. On the other hand, fresh weight and biomass can truly reflect the productivity of the crops, but taking these measurements is destructive to the plant. For morphological and physiological phenotyping of crops under stress, a conversion of stress symptoms to parameters that can be captured and digitized is needed for high-throughput automation. Commonly used methods include: 2D or 3D visible light imaging [189,190], infrared thermography [188], near-infrared imaging, spectral reflectance [191], fluorescence analysis [191,192], stable isotope analysis [193] and X-ray imaging [194]. For example, a study of wheat salt stress response suggested that the shoot area calculated from 3 digital images (2 side and 1 top images) showed a strong positive correlation with manually measured leaf area and shoot fresh weight which commonly serve as the indicators of salt tolerance in crops [195]. As a non-destructive method, the imaging system could continuously monitor the growth of the plant and distinguish its bi-phasic (osmotic stress phase and ionic stress phase) growth under salinity stress [195]. Another example is related to osmotic stresses (salinity and drought) that reduce stomatal conductance. Since the reduction in stomatal conductance will halt the cooling effect of transpiration, infrared thermal imaging can be used to monitor the degree of salinity and drought stress [188,196]. In the case of lesions on leaf surfaces caused by plant diseases, instead of measuring the lesion area on each leaf, determining the reduction in chlorophyll fluorescence is a possible alternative [197].

In addition to the physiological phenotypes, metabolite profiles in crops are also altered by both biotic and abiotic stresses [198,199]. Deleterious metabolites such as reactive oxygen species might be generated through the disruption of normal cellular processes while beneficial metabolites such as signaling molecules and osmoprotectants may be generated to alleviate the stress [200,201].

Current platforms for metabolomic analyses include various forms of liquid chromatography-coupled mass spectrometry (LC-MS), gas chromatography-coupled MS (GC-MS), capillary electrophoresis-coupled MS (CE-MS), fourier transform MS (FT-MS), fourier transform infrared spectrometry (FT-IR), and one- or two-dimension nuclear magnetic resonance (NMR). Raw spectra generated by mass spectrometry or NMR can be analyzed by making references to either in-house or online databases (Table 9). Biological multivariate data generated from metabolomic studies are commonly analyzed using principal component analysis (PCA) [202], partial least square (PLS) [203], and orthogonal projections to latent structures (O-PLS) [203,204].

TABLE 9: Mass spectra databases and other bioinformatics resources for metabolomic studies.

Database	URL	Reference
AOCS Lipid Library	http://lipidlibrary.aocs.org/index.html	-
Golm Metabolome Database	http://gmd.mpimp-golm.mpg.de/	[205]
Lipidomics Gateway	http://www.lipidmaps.org/	[206]
Madison-Qingdao Metabolomics Consortium Database	http://mmcd.nmrfam.wisc.edu/	[207]
Manchester Metabolomics Database	http://dbkgroup.org/MMD/	[208]
MassBank	http://www.massbank.jp/index.html	[209]
Metabolome Express	https://www.metabolome-express.org/	[210]
METLIN	http://metlin.scripps.edu/	[211]
NIST Chemistry WebBook	http://webbook.nist.gov/chemistry/#Notes	-
Software/Tools	URL	Reference
AMDIS	http://www.amdis.net/	-
COLMAR	http://spinportal.magnet.fsu.edu/	[212]
MetDat	http://smbl.nus.edu.sg/METDAT2/	[213]
MetaboSearch	http://omics.georgetown.edu/MetaboSearch.html	[214]

Numerous metabolomic studies have been done on crops under stress, including: salinity [215–217], drought [218–221], flooding [222], ozone treatment [223], fungal infections [224–226], bacterial infections [217,227], other infections [228], and multiple stresses [229,230].

There are two major research strategies in metabolomic studies: metabolic fingerprinting and metabolic profiling. Metabolic fingerprinting uses the mass-to-charge ratio of mass spectrometry, the peak height and/or retention time of chromatography and the strength of NMR signal as the metabolomic signature in specific samples, such that the identity of each metabolite is not necessary [231]. This helps to classify different samples into categories. For example, metabolic fingerprints have been made to differentiate between disease-resistant and susceptible varieties [217,227,228] or between salt-tolerant and sensitive varieties [232]. In one study, fourier transform infrared (FT-IR) spectroscopy was used for the metabolic fingerprinting of salt-treated tomatoes [233]. A total of 882 FT-IR spectra variables were collected between the wave number 4000 to 600 cm^{-1} for each sample [233]. Through discriminant function analysis (DFA) of the spectra variables, without knowing the identity and the quantity of each metabolite, salt-treated and control samples can be discriminated. Furthermore, key regions within the spectrum distinguishing the treated from the untreated samples were identified through genetic algorithms, and the major components were found to be amino radicals and nitrile-containing compounds [233]. Thus, disease resistance and stress tolerance of novel crop varieties can be assessed by comparing their metabolic fingerprints with those of well characterized varieties, facilitating the screening process.

On the other hand, metabolic profiling compares the metabolic compositions between samples and hence the quantitation and identification of the metabolites are required. Signal patterns must be matched to known standards or depositions in the databases in order to identify the actual compounds. For example, the accumulation of compatible solutes, such as proline, glycine-betaine, and their precursors, is usually observed in osmotically stressed crops, especially in tolerant varieties [216,220,221]. A specific example is the mitochondrial metabolic profile of flood-stressed soybean; metabolites were extracted from the roots and hypocotyls of soybean seedlings with or without submergence stress [226], and were then analyzed using capillary electrophoresis mass spectrometry. Eighty-one mitochondria-related metabolites were identified and quantified with reference to the commercially available standards [226]. There was an accumulation of TCA cycle-related metabolites, including citrate, succinate, and aconitate, but a reduction in ATP in flood-stressed plants, which can be explained by the arrest of aerobic respiration due to anoxia [222]. Following a similar logic, the accumulation of antimicrobial compounds, such as caffeic acid, phytoalexins, glycoalkaloids, and other polyphenolic compounds, are common in pathogen-infected crops compared to their uninfected counterparts [224–226,228]. Glucose oxidase secreted by a fungal pathogen, *Botrytis cinerea*, can also lead to the accumulation of gluconic acid in *Vitis vinifera* cv. Chardonnay berries [226].

5.5 FUTURE PERSPECTIVES

Sequencing throughput is no longer the major limiting factor in genomics and transcriptomics studies. The next generation sequencing platforms can actually generate enough depth for genome assembly in one or several runs [234]. However, sequence assembly and annotation for complex genomes remain challenging. The data acquisition platforms for other "-omics", on the other hand, are under rapid development to catch up with the pace of genomic research. While the data source is no longer a rate-determining step, data integration and interpretation have become the bottleneck in the research pipeline. One obstacle hindering the cross-platform analyses of different datasets is the variations in experimental designs, treatment conditions, and data formats. Drawing meaningful conclusions may sometimes be difficult when there are discrepancies between two germplasms. For example, the transcriptomic data of one germplasm may not be used effectively to explain the proteomic data of another germplasm. Researchers should therefore strategically design experiments to generate interrelated -omics data using carefully selected germplasms. The standardization of data acquisition and storage formats using strictly controlled vocabulary is also important.

With the advance of computer technology and high-throughput analysis platforms, life processes can now be captured, digitized, and stored in the hard disk of a computer. Yet, no matter how perfectly a genome is sequenced and assembled, biological data from experiments are still essential to connect the genotypes and the phenotypes. A few softwares/platforms have been developed to integrate the interactions of cellular components into networks [235,236]. For example, the VirtualPlant has been developed as a software platform for the integration and analysis of different levels of data [237]. It provides large datasets of Arabidopsis gene annotation, gene functional categories, microarray data, biochemical pathways, interaction information, and microRNA:mRNA interaction information. Users can also upload their own gene lists and microarray data for analysis, and identify coexpressed genes, interacting proteins, and metabolites associated with their genes of interest. Building a similar platform for crop plants could be extremely useful but it requires a well-coordinated effort among different research centers and groups.

REFERENCES

1. FAO, How to Feed the World in 2050; FAO: Rome, Italy, 2009; p. 35.
2. Arabidopsis Genome Initiative. The Arabidopsis genome initiative analysis of the genome sequence of the flowering plant Arabidopsis thaliana. Nature 2000, 408, 796–815.
3. BGI Library of Digital Life: Plants, Available online: http://ldl.genomics.org.cn/page/pa-plant.jsp (accessed on 5 January 2013).
4. Sayers, E.W.; Barrett, T.; Benson, D.A.; Bolton, E.; Bryant, S.H.; Canese, K.; Chetvernin, V.; Church, D.M.; DiCuccio, M.; Federhen, S.; et al. Database resources of the National Center for Biotechnology Information. Nucleic Acids Res 2012, 40, D13–D25.
5. Leinonen, R.; Akhtar, R.; Birney, E.; Bower, L.; Cerdeno-Tárraga, A.; Cheng, Y.; Cleland, I.; Faruque, N.; Goodgame, N.; Gibson, R.; et al. The European nucleotide archive. Nucleic Acids Res 2011, 39, D28–D31.
6. Miyazaki, S.; Sugawara, H.; Ikeo, K.; Gojobori, T.; Tateno, Y. DDBJ in the stream of various biological data. Nucleic Acids Res 2004, 32, D31–D34.
7. Goodstein, D.M.; Shu, S.; Howson, R.; Neupane, R.; Hayes, R.D.; Fazo, J.; Mitros, T.; Dirks, W.; Hellsten, U.; Putnam, N.; et al. Phytozome: A comparative platform for green plant genomics. Nucleic Acids Res 2012, 40, D1178–D1186.
8. Liang, C.; Jaiswal, P.; Hebbard, C.; Avraham, S.; Buckler, E.S.; Casstevens, T.; Hurwitz, B.; McCouch, S.; Ni, J.; Pujar, A.; et al. Gramene: A growing plant comparative genomics resource. Nucleic Acids Res 2008, 36, D947–D953.
9. Kanehisa, M.; Araki, M.; Goto, S.; Hattori, M.; Hirakawa, M.; Itoh, M.; Katayama, T.; Kawashima, S.; Okuda, S.; Tokimatsu, T.; et al. KEGG for linking genomes to life and the environment. Nucleic Acids Res 2008, 36, D480–D484.
10. Duvick, J.; Fu, A.; Muppirala, U.; Sabharwal, M.; Wilkerson, M.D.; Lawrence, C.J.; Lushbough, C.; Brendel, V. PlantGDB: A resource for comparative plant genomics. Nucleic Acids Res 2008, 36, D959–D965.
11. Kersey, P.J.; Staines, D.M.; Lawson, D.; Kulesha, E.; Derwent, P.; Humphrey, J.C.; Hughes, D.S.T.; Keenan, S.; Kerhornou, A.; Koscielny, G.; et al. Ensembl Genomes: An integrative resource for genome-scale data from non-vertebrate species. Nucleic Acids Res 2012, 40, D91–D97.
12. Frazer, K.A.; Pachter, L.; Poliakov, A.; Rubin, E.M.; Dubchak, I. VISTA: Computational tools for comparative genomics. Nucleic Acids Res 2004, 32, W273–W279.
13. Proost, S.; van Bel, M.; Sterck, L.; Billiau, K.; van Parys, T.; van de Peer, Y.; Vandepoele, K. PLAZA: A comparative genomics resource to study gene and genome evolution in plants. Plant Cell 2009, 21, 3718–3731.

14.	Sneddon, T.P.; Li, P.; Edmunds, S.C. GigaDB: Announcing the GigaScience database. GigaScience 2012, 1, 1–2.
15.	Bombarely, A.; Menda, N.; Tecle, I.Y.; Buels, R.M.; Strickler, S.; Fischer-York, T.; Pujar, A.; Leto, J.; Gosselin, J.; Mueller, L.A. The Sol Genomics Network (solgenomics.net): Growing tomatoes using Perl. Nucleic Acids Res 2011, 39, D1149–D1155.
16.	Carollo, V.; Matthews, D.E.; Lazo, G.R.; Blake, T.K.; Hummel, D.D.; Lui, N.; Hane, D.L.; Anderson, O.D. GrainGenes 2.0. An improved resource for the small-grains community. Plant Physiol 2005, 139, 643–651.
17.	Gonzales, M.D.; Archuleta, E.; Farmer, A.; Gajendran, K.; Grant, D.; Shoemaker, R.; Beavis, W.D.; Waugh, M.E. The Legume Information System (LIS): An integrated information resource for comparative legume biology. Nucleic Acids Res 2005, 33, D660–D665.
18.	Paterson, A.H.; Bowers, J.E.; Bruggmann, R.; Dubchak, I.; Grimwood, J.; Gundlach, H.; Haberer, G.; Hellsten, U.; Mitros, T.; Poliakov, A.; et al. The Sorghum bicolor genome and the diversification of grasses. Nature 2009, 457, 551–556.
19.	The International Barley Genome Sequencing Consortium. A physical, genetic and functional sequence assembly of the barley genome. Nature 2012, 491, 711–716.
20.	Zhang, G.; Liu, X.; Quan, Z.; Cheng, S.; Xu, X.; Pan, S.; Xie, M.; Zeng, P.; Yue, Z.; Wang, W.; et al. Genome sequence of foxtail millet (Setaria italica) provides insights into grass evolution and biofuel potential. Nat. Biotech 2012, 30, 549–554.
21.	Messmer, R.; Fracheboud, Y.; Bänziger, M.; Vargas, M.; Stamp, P.; Ribaut, J.-M. Drought stress and tropical maize: QTL-by-environment interactions and stability of QTLs across environments for yield components and secondary traits. Theor. Appl. Genet 2009, 119, 913–930.
22.	Sanseverino, W.; Hermoso, A.; D'Alessandro, R.; Vlasova, A.; Andolfo, G.; Frusciante, L.; Lowy, E.; Roma, G.; Ercolano, M.R. PRGdb 2.0: Towards a community-based database model for the analysis of R-genes in plants. Nucleic Acids Res 2013, 41, D1167–D1171.
23.	Sakai, H.; Lee, S.S.; Tanaka, T.; Numa, H.; Kim, J.; Kawahara, Y.; Wakimoto, H.; Yang, C.-C.; Iwamoto, M.; Abe, T.; et al. Rice annotation project database (RAP-DB): An integrative and interactive database for rice genomics. Plant Cell Physiol 2013, 54, e6.
24.	Schaeffer, M.L.; Harper, L.C.; Gardiner, J.M.; Andorf, C.M.; Campbell, D.A.; Cannon, E.K.S.; Sen, T.Z.; Lawrence, C.J. MaizeGDB: Curation and outreach go hand-in-hand. Database 2011, 2011, bar022.
25.	Grant, D.; Nelson, R.T.; Cannon, S.B.; Shoemaker, R.C. SoyBase, the USDA-ARS soybean genetics and genomics database. Nucleic Acids Res 2010, 38, D843–D846.
26.	Fei, Z.; Joung, J.-G.; Tang, X.; Zheng, Y.; Huang, M.; Lee, J.M.; McQuinn, R.; Tieman, D.M.; Alba, R.; Klee, H.J.; et al. Tomato functional genomics database: A comprehensive resource and analysis package for tomato functional genomics. Nucleic Acids Res 2011, 39, D1156–D1163.
27.	Gene Ontology Consortium. The Gene Ontology Website, Available online: http://www.geneontology.org/ (accessed on 10 January 2013).

28. Ashburner, M.; Ball, C.A.; Blake, J.A.; Botstein, D.; Butler, H.; Cherry, J.M.; Davis, A.P.; Dolinski, K.; Dwight, S.S.; Eppig, J.T.; et al. Gene Ontology: Tool for the unification of biology. Nat. Genet 2000, 25, 25–29.

29. Torto-Alalibo, T.; Collmer, C.; Gwinn-Giglio, M. The Plant-Associated Microbe Gene Ontology (PAMGO) Consortium: Community development of new Gene Ontology terms describing biological processes involved in microbe-host interactions. BMC Microbiol 2009, 9, 1–5.

30. Avraham, S.; Tung, C.W.; Ilic, K.; Jaiswal, P.; Kellogg, E.A.; McCouch, S.; Pujar, A.; Reiser, L.; Rhee, S.Y.; Sachs, M.M.; et al. The Plant Ontology Database: A community resource for plant structure and developmental stages controlled vocabulary and annotations. Nucleic Acids Res 2008, 36, D449–D454.

31. Eilbeck, K.; Lewis, S.E.; Mungall, C.J.; Yandell, M.; Stein, L.; Durbin, R.; Ashburner, M. The Sequence Ontology: A tool for the unification of genome annotations. Genome Biol 2005, 6, R44.

32. The Catalogue of Life, Available online: http://www.catalogueoflife.org/ (accessed on 8 January 2013).

33. Integrated Taxonomic Information System. Available online: http://www.itis.gov/ (accessed on 8 January 2013).

34. Wang, K.; Wang, Z.; Li, F.; Ye, W.; Wang, J.; Song, G.; Yue, Z.; Cong, L.; Shang, H.; Zhu, S.; et al. The draft genome of a diploid cotton Gossypium raimondii. Nat. Genet 2012, 44, 1098–1103.

35. Shulaev, V.; Korban, S.S.; Sosinski, B.; Abbott, A.G.; Aldwinckle, H.S.; Folta, K.M.; Iezzoni, A.; Main, D.; Arús, P.; Dandekar, A.M.; et al. Multiple models for rosaceae genomics. Plant Physiol 2008, 147, 985–1003.

36. Shulaev, V.; Sargent, D.J.; Crowhurst, R.N.; Mockler, T.C.; Folkerts, O.; Delcher, A.L.; Jaiswal, P.; Mockaitis, K.; Liston, A.; Mane, S.P.; et al. The genome of woodland strawberry (Fragaria vesca). Nat. Genet 2011, 43, 109–116.

37. Brenchley, R.; Spannagl, M.; Pfeifer, M.; Barker, G.L.A.; D'Amore, R.; Allen, A.M.; McKenzie, N.; Kramer, M.; Kerhornou, A.; Bolser, D.; et al. Analysis of the bread wheat genome using whole-genome shotgun sequencing. Nature 2012, 491, 705–710.

38. Jia, J.; Zhao, S.; Kong, X.; Li, Y.; Zhao, G.; He, W.; Appels, R.; Pfeifer, M.; Tao, Y.; Zhang, X.; et al. Aegilops tauschii draft genome sequence reveals a gene repertoire for wheat adaptation. Nature 2013, 496, 91–95.

39. Ling, H.-Q.; Zhao, S.; Liu, D.; Wang, J.; Sun, H.; Zhang, C.; Fan, H.; Li, D.; Dong, L.; Tao, Y.; et al. Draft genome of the wheat A-genome progenitor Triticum urartu. Nature 2013, 496, 87–90.

40. Lam, H.M.; Xu, X.; Liu, X.; Chen, W.B.; Yang, G.H.; Wong, F.L.; Li, M.W.; He, W.M.; Qin, N.; Wang, B.; et al. Resequencing of 31 wild and cultivated soybean genomes identifies patterns of genetic diversity and selection. Nat. Genet 2010, 42, 1053–1059.

41. Xu, X.; Liu, X.; Ge, S.; Jensen, J.D.; Hu, F.; Li, X.; Dong, Y.; Gutenkunst, R.N.; Fang, L.; Huang, L.; et al. Resequencing 50 accessions of cultivated and wild rice yields markers for identifying agronomically important genes. Nat. Biotechnol 2012, 30, 105–111.

42. Hufford, M.B.; Xu, X.; van Heerwaarden, J.; Pyhajarvi, T.; Chia, J.-M.; Cartwright, R.A.; Elshire, R.J.; Glaubitz, J.C.; Guill, K.E.; Kaeppler, S.M.; et al. Comparative population genomics of maize domestication and improvement. Nat. Genet 2012, 44, 808–811.

43. Kim, M.Y.; Lee, S.; Van, K.; Kim, T.-H.; Jeong, S.-C.; Choi, I.-Y.; Kim, D.-S.; Lee, Y.-S.; Park, D.; Ma, J.; et al. Whole-genome sequencing and intensive analysis of the undomesticated soybean (Glycine soja Sieb. and Zucc.) genome. Proc. Natl. Acad. Sci. USA 2010, 107, 22032–22037.

44. Lai, J.; Li, R.; Xu, X.; Jin, W.; Xu, M.; Zhao, H.; Xiang, Z.; Song, W.; Ying, K.; Zhang, M.; et al. Genome-wide patterns of genetic variation among elite maize inbred lines. Nat. Genet 2010, 42, 1027–1030.

45. Garcia-Mas, J.; Benjak, A.; Sanseverino, W.; Bourgeois, M.; Mir, G.; Gonzalez, V.M.; Henaff, E.; Camara, F.; Cozzuto, L.; Lowy, E.; et al. The genome of melon (Cucumis melo L.). Proc. Natl. Acad. Sci. USA 2012, 109, 11872–11877.

46. Dangl, J.L.; Jones, J.D.G. Plant pathogens and integrated defence responses to infection. Nature 2001, 411, 826–833.

47. Shanmugam, V. Role of extracytoplasmic leucine rich repeat proteins in plant defence mechanisms. Microbiol. Res 2005, 160, 83–94.

48. Torii, K.U. Leucine-rich repeat receptor kinases in plants: Structure, function, and signal transduction pathways. Int. Rev. Cytol 2004, 234, 1–46.

49. Van Ooijen, G.; Mayr, G.; Kasiem, M.M.A.; Albrecht, M.; Cornelissen, B.J.C.; Takken, F.L.W. Structure—Function analysis of the NB-ARC domain of plant disease resistance proteins. J. Exp. Bot 2008, 59, 1383–1397.

50. McNally, K.L.; Childs, K.L.; Bohnert, R.; Davidson, R.M.; Zhao, K.; Ulat, V.J.; Zeller, G.; Clark, R.M.; Hoen, D.R.; Bureau, T.E.; et al. Genomewide SNP variation reveals relationships among landraces and modern varieties of rice. Proc. Natl. Acad. Sci. USA 2009, 106, 12273–12278.

51. Zheng, L.-Y.; Guo, X.-S.; He, B.; Sun, L.-J.; Peng, Y.; Dong, S.-S.; Liu, T.-F.; Jiang, S.; Ramachandran, S.; Liu, C.-M.; et al. Genome-wide patterns of genetic variation in sweet and grain sorghum (Sorghum bicolor). Genome Biol 2011, 12, R114.

52. Dodds, P.N.; Rathjen, J.P. Plant immunity: Towards an integrated view of plant-pathogen interactions. Nat. Rev. Genet 2010, 11, 539–548.

53. D'Hont, A.; Denoeud, F.; Aury, J.-M.; Baurens, F.-C.; Carreel, F.; Garsmeur, O.; Noel, B.; Bocs, S.; Droc, G.; Rouard, M.; et al. The banana (Musa acuminata) genome and the evolution of monocotyledonous plants. Nature 2012, 488, 213–217.

54. Singh, K.B.; Foley, R.C.; Oñate-Sánchez, L. Transcription factors in plant defense and stress responses. Curr. Opin. Plant Biol 2002, 5, 430–436.

55. Collard, B.C.Y.; Jahufer, M.Z.Z.; Brouwer, J.B.; Pang, E.C.K. An introduction to markers, quantitative trait loci (QTL) mapping and marker-assisted selection for crop improvement: The basic concepts. Euphytica 2005, 142, 169–196.

56. Rafalski, J.A. Association genetics in crop improvement. Curr. Opin. Plant Biol 2010, 13, 174–180.

57. Huang, S.; Spielmeyer, W.; Lagudah, E.S.; James, R.A.; Platten, J.D.; Dennis, E.S.; Munns, R. A sodium transporter (HKT7) is a candidate for Nax1, a gene for salt tolerance in durum wheat. Plant Physiol 2006, 142, 1718–1727.

58. Ren, Z.-H.; Gao, J.-P.; Li, L.-G.; Cai, X.-L.; Huang, W.; Chao, D.-Y.; Zhu, M.-Z.; Wang, Z.-Y.; Luan, S.; Lin, H.-X. A rice quantitative trait locus for salt tolerance encodes a sodium transporter. Nat. Genet 2005, 37, 1141–1146.

59. Sutton, T.; Baumann, U.; Hayes, J.; Collins, N.C.; Shi, B.-J.; Schnurbusch, T.; Hay, A.; Mayo, G.; Pallotta, M.; Tester, M.; et al. Boron-toxicity tolerance in barley arising from efflux transporter amplification. Science 2007, 318, 1446–1449.

60. Huang, X.; Feng, Q.; Qian, Q.; Zhao, Q.; Wang, L.; Wang, A.; Guan, J.; Fan, D.; Weng, Q.; Huang, T.; et al. High-throughput genotyping by whole-genome resequencing. Genome Res 2009, 19, 1068–1076.

61. Huang, X.; Wei, X.; Sang, T.; Zhao, Q.; Feng, Q.; Zhao, Y.; Li, C.; Zhu, C.; Lu, T.; Zhang, Z.; et al. Genome-wide association studies of 14 agronomic traits in rice landraces. Nat. Genet 2010, 42, U961–U976.

62. Yu, H.; Xie, W.; Wang, J.; Xing, Y.; Xu, C.; Li, X.; Xiao, J.; Zhang, Q. Gains in QTL detection using an ultra-high density SNP map based on population sequencing relative to traditional RFLP/SSR markers. PLoS One 2011, 6, e17595.

63. Zou, G.; Zhai, G.; Feng, Q.; Yan, S.; Wang, A.; Zhao, Q.; Shao, J.; Zhang, Z.; Zou, J.; Han, B.; et al. Identification of QTLs for eight agronomically important traits using an ultra-high-density map based on SNPs generated from high-throughput sequencing in sorghum under contrasting photoperiods. J. Exp. Bot 2012, 63, 5451–5462.

64. Li, H.; Durbin, R. Fast and accurate long-read alignment with Burrows–Wheeler transform. Bioinformatics 2010, 26, 589–595.

65. Wetterstrand, K. DNA Sequencing Costs: Data from the NHGRI Genome Sequencing Program (GSP), Available online: http://www.genome.gov/sequencingcosts (accessed on 3 January 2013).

66. Heffner, E.L.; Sorrells, M.E.; Jannink, J.-L. Genomic selection for crop improvement. Crop Sci 2009, 49, 1–12.

67. Meuwissen, T.H.E.; Hayes, B.J.; Goddard, M.E. Prediction of total genetic value using genome-wide dense marker maps. Genetics 2001, 157, 1819–1829.

68. Zhao, Y.; Gowda, M.; Liu, W.; Würschum, T.; Maurer, H.P.; Longin, F.H.; Ranc, N.; Reif, J.C. Accuracy of genomic selection in European maize elite breeding populations. Theor. Appl. Genet 2012, 124, 769–776.

69. Heffner, E.L.; Jannink, J.-L.; Iwata, H.; Souza, E.; Sorrells, M.E. Genomic selection accuracy for grain quality traits in biparental wheat populations. Crop Sci 2011, 51, 2597–2606.

70. Heffner, E.L.; Jannink, J.-L.; Sorrells, M.E. Genomic selection accuracy using multifamily prediction models in a wheat breeding program. Plant Gen 2011, 4, 65–75.

71. Zhong, S.Q.; Dekkers, J.C.M.; Fernando, R.L.; Jannink, J.L. Factors affecting accuracy from genomic selection in populations derived from multiple inbred lines: A barley case study. Genetics 2009, 182, 355–364.

72. Oliveira, E.; Resende, M.; Silva Santos, V.; Ferreira, C.; Oliveira, G.; Silva, M.; Oliveira, L.; Aguilar-Vildoso, C. Genome-wide selection in cassava. Euphytica 2012, 187, 263–276.

73. Xu, S. Estimating polygenic effects using markers of the entire genome. Genetics 2003, 163, 789–801.

74. Nakaya, A.; Isobe, S.N. Will genomic selection be a practical method for plant breeding? Ann. Bot 2012, 110, 1303–1316.

75. Jannink, J.-L.; Lorenz, A.J.; Iwata, H. Genomic selection in plant breeding: From theory to practice. Brief. Funct. Genomics 2010, 9, 166–177.
76. Rutkoski, J.; Heffner, E.; Sorrells, M. Genomic selection for durable stem rust resistance in wheat. Euphytica 2011, 179, 161–173.
77. Salse, J. In silico archeogenomics unveils modern plant genome organisation, regulation and evolution. Curr. Opin. Plant Biol 2012, 15, 122–130.
78. Salamov, A.A.; Solovyev, V.V. Ab initio gene finding in drosophila genomic DNA. Genome Res 2000, 10, 516–522.
79. Howe, K.L.; Chothia, T.; Durbin, R. GAZE: A generic framework for the integration of gene-prediction data by dynamic programming. Genome Res 2002, 12, 1418–1427.
80. Allen, J.E.; Salzberg, S.L. JIGSAW: Integration of multiple sources of evidence for gene prediction. Bioinformatics 2005, 21, 3596–3603.
81. Finn, R.D.; Mistry, J.; Tate, J.; Coggill, P.; Heger, A.; Pollington, J.E.; Gavin, O.L.; Gunasekaran, P.; Ceric, G.; Forslund, K.; et al. The Pfam protein families database. Nucleic Acids Res 2010, 38, D211–D222.
82. Hunter, S.; Jones, P.; Mitchell, A.; Apweiler, R.; Attwood, T.K.; Bateman, A.; Bernard, T.; Binns, D.; Bork, P.; Burge, S.; et al. InterPro in 2011: New developments in the family and domain prediction database. Nucleic Acids Res 2012, 40, D306–D312.
83. Kanehisa, M.; Goto, S.; Sato, Y.; Furumichi, M.; Tanabe, M. KEGG for integration and interpretation of large-scale molecular data sets. Nucleic Acids Res 2012, 40, D109–D114.
84. Altschul, S.F.; Gish, W.; Miller, W.; Myers, E.W.; Lipman, D.J. Basic local alignment search tool. J. Mol. Biol 1990, 215, 403–410.
85. Letunic, I.; Doerks, T.; Bork, P. SMART 7: Recent updates to the protein domain annotation resource. Nucleic Acids Res 2012, 40, D302–D305.
86. Finn, R.D.; Clements, J.; Eddy, S.R. HMMER web server: Interactive sequence similarity searching. Nucleic Acids Res 2011, 39, W29–W37.
87. Sharoni, A.M.; Nuruzzaman, M.; Satoh, K.; Shimizu, T.; Kondoh, H.; Sasaya, T.; Choi, I.-R.; Omura, T.; Kikuchi, S. Gene structures, classification and expression models of the AP2/EREBP transcription factor family in rice. Plant Cell Physiol 2011, 52, 344–360.
88. Schena, M.; Shalon, D.; Davis, R.W.; Brown, P.O. Quantitative monitoring of gene expression patterns with a complementary DNA microarray. Science 1995, 270, 467–470.
89. Ozsolak, F.; Milos, P.M. RNA sequencing: Advances, challenges and opportunities. Nat. Rev. Genet 2011, 12, 87–98.
90. Rustici, G.; Kolesnikov, N.; Brandizi, M.; Burdett, T.; Dylag, M.; Emam, I.; Farne, A.; Hastings, E.; Ison, J.; Keays, M.; et al. ArrayExpress update—Trends in database growth and links to data analysis tools. Nucleic Acids Res 2013, 41, D987–D990.
91. Lou, S.-K.; Ni, B.; Lo, L.-Y.; Tsui, S.K.-W.; Chan, T.-F.; Leung, K.-S. ABMapper: A suffix array-based tool for multi-location searching and splice-junction mapping. Bioinformatics 2011, 27, 421–422.
92. Langmead, B.; Salzberg, S.L. Fast gapped-read alignment with Bowtie 2. Nat. Meth 2012, 9, 357–359.

93. Trapnell, C.; Roberts, A.; Goff, L.; Pertea, G.; Kim, D.; Kelley, D.R.; Pimentel, H.; Salzberg, S.L.; Rinn, J.L.; Pachter, L. Differential gene and transcript expression analysis of RNA-seq experiments with TopHat and Cufflinks. Nat. Protoc 2012, 7, 562–578.

94. Wang, L.; Feng, Z.; Wang, X.; Wang, X.; Zhang, X. DEGseq: An R package for identifying differentially expressed genes from RNA-seq data. Bioinformatics 2010, 26, 136–138.

95. Nawrocki, E.P.; Kolbe, D.L.; Eddy, S.R. Infernal 1.0: Inference of RNA alignments. Bioinformatics 2009, 25, 1335–1337.

96. Schulz, M.H.; Zerbino, D.R.; Vingron, M.; Birney, E. Oases: Robust de novo RNA-seq assembly across the dynamic range of expression levels. Bioinformatics 2012, 28, 1086–1092.

97. Trapnell, C.; Pachter, L.; Salzberg, S.L. TopHat: Discovering splice junctions with RNA-Seq. Bioinformatics 2009, 25, 1105–1111.

98. Robertson, G.; Schein, J.; Chiu, R.; Corbett, R.; Field, M.; Jackman, S.D.; Mungall, K.; Lee, S.; Okada, H.M.; Qian, J.Q.; et al. De novo assembly and analysis of RNA-seq data. Nat. Meth 2010, 7, 909–912.

99. Grabherr, M.G.; Haas, B.J.; Yassour, M.; Levin, J.Z.; Thompson, D.A.; Amit, I.; Adiconis, X.; Fan, L.; Raychowdhury, R.; Zeng, Q.; et al. Full-length transcriptome assembly from RNA-Seq data without a reference genome. Nat. Biotechnol 2011, 29, 644–652.

100. Sekhon, R.S.; Lin, H.; Childs, K.L.; Hansey, C.N.; Buell, C.R.; de Leon, N.; Kaeppler, S.M. Genome-wide atlas of transcription during maize development. Plant J 2011, 66, 553–563.

101. Severin, A.; Woody, J.; Bolon, Y.-T.; Joseph, B.; Diers, B.; Farmer, A.; Muehlbauer, G.; Nelson, R.; Grant, D.; Specht, J.; et al. RNA-Seq Atlas of Glycine max: A guide to the soybean transcriptome. BMC Plant Biol 2010, 10, 160.

102. Ge, Y.; Li, Y.; Zhu, Y.-M.; Bai, X.; Lv, D.-K.; Guo, D.; Ji, W.; Cai, H. Global transcriptome profiling of wild soybean (Glycine soja) roots under NaHCO3 treatment. BMC Plant Biol 2010, 10, 153.

103. Ma, T.-L.; Wu, W.-H.; Wang, Y. Transcriptome analysis of rice root responses to potassium deficiency. BMC Plant Biol 2012, 12, 161.

104. An, D.; Yang, J.; Zhang, P. Transcriptome profiling of low temperature-treated cassava apical shoots showed dynamic responses of tropical plant to cold stress. BMC Genomics 2012, 13, 64.

105. Zabala, G.; Zou, J.; Tuteja, J.; Gonzalez, D.; Clough, S.; Vodkin, L. Transcriptome changes in the phenylpropanoid pathway of Glycine max in response to Pseudomonas syringae infection. BMC Plant Biol 2006, 6, 26.

106. Wang, C.; Chen, H.; Hao, Q.; Sha, A.; Shan, Z.; Chen, L.; Zhou, R.; Zhi, H.; Zhou, X. Transcript profile of the response of two soybean genotypes to potassium deficiency. PLoS One 2012, 7, e39856.

107. Lenka, S.K.; Katiyar, A.; Chinnusamy, V.; Bansal, K.C. Comparative analysis of drought-responsive transcriptome in Indica rice genotypes with contrasting drought tolerance. Plant Biotechnol. J 2011, 9, 315–327.

108. Zhang, T.; Zhao, X.; Wang, W.; Pan, Y.; Huang, L.; Liu, X.; Zong, Y.; Zhu, L.; Yang, D.; Fu, B. Comparative transcriptome profiling of chilling stress responsiveness in two contrasting rice genotypes. PLoS One 2012, 7, e43274.

109. Li, Y.C.; Meng, F.R.; Zhang, C.Y.; Zhang, N.; Sun, M.S.; Ren, J.P.; Niu, H.B.; Wang, X.; Yin, J. Comparative analysis of water stress-responsive transcriptomes in drought-susceptible and -tolerant wheat (Triticum aestivum L.). J. Plant Biol 2012, 55, 349–360.

110. Zahaf, O.; Blanchet, S.; de Zelicourt, A.; Alunni, B.; Plet, J.; Laffont, C.; de Lorenzo, L.; Imbeaud, S.; Ichante, J.-L.; Diet, A.; et al. Comparative transcriptomic analysis of salt adaptation in roots of contrasting Medicago truncatula genotypes. Mol. Plant 2012, 5, 1068–1081.

111. Puranik, S.; Jha, S.; Srivastava, P.S.; Sreenivasulu, N.; Prasad, M. Comparative transcriptome analysis of contrasting foxtail millet cultivars in response to short-term salinity stress. J. Plant Physiol 2011, 168, 280–287.

112. Delker, C.; Quint, M. Expression level polymorphisms: Heritable traits shaping natural variation. Trends Plant Sci 2011, 16, 481–488.

113. Holloway, B.; Luck, S.; Beatty, M.; Rafalski, J.-A.; Li, B. Genome-wide expression quantitative trait loci (eQTL) analysis in maize. BMC Genomics 2011, 12, 336.

114. Wang, J.; Yu, H.; Xie, W.; Xing, Y.; Yu, S.; Xu, C.; Li, X.; Xiao, J.; Zhang, Q. A global analysis of QTLs for expression variations in rice shoots at the early seedling stage. Plant J 2010, 63, 1063–1074.

115. Chen, X.; Hackett, C.A.; Niks, R.E.; Hedley, P.E.; Booth, C.; Druka, A.; Marcel, T.C.; Vels, A.; Bayer, M.; Milne, I.; et al. An eQTL Analysis of partial resistance to Puccinia hordei in barley. PLoS One 2010, 5, e8598.

116. Mann, M. Can proteomics retire the western blot? J. Proteome Res 2008, 7, 3065–3065.

117. Gygi, S.P.; Rist, B.; Gerber, S.A.; Turecek, F.; Gelb, M.H.; Aebersold, R. Quantitative analysis of complex protein mixtures using isotope-coded affinity tags. Nat. Biotechnol 1999, 17, 994–999.

118. Ong, S.E.; Blagoev, B.; Kratchmarova, I.; Kristensen, D.B.; Steen, H.; Pandey, A.; Mann, M. Stable isotope labeling by amino acids in cell culture, SILAC, as a simple and accurate approach to expression proteomics. Mol. Cell. Proteomics 2002, 1, 376–386.

119. Wiese, S.; Reidegeld, K.A.; Meyer, H.E.; Warscheid, B. Protein labeling by iTRAQ: A new tool for quantitative mass spectrometry in proteome research. Proteomics 2007, 7, 340–350.

120. Cox, J.; Mann, M. MaxQuant enables high peptide identification rates, individualized p.p.b.-range mass accuracies and proteome-wide protein quantification. Nat. Biotechnol 2008, 26, 1367–1372.

121. Old, W.M.; Meyer-Arendt, K.; Aveline-Wolf, L.; Pierce, K.G.; Mendoza, A.; Sevinsky, J.R.; Resing, K.A.; Ahn, N.G. Comparison of label-free methods for quantifying human proteins by shotgun proteomics. Mol. Cell. Proteomics 2005, 4, 1487–1502.

122. Jorrín, J.V.; Maldonado, A.M.; Castillejo, M.A. Plant proteome analysis: A 2006 update. Proteomics 2007, 7, 2947–2962.

123. Jorrin-Novo, J.V.; Maldonado, A.M.; Echevarria-Zomeno, S.; Valledor, L.; Castillejo, M.A.; Curto, M.; Valero, J.; Sghaier, B.; Donoso, G.; Redondo, I. Plant proteomics update (2007–2008): Second-generation proteomic techniques, an appropriate experimental design, and data analysis to fulfill MIAPE standards, increase

plant proteome coverage and expand biological knowledge. J. Proteomics 2009, 72, 285–314.

124. Kamal, A.H.M.; Cho, K.; Kim, D.-E.; Uozumi, N.; Chung, K.-Y.; Lee, S.Y.; Choi, J.-S.; Cho, S.-W.; Shin, C.-S.; Woo, S.H. Changes in physiology and protein abundance in salt-stressed wheat chloroplasts. Mol. Biol. Rep 2012, 39, 9059–9074.

125. Ahsan, N.; Nakamura, T.; Komatsu, S. Differential responses of microsomal proteins and metabolites in two contrasting cadmium (Cd)-accumulating soybean cultivars under Cd stress. Amino Acids 2012, 42, 317–327.

126. Wang, H.; Wang, S.; Xia, Y. Identification and verification of redox-sensitive proteins in Arabidopsis thaliana. Methods Mol. Biol 2012, 876, 83–94.

127. Galant, A.; Koester, R.P.; Ainsworth, E.A.; Hicks, L.M.; Jez, J.M. From climate change to molecular response: Redox proteomics of ozone-induced responses in soybean. New Phytol 2012, 194, 220–229.

128. Nakagami, H.; Sugiyama, N.; Mochida, K.; Daudi, A.; Yoshida, Y.; Toyoda, T.; Tomita, M.; Ishihama, Y.; Shirasu, K. Large-scale comparative phosphoproteomics identifies conserved phosphorylation sites in plants. Plant Physiol 2010, 153, 1161–1174.

129. Agrawal, G.K.; Jwa, N.-S.; Lebrun, M.-H.; Job, D.; Rakwal, R. Plant secretome: Unlocking secrets of the secreted proteins. Proteomics 2010, 10, 799–827.

130. Alexandersson, E.; Ashfaq, A.; Resjö, S.; Andreasson, E. Plant secretome proteomics. Front. Plant Sci 2013, 4, 9.

131. Catalá, C.; Howe, K.J.; Hucko, S.; Rose, J.K.C.; Thannhauser, T.W. Towards characterization of the glycoproteome of tomato (Solanum lycopersicum) fruit using Concanavalin A lectin affinity chromatography and LC-MALDI-MS/MS analysis. Proteomics 2011, 11, 1530–1544.

132. Ruiz-May, E.; Kim, S.J.; Brandizzi, F.; Rose, J.K.C. The secreted plant n-glycoproteome and associated secretory pathways. Front. Plant Sci 2012, 3, 117.

133. Pawson, T.; Nash, P. Protein-protein interactions define specificity in signal transduction. Genes Dev 2000, 14, 1027–1047.

134. Zhang, Y.; Gao, P.; Yuan, J.S. Plant protein-protein interaction network and interactome. Curr. Genomics 2010, 11, 40–46.

135. Wittig, I.; Braun, H.-P.; Schagger, H. Blue native PAGE. Nat. Protoc 2006, 1, 418–428.

136. Hue, M.; Riffle, M.; Vert, J.-P.; Noble, W. Large-scale prediction of protein-protein interactions from structures. BMC Bioinforma 2010, 11, 144.

137. Moal, I.H.; Agius, R.; Bates, P.A. Protein-protein binding affinity prediction on a diverse set of structures. Bioinformatics 2011, 27, 3002–3009.

138. Cusick, M.E.; Yu, H.; Smolyar, A.; Venkatesan, K.; Carvunis, A.-R.; Simonis, N.; Rual, J.-F.; Borick, H.; Braun, P.; Dreze, M.; et al. Literature-curated protein interaction datasets. Nat. Meth 2009, 6, 39–46.

139. Ho, C.-L.; Wu, Y.; Shen, H.-B.; Provart, N.; Geisler, M. A predicted protein interactome for rice. Rice 2012, 5, 15.

140. Cui, J.; Li, P.; Li, G.; Xu, F.; Zhao, C.; Li, Y.H.; Yang, Z.N.; Wang, G.; Yu, Q.B.; Li, Y.X.; et al. AtPID: *Arabidopsis thaliana* protein interactome database—An integrative platform for plant systems biology. Nucleic Acids Res 2008, 36, D999–D1008.

141. Arabidopsis interactome mapping consortium. Evidence for network evolution in an Arabidopsis interactome map. Science 2011, 333, 601–607.

142. Seo, Y.-S.; Chern, M.; Bartley, L.E.; Han, M.; Jung, K.-H.; Lee, I.; Walia, H.; Richter, T.; Xu, X.; Cao, P.; et al. Towards establishment of a rice stress response interactome. PLoS Genet 2011, 7, e1002020.

143. Gu, H.; Zhu, P.; Jiao, Y.; Meng, Y.; Chen, M. PRIN: A predicted rice interactome network. BMC Bioinforma 2011, 12, 161.

144. Chatr-aryamontri, A.; Breitkreutz, B.-J.; Heinicke, S.; Boucher, L.; Winter, A.; Stark, C.; Nixon, J.; Ramage, L.; Kolas, N.; O'Donnell, L.; et al. The BioGRID interaction database: 2013 update. Nucleic Acids Res 2013, 41, D816–D823.

145. Xenarios, I.; Salwínski, L.; Duan, X.J.; Higney, P.; Kim, S.-M.; Eisenberg, D. DIP, the Database of Interacting Proteins: A research tool for studying cellular networks of protein interactions. Nucleic Acids Res 2002, 30, 303–305.

146. Mingwei, M.; Haoyang, C.; Wen, Z.; Zhirui, Y.; Xiao, L.; Xinjian, F.; Quansheng, F. PlaPID: A Database of Protein-Protein Interactions in Plants. Proceedings of the 4th International Conference on Bioinformatics and Biomedical Engineering (iCBBE), Chengdu, China, 18–20 June 2010; pp. 1–4.

147. Kerrien, S.; Aranda, B.; Breuza, L.; Bridge, A.; Broackes-Carter, F.; Chen, C.; Duesbury, M.; Dumousseau, M.; Feuermann, M.; Hinz, U.; et al. The IntAct molecular interaction database in 2012. Nucleic Acids Res 2012, 40, D841–D846.

148. Cooper, B.; Clarke, J.D.; Budworth, P.; Kreps, J.; Hutchison, D.; Park, S.; Guimil, S.; Dunn, M.; Luginbühl, P.; Ellero, C.; et al. A network of rice genes associated with stress response and seed development. Proc. Natl. Acad. Sci. USA 2003, 100, 4945–4950.

149. Tardif, G.; Kane, N.; Adam, H.; Labrie, L.; Major, G.; Gulick, P.; Sarhan, F.; Laliberté, J.-F. Interaction network of proteins associated with abiotic stress response and development in wheat. Plant Mol. Biol 2007, 63, 703–718.

150. Afzal, A.J.; Natarajan, A.; Saini, N.; Iqbal, M.J.; Geisler, M.; El Shemy, H.A.; Mungur, R.; Willmitzer, L.; Lightfoot, D.A. The nematode resistance allele at the rhg1 locus alters the proteome and primary metabolism of soybean roots. Plant Physiol 2009, 151, 1264–1280.

151. Gendler, K.; Paulsen, T.; Napoli, C. ChromDB: The chromatin database. Nucleic Acids Res 2008, 36, D298–D302.

152. Morison, I.M.; Ramsay, J.P.; Spencer, H.G. A census of mammalian imprinting. Trends Genet 2005, 21, 457–465.

153. Tsukahara, S.; Kobayashi, A.; Kawabe, A.; Mathieu, O.; Miura, A.; Kakutani, T. Bursts of retrotransposition reproduced in Arabidopsis. Nature 2009, 461, 423–426.

154. Zhang, X.Y.; Yazaki, J.; Sundaresan, A.; Cokus, S.; Chan, S.W.L.; Chen, H.M.; Henderson, I.R.; Shinn, P.; Pellegrini, M.; Jacobsen, S.E.; et al. Genome-wide high-resolution mapping and functional analysis of DNA methylation in Arabidopsis. Cell 2006, 126, 1189–1201.

155. Zilberman, D.; Gehring, M.; Tran, R.K.; Ballinger, T.; Henikoff, S. Genome-wide analysis of *Arabidopsis thaliana* DNA methylation uncovers an interdependence between methylation and transcription. Nat. Genet 2007, 39, 61–69.

156. Seifert, M.; Cortijo, S.; Colome-Tatche, M.; Johannes, F.; Roudier, F.; Colot, V. Me-
     DIP-HMM: Genome-wide identification of distinct DNA methylation states from
     high-density tiling arrays. Bioinformatics 2012, 28, 2930–2939.
157. Cokus, S.J.; Feng, S.H.; Zhang, X.Y.; Chen, Z.G.; Merriman, B.; Haudenschild,
     C.D.; Pradhan, S.; Nelson, S.F.; Pellegrini, M.; Jacobsen, S.E. Shotgun bisulphite
     sequencing of the Arabidopsis genome reveals DNA methylation patterning. Nature
     2008, 452, 215–219.
158. Bibikova, M.; Barnes, B.; Tsan, C.; Ho, V.; Klotzle, B.; Le, J.M.; Delano, D.; Zhang,
     L.; Schroth, G.P.; Gunderson, K.L.; et al. High density DNA methylation array with
     single CpG site resolution. Genomics 2011, 98, 288–295.
159. Meissner, A.; Gnirke, A.; Bell, G.W.; Ramsahoye, B.; Lander, E.S.; Jaenisch, R.
     Reduced representation bisulfite sequencing for comparative high-resolution DNA
     methylation analysis. Nucleic Acids Res 2005, 33, 5868–5877.
160. Dowen, R.H.; Pelizzola, M.; Schmitz, R.J.; Lister, R.; Dowen, J.M.; Nery, J.R.;
     Dixon, J.E.; Ecker, J.R. Widespread dynamic DNA methylation in response to biotic
     stress. Proc. Natl. Acad. Sci. USA 2012, 109, E2183–E2191.
161. Wang, W.S.; Pan, Y.J.; Zhao, X.Q.; Dwivedi, D.; Zhu, L.H.; Ali, J.; Fu, B.Y.; Li, Z.K.
     Drought-induced site-specific DNA methylation and its association with drought
     tolerance in rice (Oryza sativa L.). J. Exp. Bot 2011, 62, 1951–1960.
162. Zhong, L.; Xu, Y.H.; Wang, J.B. DNA-methylation changes induced by salt stress in
     wheat Triticum aestivum. Afr. J. Biotechnol 2009, 8, 6201–6207.
163. Calarco, J.P.; Borges, F.; Donoghue, M.T.; van Ex, F.; Jullien, P.E.; Lopes, T.; Gard-
     ner, R.; Berger, F.; Feijo, J.A.; Becker, J.D.; et al. Reprogramming of DNA methyla-
     tion in pollen guides epigenetic inheritance via small RNA. Cell 2012, 151, 194–205.
164. Holeski, L.M.; Jander, G.; Agrawal, A.A. Transgenerational defense induction and
     epigenetic inheritance in plants. Trends Ecol. Evol 2012, 27, 618–626.
165. Kou, H.P.; Li, Y.; Song, X.X.; Ou, X.F.; Xing, S.C.; Ma, J.; von Wettstein, D.; Liu,
     B. Heritable alteration in DNA methylation induced by nitrogen-deficiency stress
     accompanies enhanced tolerance by progenies to the stress in rice (Oryza sativa L.).
     J. Plant Physiol 2011, 168, 1685–1693.
166. Lutsik, P.; Feuerbach, L.; Arand, J.; Lengauer, T.; Walter, J.; Bock, C. BiQ Analyzer
     HT: Locus-specific analysis of DNA methylation by high-throughput bisulfite se-
     quencing. Nucleic Acids Res 2011, 39, W551–W556.
167. Krueger, F.; Andrews, S.R. Bismark: A flexible aligner and methylation caller for
     Bisulfite-Seq applications. Bioinformatics 2011, 27, 1571–1572.
168. Harris, E.Y.; Ponts, N.; Le Roch, K.G.; Lonardi, S. BRAT-BW: Efficient and accu-
     rate mapping of bisulfite-treated reads. Bioinformatics 2012, 28, 1795–1796.
169. Hansen, K.D.; Langmead, B.; Irizarry, R.A. BSmooth: From whole genome bisulfite
     sequencing reads to differentially methylated regions. Genome Biol 2012, 13, R83.
170. Chen, P.Y.; Cokus, S.J.; Pellegrini, M. BS Seeker: Precise mapping for bisulfite se-
     quencing. BMC Bioinforma 2010, 11, 203.
171. Xi, Y.; Li, W. BSMAP: Whole genome bisulfite sequence MAPping program. BMC
     Bioinforma 2009, 10, 232.
172. Su, J.Z.; Yan, H.D.; Wei, Y.J.; Liu, H.B.; Liu, H.; Wang, F.; Lv, J.; Wu, Q.; Zhang,
     Y. CpG_MPs: Identification of CpG methylation patterns of genomic regions from
     high-throughput bisulfite sequencing data. Nucleic Acids Res 2013, 41, e4.

173. Benoukraf, T.; Wongphayak, S.; Hadi, L.H.; Wu, M.; Soong, R. GBSA: A comprehensive software for analysing whole genome bisulfite sequencing data. Nucleic Acids Res. 2012. [CrossRef]
174. Gruntman, E.; Qi, Y.J.; Slotkin, R.K.; Roeder, T.; Martienssen, R.A.; Sachidanandam, R. Kismeth: Analyzer of plant methylation states through bisulfite sequencing. BMC Bioinforma 2008, 9, 371.
175. Kumaki, Y.; Oda, M.; Okano, M. QUMA: Quantification tool for methylation analysis. Nucleic Acids Res 2008, 36, W170–W175.
176. Xi, Y.X.; Bock, C.; Muller, F.; Sun, D.Q.; Meissner, A.; Li, W. RRBSMAP: A fast, accurate and user-friendly alignment tool for reduced representation bisulfite sequencing. Bioinformatics 2012, 28, 430–432.
177. Strahl, B.D.; Allis, C.D. The language of covalent histone modifications. Nature 2000, 403, 41–45.
178. Marino-Ramirez, L.; Levine, K.M.; Morales, M.; Zhang, S.Y.; Moreland, R.T.; Baxevanis, A.D.; Landsman, D. The histone database: An integrated resource for histones and histone fold-containing proteins. Database-Oxford 2011. [CrossRef]
179. Lee, K.K.; Workman, J.L. Histone acetyltransferase complexes: One size doesn't fit all. Nat. Rev. Mol. Cell Biol 2007, 8, 284–295.
180. Barski, A.; Cuddapah, S.; Cui, K.R.; Roh, T.Y.; Schones, D.E.; Wang, Z.B.; Wei, G.; Chepelev, I.; Zhao, K.J. High-resolution profiling of histone methylations in the human genome. Cell 2007, 129, 823–837.
181. Oki, M.; Aihara, H.; Ito, T. Role Of Histone Phosphorylation In Chromatin Dynamics And Its Implications in Diseases. In Chromatin and Disease; Kundu, T., Bittman, R., Dasgupta, D., Engelhardt, H., Flohe, L., Herrmann, H., Holzenburg, A., Nasheuer, H.P., Rottem, S., Wyss, M., Zwickl, P., Eds.; Springer: London, UK, 2007; Volume 41, pp. 323–340.
182. Shivaswamy, S.; Iyer, V.R. Genome-wide analysis of chromatin status using tiling microarrays. Methods 2007, 41, 304–311.
183. Johnson, D.S.; Mortazavi, A.; Myers, R.M.; Wold, B. Genome-wide mapping of in vivo protein-DNA interactions. Science 2007, 316, 1497–1502.
184. Papaefthimiou, D.; Tsaftaris, A.S. Characterization of a drought inducible trithorax-like H3K4 methyltransferase from barley. Biol. Plant 2012, 56, 683–692.
185. Ding, B.; Bellizzi, M.D.R.; Ning, Y.; Meyers, B.C.; Wang, G.-L. HDT701, a Histone H4 deacetylase, negatively regulates plant innate immunity by modulating histone H4 acetylation of defense-related genes in rice. Plant Cell 2012, 24, 3783–3794.
186. Song, Y.G.; Ji, D.D.; Li, S.; Wang, P.; Li, Q.; Xiang, F.N. The dynamic changes of DNA methylation and histone modifications of salt responsive transcription factor genes in soybean. PLoS One 2012, 7, e41274.
187. Zong, W.; Zhong, X.; You, J.; Xiong, L. Genome-wide profiling of histone H3K4-tri-methylation and gene expression in rice under drought stress. Plant Mol. Biol 2013, 81, 175–188.
188. Munns, R.; James, R.A.; Sirault, X.R.R.; Furbank, R.T.; Jones, H.G. New phenotyping methods for screening wheat and barley for beneficial responses to water deficit. J. Exp. Bot 2010, 61, 3499–3507.

189. Golzarian, M.; Frick, R.; Rajendran, K.; Berger, B.; Roy, S.; Tester, M.; Lun, D. Accurate inference of shoot biomass from high-throughput images of cereal plants. Plant Methods 2011, 7, 2.
190. Paproki, A.; Sirault, X.; Berry, S.; Furbank, R.; Fripp, J. A novel mesh processing based technique for 3D plant analysis. BMC Plant Biol 2012, 12, 63.
191. Spielbauer, G.; Armstrong, P.; Baier, J.W.; Allen, W.B.; Richardson, K.; Shen, B.; Settles, A.M. High-throughput near-infrared reflectance spectroscopy for predicting quantitative and qualitative composition phenotypes of individual maize kernels. Cereal Chem 2009, 86, 556–564.
192. Baker, N.R. Chlorophyll fluorescence: A probe of photosynthesis in vivo. Annu. Rev. Plant Biol 2008, 59, 89–113.
193. Condon, A.G.; Richards, R.A.; Rebetzke, G.J.; Farquhar, G.D. Breeding for high water-use efficiency. J. Exp. Bot 2004, 55, 2447–2460.
194. Hargreaves, C.; Gregory, P.; Bengough, A.G. Measuring root traits in barley (Hordeum vulgare ssp. vulgare and ssp. spontaneum) seedlings using gel chambers, soil sacs and X-ray microtomography. Plant Soil 2009, 316, 285–297.
195. Rajendran, K.; Tester, M.; Roy, S.J. Quantifying the three main components of salinity tolerance in cereals. Plant Cell Environ 2009, 32, 237–249.
196. Jones, H.G.; Serraj, R.; Loveys, B.R.; Xiong, L.; Wheaton, A.; Price, A.H. Thermal infrared imaging of crop canopies for the remote diagnosis and quantification of plant responses to water stress in the field. Funct. Plant Biol 2009, 36, 978–989.
197. Moshou, D.; Bravo, C.; Oberti, R.; West, J.; Bodria, L.; McCartney, A.; Ramon, H. Plant disease detection based on data fusion of hyper-spectral and multi-spectral fluorescence imaging using Kohonen maps. Real-Time Imaging 2005, 11, 75–83.
198. Genga, A.; Mattana, M.; Coraggio, I.; Locatelli, F.; Piffanelli, P.; Consonni, R. Plant Metabolomics: A Characterisation of Plant Responses to Abiotic Stresses. In Abiotic Stress in Plants—Mechanisms and Adaptations; Shanker, A., Venkateswarlu, B., Eds.; InTech: Rijeka, Croatia, 2011.
199. Kooke, R.; Keurentjes, J.J.B. Multi-dimensional regulation of metabolic networks shaping plant development and performance. J. Exp. Bot. 2011. [CrossRef]
200. Cramer, G.R.; Urano, K.; Delrot, S.; Pezzotti, M.; Shinozaki, K. Effects of abiotic stress on plants: A systems biology perspective. BMC Plant Biol 2011, 11, 163.
201. Huang, G.-T.; Ma, S.-L.; Bai, L.-P.; Zhang, L.; Ma, H.; Jia, P.; Liu, J.; Zhong, M.; Guo, Z.-F. Signal transduction during cold, salt, and drought stresses in plants. Mol. Biol. Rep 2012, 39, 969–987.
202. Liland, K.H. Multivariate methods in metabolomics—From pre-processing to dimension reduction and statistical analysis. Trends Anal. Chem 2011, 30, 827–841.
203. Stenlund, H.; Gorzsas, A.; Persson, P.; Sundberg, B.; Trygg, J. Orthogonal projections to latent structures discriminant analysis modeling on in situ FT-IR spectral imaging of liver tissue for identifying sources of variability. Anal. Chem 2008, 80, 6898–6906.
204. Trygg, J.; Wold, S. Orthogonal projections to latent structures (O-PLS). J. Chemom 2002, 16, 119–128.
205. Kopka, J.; Schauer, N.; Krueger, S.; Birkemeyer, C.; Usadel, B.; Bergmüller, E.; Dörmann, P.; Weckwerth, W.; Gibon, Y.; Stitt, M.; et al. GMD@CSB.DB: The golm metabolome database. Bioinformatics 2005, 21, 1635–1638.

206. Fahy, E.; Sud, M.; Cotter, D.; Subramaniam, S. LIPID MAPS online tools for lipid research. Nucleic Acids Res 2007, 35, W606–W612.
207. Cui, Q.; Lewis, I.A.; Hegeman, A.D.; Anderson, M.E.; Li, J.; Schulte, C.F.; Westler, W.M.; Eghbalnia, H.R.; Sussman, M.R.; Markley, J.L. Metabolite identification via the Madison Metabolomics Consortium Database. Nat. Biotechnol 2008, 26, 162–164.
208. Brown, M.; Dunn, W.B.; Dobson, P.; Patel, Y.; Winder, C.L.; Francis-McIntyre, S.; Begley, P.; Carroll, K.; Broadhurst, D.; Tseng, A.; et al. Mass spectrometry tools and metabolite-specific databases for molecular identification in metabolomics. Analyst 2009, 134, 1322–1332.
209. Horai, H.; Arita, M.; Kanaya, S.; Nihei, Y.; Ikeda, T.; Suwa, K.; Ojima, Y.; Tanaka, K.; Tanaka, S.; Aoshima, K.; et al. MassBank: A public repository for sharing mass spectral data for life sciences. J. Mass Spectrom 2010, 45, 703–714.
210. Carroll, A.; Badger, M.; Harvey Millar, A. The MetabolomeExpress Project: Enabling web-based processing, analysis and transparent dissemination of GC/MS metabolomics datasets. BMC Bioinforma 2010, 11, 376.
211. Tautenhahn, R.; Cho, K.; Uritboonthai, W.; Zhu, Z.; Patti, G.J.; Siuzdak, G. An accelerated workflow for untargeted metabolomics using the METLIN database. Nat. Biotechnol 2012, 30, 826–828.
212. Zhang, F.; Robinette, S.L.; Bruschweiler-Li, L.; Brüschweiler, R. Web server suite for complex mixture analysis by covariance NMR. Magn. Reson. Chem 2009, 47, S118–S122.
213. Biswas, A.; Mynampati, K.C.; Umashankar, S.; Reuben, S.; Parab, G.; Rao, R.; Kannan, V.S.; Swarup, S. MetDAT: A modular and workflow-based free online pipeline for mass spectrometry data processing, analysis and interpretation. Bioinformatics 2010, 26, 2639–2640.
214. Zhou, B.; Wang, J.; Ressom, H.W. MetaboSearch: Tool for mass-based metabolite identification using multiple databases. PLoS One 2012, 7, e40096.
215. Gavaghan, C.L.; Li, J.V.; Hadfield, S.T.; Hole, S.; Nicholson, J.K.; Wilson, I.D.; Howe, P.W.A.; Stanley, P.D.; Holmes, E. Application of NMR-based Metabolomics to the Investigation of Salt Stress in Maize (Zea mays). Phytochem. Anal 2011, 22, 214–224.
216. Widodo Patterson, J.H.; Newbigin, E.; Tester, M.; Bacic, A.; Roessner, U. Metabolic responses to salt stress of barley (Hordeum vulgare L.) cultivars, Sahara and Clipper, which differ in salinity tolerance. J. Exp. Bot. 2009, 60, 4089–4103.
217. Wu, J.; Yu, H.; Dai, H.; Mei, W.; Huang, X.; Zhu, S.; Peng, M. Metabolite profiles of rice cultivars containing bacterial blight-resistant genes are distinctive from susceptible rice. Acta Biochim. Biophys. Sin 2012, 44, 650–659.
218. Levi, A.; Paterson, A.H.; Cakmak, I.; Saranga, Y. Metabolite and mineral analyses of cotton near-isogenic lines introgressed with QTLs for productivity and drought-related traits. Physiol. Plant 2011, 141, 265–275.
219. Semel, Y.; Schauer, N.; Roessner, U.; Zamir, D.; Fernie, A.R. Metabolite analysis for the comparison of irrigated and non-irrigated field grown tomato of varying genotype. Metabolomics 2007, 3, 289–295.
220. Silvente, S.; Sobolev, A.P.; Lara, M. Metabolite adjustments in drought tolerant and sensitive soybean genotypes in response to water stress. PLoS One 2012, 7, e38554.

221. Witt, S.; Galicia, L.; Lisec, J.; Cairns, J.; Tiessen, A.; Luis Araus, J.; Palacios-Rojas, N.; Fernie, A.R. Metabolic and phenotypic responses of greenhouse-grown maize hybrids to experimentally controlled drought stress. Mol. Plant 2012, 5, 401–417.

222. Komatsu, S.; Yamamoto, A.; Nakamura, T.; Nouri, M.-Z.; Nanjo, Y.; Nishizawa, K.; Furukawa, K. Comprehensive analysis of mitochondria in roots and hypocotyls of soybean under flooding stress using proteomics and metabolomics techniques. J. Proteome Res 2011, 10, 3993–4004.

223. Cho, K.; Shibato, J.; Agrawal, G.K.; Jung, Y.-H.; Kubo, A.; Jwa, N.-S.; Tamogami, S.; Satoh, K.; Kikuchi, S.; Higashi, T.; et al. Integrated transcriptomics, proteomics, and metabolomics analyses to survey ozone responses in the leaves of rice seedling. J. Proteome Res 2008, 7, 2980–2998.

224. Aliferis, K.A.; Jabaji, S. FT-ICR/MS and GC-EI/MS metabolomics networking unravels global potato sprout's responses to Rhizoctonia solani infection. PLoS One 2012, 7, e42576.

225. Figueiredo, A.; Fortes, A.M.; Ferreira, S.; Sebastiana, M.; Choi, Y.H.; Sousa, L.; Acioli-Santos, B.; Pessoa, F.; Verpoorte, R.; Pais, M.S. Transcriptional and metabolic profiling of grape (Vitis vinifera L.) leaves unravel possible innate resistance against pathogenic fungi. J. Exp. Bot 2008, 59, 3371–3381.

226. Hong, Y.-S.; Martinez, A.; Liger-Belair, G.; Jeandet, P.; Nuzillard, J.-M.; Cilindre, C. Metabolomics reveals simultaneous influences of plant defence system and fungal growth in Botrytis cinerea-infected Vitis vinifera cv. Chardonnay berries. J. Exp. Bot 2012, 63, 5773–5785.

227. Cevallos-Cevallos, J.M.; Futch, D.B.; Shilts, T.; Folimonova, S.Y.; Reyes-De-Corcuera, J.I. GC-MS metabolomic differentiation of selected citrus varieties with different sensitivity to citrus huanglongbing. Plant Physiol. Biochem 2012, 53, 69–76.

228. Ali, K.; Maltese, F.; Figueiredo, A.; Rex, M.; Fortes, A.M.; Zyprian, E.; Pais, M.S.; Verpoorte, R.; Choi, Y.H. Alterations in grapevine leaf metabolism upon inoculation with Plasmopara viticola in different time-points. Plant Sci 2012, 191, 100–107.

229. Fumagalli, E.; Baldoni, E.; Abbruscato, P.; Piffanelli, P.; Genga, A.; Lamanna, R.; Consonni, R. NMR techniques coupled with multivariate statistical analysis: Tools to analyse Oryza sativa metabolic content under stress conditions. J. Agron. Crop Sci 2009, 195, 77–88.

230. Rose, M.T.; Rose, T.J.; Pariasca-Tanaka, J.; Yoshihashi, T.; Neuweger, H.; Goesmann, A.; Frei, M.; Wissuwa, M. Root metabolic response of rice (Oryza sativa L.) genotypes with contrasting tolerance to zinc deficiency and bicarbonate excess. Planta 2012, 236, 959–973.

231. Shulaev, V. Metabolomics technology and bioinformatics. Brief. Bioinforma 2006, 7, 128–139.

232. Wu, W.; Zhang, Q.; Zhu, Y.; Lam, H.-M.; Cai, Z.; Guo, D. Comparative metabolic profiling reveals secondary metabolites correlated with soybean salt tolerance. J. Agric. Food Chem 2008, 56, 11132–11138.

233. Johnson, H.E.; Broadhurst, D.; Goodacre, R.; Smith, A.R. Metabolic fingerprinting of salt-stressed tomatoes. Phytochemistry 2003, 62, 919–928.

234. Liu, L.; Li, Y.H.; Li, S.L.; Hu, N.; He, Y.M.; Pong, R.; Lin, D.N.; Lu, L.H.; Law, M. Comparison of next-generation sequencing systems. J. Biomed. Biotechnol. 2012. [CrossRef]

235. Kao, H.-L.; Gunsalus, K.C. Browsing Multidimensional Molecular Networks with the Generic Network Browser (N-Browse). In Current Protocols in Bioinformatics; John Wiley & Sons, Inc: Hoboken, NJ, USA, 2002.

236. Smoot, M.E.; Ono, K.; Ruscheinski, J.; Wang, P.-L.; Ideker, T. Cytoscape 2.8: New features for data integration and network visualization. Bioinformatics 2011, 27, 431–432.

237. Katari, M.S.; Nowicki, S.D.; Aceituno, F.F.; Nero, D.; Kelfer, J.; Thompson, L.P.; Cabello, J.M.; Davidson, R.S.; Goldberg, A.P.; Shasha, D.E.; et al. VirtualPlant: A software platform to support systems biology research. Plant Physiol 2010, 152, 500–515.

238. Jami, S.K.; Clark, G.B.; Ayele, B.T.; Ashe, P.; Kirti, P.B. Genome-wide comparative analysis of annexin superfamily in plants. PLoS One 2012, 7, e47801.

239. Wan, H.; Yuan, W.; Bo, K.; Shen, J.; Pang, X.; Chen, J. Genome-wide analysis of NBS-encoding disease resistance genes in Cucumis sativus and phylogenetic study of NBS-encoding genes in Cucurbitaceae crops. BMC Genomics 2013, 14, 109.

240. Hu, L.; Liu, S. Genome-wide identification and phylogenetic analysis of the ERF gene family in cucumbers. Genet. Mol. Biol 2011, 34, 624–633.

241. Li, Q.; Zhang, C.; Li, J.; Wang, L.; Ren, Z. Genome-wide identification and characterization of R2R3MYB family in Cucumis sativus. PLoS One 2012, 7, e47576.

242. Ling, J.; Jiang, W.; Zhang, Y.; Yu, H.; Mao, Z.; Gu, X.; Huang, S.; Xie, B. Genome-wide analysis of WRKY gene family in Cucumis sativus. BMC Genomics 2011, 12, 471.

243. Kang, Y.; Kim, K.; Shim, S.; Yoon, M.; Sun, S.; Kim, M.; Van, K.; Lee, S.-H. Genome-wide mapping of NBS-LRR genes and their association with disease resistance in soybean. BMC Plant Biol 2012, 12, 139.

244. Du, H.; Yang, S.-S.; Liang, Z.; Feng, B.-R.; Liu, L.; Huang, Y.-B.; Tang, Y.-X. Genome-wide analysis of the MYB transcription factor superfamily in soybean. BMC Plant Biol 2012, 12, 106.

245. Dung Tien, L.; Nishiyama, R.; Watanabe, Y.; Mochida, K.; Yamaguchi-Shinozaki, K.; Shinozaki, K.; Lam-Son Phan, T. Genome-wide survey and expression analysis of the plant-specific NAC transcription factor family in soybean during development and dehydration stress. DNA Res 2011, 18, 263–276.

246. Osorio, M.B.; Buecker-Neto, L.; Castilhos, G.; Turchetto-Zolet, A.C.; Wiebke-Strohm, B.; Bodanese-Zanettini, M.H.; Margis-Pinheiro, M. Identification and in silico characterization of soybean trihelix-GT and bHLH transcription factors involved in stress responses. Genet. Mol. Biol 2012, 35, 233–246.

247. Tran, L.-S.P.; Quach, T.N.; Guttikonda, S.K.; Aldrich, D.L.; Kumar, R.; Neelakandan, A.; Valliyodan, B.; Nguyen, H.T. Molecular characterization of stress-inducible GmNAC genes in soybean. Mol. Genet. Genomics 2009, 281, 647–664.

248. Zhou, Q.-Y.; Tian, A.-G.; Zou, H.-F.; Xie, Z.-M.; Lei, G.; Huang, J.; Wang, C.-M.; Wang, H.-W.; Zhang, J.-S.; Chen, S.-Y. Soybean WRKY-type transcription factor genes, GmWRKY13, GmWRKY21, and GmWRKY54, confer differential tolerance to abiotic stresses in transgenic Arabidopsis plants. Plant Biotechnol. J 2008, 6, 486–503.

249. Liang, D.; Xia, H.; Wu, S.; Ma, F. Genome-wide identification and expression profiling of dehydrin gene family in Malus domestica. Mol. Biol. Rep 2012, 39, 10759–10768.

250. Zhao, T.; Liang, D.; Wang, P.; Liu, J.; Ma, F. Genome-wide analysis and expression profiling of the DREB transcription factor gene family in Malus under abiotic stress. Mol. Genet. Genomics 2012, 287, 423–436.

251. Agalou, A.; Purwantomo, S.; Oevernaes, E.; Johannesson, H.; Zhu, X.; Estiati, A.; de Kam, R.J.; Engstroem, P.; Slamet-Loedin, I.H.; Zhu, Z.; et al. A genome-wide survey of HD-Zip genes in rice and analysis of drought-responsive family members. Plant Mol. Biol 2008, 66, 87–103.

252. Agarwal, P.; Arora, R.; Ray, S.; Singh, A.K.; Singh, V.P.; Takatsuji, H.; Kapoor, S.; Tyagi, A.K. Genome-wide identification of C2H2 zinc-finger gene family in rice and their phylogeny and expression analysis. Plant Mol. Biol 2007, 65, 467–485.

253. Amrutha, R.N.; Sekhar, P.N.; Varshney, R.K.; Kishor, P.B.K. Genome-wide analysis and identification of genes related to potassium transporter families in rice (Oryza sativa L.). Plant Sci 2007, 172, 708–721.

254. Chen, R.; Jiang, Y.; Dong, J.; Zhang, X.; Xiao, H.; Xu, Z.; Gao, X. Genome-wide analysis and environmental response profiling of SOT family genes in rice (Oryza sativa). Genes Genomics 2012, 34, 549–560.

255. Ding, X.; Hou, X.; Xie, K.; Xiong, L. Genome-wide identification of BURP domain-containing genes in rice reveals a gene family with diverse structures and responses to abiotic stresses. Planta 2009, 230, 149–163.

256. Gollan, P.J.; Bhave, M. Genome-wide analysis of genes encoding FK506-binding proteins in rice. Plant Mol. Biol 2010, 72, 1–16.

257. Huang, J.; Zhao, X.; Yu, H.; Ouyang, Y.; Wang, L.; Zhang, Q. The ankyrin repeat gene family in rice: Genome-wide identification, classification and expression profiling. Plant Mol. Biol 2009, 71, 207–226.

258. Jain, M.; Tyagi, A.K.; Khurana, J.P. Genome-wide identification, classification, evolutionary expansion and expression analyses of homeobox genes in rice. FEBS J 2008, 275, 2845–2861.

259. Jiang, S.-Y.; Ramamoorthy, R.; Bhalla, R.; Luan, H.-F.; Venkatesh, P.N.; Cai, M.; Ramachandran, S. Genome-wide survey of the RIP domain family in Oryza sativa and their expression profiles under various abiotic and biotic stresses. Plant Mol. Biol 2008, 67, 603–614.

260. Nuruzzaman, M.; Manimekalai, R.; Sharoni, A.M.; Satoh, K.; Kondoh, H.; Ooka, H.; Kikuchi, S. Genome-wide analysis of NAC transcription factor family in rice. Gene 2010, 465, 30–44.

261. Nuruzzaman, M.; Sharoni, A.M.; Satoh, K.; Al-Shammari, T.; Shimizu, T.; Sasaya, T.; Omura, T.; Kikuchi, S. The thioredoxin gene family in rice: Genome-wide identification and expression profiling under different biotic and abiotic treatments. Biochem. Biophys. Res. Commun 2012, 423, 417–423.

262. Ouyang, Y.; Chen, J.; Xie, W.; Wang, L.; Zhang, Q. Comprehensive sequence and expression profile analysis of Hsp20 gene family in rice. Plant. Mol. Biol 2009, 70, 341–357.

263. Vij, S.; Giri, J.; Dansana, P.K.; Kapoor, S.; Tyagi, A.K. The receptor-like cytoplasmic kinase (OsRLCK) gene family in rice: Organization, phylogenetic relationship, and expression during development and stress. Mol. Plant 2008, 1, 732–750.

264. Wang, D.; Guo, Y.; Wu, C.; Yang, G.; Li, Y.; Zheng, C. Genome-wide analysis of CCCH zinc finger family in Arabidopsis and rice. BMC Genomics 2008, 9, 44.

265. Zhao, H.; Ma, H.; Yu, L.; Wang, X.; Zhao, J. Genome-wide survey and expression analysis of amino acid transporter gene family in rice (Oryza sativa L.). PLoS One 2012, 7, e49210.

266. Wu, J.; Peng, Z.; Liu, S.; He, Y.; Cheng, L.; Kong, F.; Wang, J.; Lu, G. Genome-wide analysis of Aux/IAA gene family in Solanaceae species using tomato as a model. Mol. Genet. Genomics 2012, 287, 295–311.

267. Bai, M.; Yang, G.-S.; Chen, W.-T.; Mao, Z.-C.; Kang, H.-X.; Chen, G.-H.; Yang, Y.-H.; Xie, B.-Y. Genome-wide identification of Dicer-like, Argonaute and RNA-dependent RNA polymerase gene families and their expression analyses in response to viral infection and abiotic stresses in Solanum lycopersicum. Gene 2012, 501, 52–62.

268. Huang, S.; Gao, Y.; Liu, J.; Peng, X.; Niu, X.; Fei, Z.; Cao, S.; Liu, Y. Genome-wide analysis of WRKY transcription factors in Solanum lycopersicum. Mol. Genet. Genomics 2012, 287, 495–513.

269. Kong, F.; Wang, J.; Cheng, L.; Liu, S.; Wu, J.; Peng, Z.; Lu, G. Genome-wide analysis of the mitogen-activated protein kinase gene family in Solanum lycopersicum. Gene 2012, 499, 108–120.

270. Gan, D.; Jiang, H.; Zhang, J.; Zhao, Y.; Zhu, S.; Cheng, B. Genome-wide analysis of BURP domain-containing genes in Maize and Sorghum. Mol. Biol. Rep 2011, 38, 4553–4563.

271. Vannozzi, A.; Dry, I.B.; Fasoli, M.; Zenoni, S.; Lucchin, M. Genome-wide analysis of the grapevine stilbene synthase multigenic family: Genomic organization and expression profiles upon biotic and abiotic stresses. BMC Plant Biol 2012, 12, 130.

272. Zhuang, J.; Peng, R.-H.; Cheng, Z.-M.; Zhang, J.; Cai, B.; Zhang, Z.; Gao, F.; Zhu, B.; Fu, X.-Y.; Jin, X.-F.; et al. Genome-wide analysis of the putative AP2/ERF family genes in Vitis vinifera. Sci. Hortic 2009, 123, 73–81.

273. Cheng, Y.; Li, X.; Jiang, H.; Ma, W.; Miao, W.; Yamada, T.; Zhang, M. Systematic analysis and comparison of nucleotide-binding site disease resistance genes in maize. FEBS J 2012, 279, 2431–2443.

274. Gomez-Anduro, G.; Adriana Ceniceros-Ojeda, E.; Edith Casados-Vazquez, L.; Bencivenni, C.; Sierra-Beltran, A.; Murillo-Amador, B.; Tiessen, A. Genome-wide analysis of the beta-glucosidase gene family in maize (Zea mays L. var B73). Plant Mol. Biol 2011, 77, 159–183.

275. Lin, Y.-X.; Jiang, H.-Y.; Chu, Z.-X.; Tang, X.-L.; Zhu, S.-W.; Cheng, B.-J. Genome-wide identification, classification and analysis of heat shock transcription factor family in maize. BMC Genomics 2011, 12, 76.

276. Peng, X.; Zhao, Y.; Cao, J.; Zhang, W.; Jiang, H.; Li, X.; Ma, Q.; Zhu, S.; Cheng, B. CCCH-type zinc finger family in Maize: Genome-wide identification, classification and expression profiling under abscisic acid and drought treatments. PLoS One 2012, 7, e40120.

277. Wang, W.W.; Ma, Q.; Xiang, Y.; Zhu, S.W.; Cheng, B.J. Genome-wide analysis of immunophilin FKBP genes and expression patterns in Zea mays. Genet. Mol. Res 2012, 11, 1690–1700.

278. Wei, K.-F.; Chen, J.; Chen, Y.-F.; Wu, L.-J.; Xie, D.-X. Molecular phylogenetic and expression analysis of the complete WRKY transcription factor family in Maize. DNA Res 2012, 19, 153–164.

279. Zhang, Z.; Zhang, J.; Chen, Y.; Li, R.; Wang, H.; Wei, J. Genome-wide analysis and identification of HAK potassium transporter gene family in maize (Zea mays L.). Mol. Biol. Rep 2012, 39, 8465–8473.

280. Zhou, M.-L.; Zhang, Q.; Zhou, M.; Sun, Z.-M.; Zhu, X.-M.; Shao, J.-R.; Tang, Y.-X.; Wu, Y.-M. Genome-wide identification of genes involved in raffinose metabolism in Maize. Glycobiology 2012, 22, 1775–1785.

281. Wendelboe-Nelson, C.; Morris, P.C. Proteins linked to drought tolerance revealed by DIGE analysis of drought resistant and susceptible barley varieties. Proteomics 2012, 12, 3374–3385.

282. Fatehi, F.; Hosseinzadeh, A.; Alizadeh, H.; Brimavandi, T.; Struik, P.C. The proteome response of salt-resistant and salt-sensitive barley genotypes to long-term salinity stress. Mol. Biol. Rep 2012, 39, 6387–6397.

283. Cheng, Z.Y.; Woody, O.Z.; McConkey, B.J.; Glick, B.R. Combined effects of the plant growth-promoting bacterium Pseudomonas putida UW4 and salinity stress on the Brassica napus proteome. Appl. Soil Ecol 2012, 61, 255–263.

284. Louarn, S.; Nawrocki, A.; Edelenbos, M.; Jensen, D.F.; Jensen, O.N.; Collinge, D.B.; Jensen, B. The influence of the fungal pathogen Mycocentrospora acerina on the proteome and polyacetylenes and 6-methoxymellein in organic and conventionally cultivated carrots (Daucus carota) during post harvest storage. J. Proteomics 2012, 75, 962–977.

285. Deeba, F.; Pandey, A.K.; Ranjan, S.; Mishra, A.; Singh, R.; Sharma, Y.K.; Shirke, P.A.; Pandey, V. Physiological and proteomic responses of cotton (Gossypium herbaceum L.) to drought stress. Plant Physiol. Biochnol 2012, 53, 6–18.

286. Zheng, M.; Wang, Y.H.; Liu, K.; Shu, H.M.; Zhou, Z.G. Protein expression changes during cotton fiber elongation in response to low temperature stress. J. Plant Physiol 2012, 169, 399–409.

287. Wang, Y.H.; Zheng, M.; Gao, X.B.; Zhou, Z.G. Protein differential expression in the elongating cotton (Gossypium hirsutum L.) fiber under nitrogen stress. Sci. China Life Sci 2012, 55, 984–992.

288. Li, J.; Sun, J.; Yang, Y.J.; Guo, S.R.; Glick, B.R. Identification of hypoxic-responsive proteins in cucumber roots using a proteomic approach. Plant Physiol. Biochnol 2012, 51, 74–80.

289. Palmieri, M.C.; Perazzolli, M.; Matafora, V.; Moretto, M.; Bachi, A.; Pertot, I. Proteomic analysis of grapevine resistance induced by Trichoderma harzianum T39 reveals specific defence pathways activated against downy mildew. J. Exp. Bot 2012, 63, 6237–6251.

290. Wang, Z.; Zhao, F.X.; Zhao, X.; Ge, H.; Chai, L.J.; Chen, S.W.; Perl, A.; Ma, H.Q. Proteomic analysis of berry-sizing effect of GA3 on seedless Vitis vinifera L. Proteomics 2012, 12, 86–94.

291. Minas, I.S.; Tanou, G.; Belghazi, M.; Job, D.; Manganaris, G.A.; Molassiotis, A.; Vasilakakis, M. Physiological and proteomic approaches to address the active role of ozone in kiwifruit post-harvest ripening. J. Exp. Bot 2012, 63, 2449–2464.

292. Huang, H.; Moller, I.M.; Song, S.Q. Proteomics of desiccation tolerance during development and germination of maize embryos. J. Proteomics 2012, 75, 1247–1262.

293. Benesova, M.; Hola, D.; Fischer, L.; Jedelsky, P.L.; Hnilicka, F.; Wilhelmova, N.; Rothova, O.; Kocova, M.; Prochazkova, D.; Honnerova, J.; et al. The physiology

and proteomics of drought tolerance in Maize: Early stomatal closure as a cause of lower tolerance to short-term dehydration? PLoS One 2012, 7, e38017.

294. Fristedt, R.; Wasilewska, W.; Romanowska, E.; Vener, A.V. Differential phosphorylation of thylakoid proteins in mesophyll and bundle sheath chloroplasts from maize plants grown under low or high light. Proteomics 2012, 12, 2852–2861.

295. Muneer, S.; Kim, T.H.; Qureshi, M.I. Fe modulates Cd-induced oxidative stress and the expression of stress responsive proteins in the nodules of Vigna radiata. Plant Growth Regul 2012, 68, 421–433.

296. Rodrigues, S.P.; Ventura, J.A.; Aguilar, C.; Nakayasu, E.S.; Choi, H.; Sobreira, T.J.P.; Nohara, L.L.; Wermelinger, L.S.; Almeida, I.C.; Zingali, R.B.; et al. Label-free quantitative proteomics reveals differentially regulated proteins in the latex of sticky diseased Carica papaya L. plants. J. Proteomics 2012, 75, 3191–3198.

297. Mohammadi, P.P.; Moieni, A.; Komatsu, S. Comparative proteome analysis of drought-sensitive and drought-tolerant rapeseed roots and their hybrid F1 line under drought stress. Amino Acids 2012, 43, 2137–2152.

298. Zhu, M.M.; Dai, S.J.; Zhu, N.; Booy, A.; Simons, B.; Yi, S.; Chen, S.X. Methyl jasmonate responsive proteins in Brassica napus guard cells revealed by iTRAQ-based quantitative proteomics. J. Proteome Res 2012, 11, 3728–3742.

299. Chen, J.H.; Tian, L.; Xu, H.F.; Tian, D.G.; Luo, Y.M.; Ren, C.M.; Yang, L.M.; Shi, J.S. Cold-induced changes of protein and phosphoprotein expression patterns from rice roots as revealed by multiplex proteomic analysis. Plant Omics 2012, 5, 194–199.

300. Ji, K.X.; Wang, Y.Y.; Sun, W.N.; Lou, Q.J.; Mei, H.W.; Shen, S.H.; Chen, H. Drought-responsive mechanisms in rice genotypes with contrasting drought tolerance during reproductive stage. J. Plant Physiol 2012, 169, 336–344.

301. Mirzaei, M.; Soltani, N.; Sarhadi, E.; Pascovici, D.; Keighley, T.; Salekdeh, G.H.; Haynes, P.A.; Atwell, B.J. Shotgun proteomic analysis of long-distance drought signaling in rice roots. J. Proteome Res 2012, 11, 348–358.

302. Koga, H.; Dohi, K.; Nishiuchi, T.; Kato, T.; Takahara, H.; Mori, M.; Komatsu, S. Proteomic analysis of susceptible rice plants expressing the whole plant-specific resistance against Magnaporthe oryzae: Involvement of a thaumatin-like protein. Physiol. Mol. Plant P 2012, 77, 60–66.

303. Li, Y.F.; Zhang, Z.H.; Nie, Y.F.; Zhang, L.H.; Wang, Z.Z. Proteomic analysis of salicylic acid-induced resistance to Magnaporthe oryzae in susceptible and resistant rice. Proteomics 2012, 12, 2340–2354.

304. Hakeem, K.R.; Chandna, R.; Ahmad, A.; Qureshi, M.I.; Iqbal, M. Proteomic analysis for low and high nitrogen-responsive proteins in the leaves of rice genotypes grown at three nitrogen levels. Appl. Biochem. Biotechnol 2012, 168, 834–850.

305. Sawada, H.; Komatsu, S.; Nanjo, Y.; Khan, N.A.; Kohno, Y. Proteomic analysis of rice response involved in reduction of grain yield under elevated ozone stress. Environ. Exp. Bot 2012, 77, 108–116.

306. Wang, Y.D.; Wang, X.; Wong, Y.S. Proteomics analysis reveals multiple regulatory mechanisms in response to selenium in rice. J. Proteomics 2012, 75, 1849–1866.

307. Li, D.X.; Wang, L.J.; Teng, S.L.; Zhang, G.G.; Guo, L.J.; Mao, Q.; Wang, W.; Li, M.; Chen, L. Proteomics analysis of rice proteins up-regulated in response to bacterial leaf streak disease. J. Plant Biol 2012, 55, 316–324.

308. Ngara, R.; Ndimba, R.; Borch-Jensen, J.; Jensen, O.N.; Ndimba, B. Identification and profiling of salinity stress-responsive proteins in Sorghum bicolor seedlings. J. Proteomics 2012, 75, 4139–4150.

309. Hossain, Z.; Hajika, M.; Komatsu, S. Comparative proteome analysis of high and low cadmium accumulating soybeans under cadmium stress. Amino Acids 2012, 43, 2393–2416.

310. Mohammadi, P.P.; Moieni, A.; Hiraga, S.; Komatsu, S. Organ-specific proteomic analysis of drought-stressed soybean seedlings. J. Proteomics 2012, 75, 1906–1923.

311. Salavati, A.; Khatoon, A.; Nanjo, Y.; Komatsu, S. Analysis of proteomic changes in roots of soybean seedlings during recovery after flooding. J. Proteomics 2012, 75, 878–893.

312. Yanagawa, Y.; Komatsu, S. Ubiquitin/proteasome-mediated proteolysis is involved in the response to flooding stress in soybean roots, independent of oxygen limitation. Plant Sci 2012, 185, 250–258.

313. Khatoon, A.; Rehman, S.; Salavati, A.; Komatsu, S. A comparative proteomics analysis in roots of soybean to compatible symbiotic bacteria under flooding stress. Amino Acids 2012, 43, 2513–2525.

314. Wang, L.Q.; Ma, H.; Song, L.R.; Shu, Y.J.; Gu, W.H. Comparative proteomics analysis reveals the mechanism of pre-harvest seed deterioration of soybean under high temperature and humidity stress. J. Proteomics 2012, 75, 2109–2127.

315. Wang, Y.; Yuan, X.Z.; Hu, H.; Liu, Y.; Sun, W.H.; Shan, Z.H.; Zhou, X.A. Proteomic analysis of differentially expressed proteins in resistant soybean leaves after Phakopsora pachyrhizi infection. J. Phytopathol 2012, 160, 554–560.

316. Ma, H.; Song, L.; Shu, Y.; Wang, S.; Niu, J.; Wang, Z.; Yu, T.; Gu, W.; Ma, H. Comparative proteomic analysis of seedling leaves of different salt tolerant soybean genotypes. J. Proteomics 2012, 75, 1529–1546.

317. Khatoon, A.; Rehman, S.; Hiraga, S.; Makino, T.; Komatsu, S. Organ-specific proteomics analysis for identification of response mechanism in soybean seedlings under flooding stress. J. Proteomics 2012, 75, 5706–5723.

318. Nanjo, Y.; Skultety, L.; Uvackova, L.; Kubicova, K.; Hajduch, M.; Komatsu, S. Mass spectrometry-based analysis of proteomic changes in the root tips of flooded soybean seedlings. J. Proteome Res 2012, 11, 372–385.

319. Khatoon, A.; Rehman, S.; Oh, M.W.; Woo, S.H.; Komatsu, S. Analysis of response mechanism in soybean under low oxygen and flooding stresses using gel-base proteomics technique. Mol. Biol. Rep 2012, 39, 10581–10594.

320. Koehler, G.; Wilson, R.C.; Goodpaster, J.V.; Sonsteby, A.; Lai, X.; Witzmann, F.A.; You, J.S.; Rohloff, J.; Randall, S.K.; Alsheikh, M. Proteomic study of low-temperature responses in strawberry cultivars (Fragaria x ananassa) that differ in cold tolerance. Plant Physiol 2012, 159, 1787–1805.

321. Fang, X.P.; Chen, W.Y.; Xin, Y.; Zhang, H.M.; Yan, C.Q.; Yu, H.; Liu, H.; Xiao, W.F.; Wang, S.Z.; Zheng, G.Z.; et al. Proteomic analysis of strawberry leaves infected with Colletotrichum fragariae. J. Proteomics 2012, 75, 4074–4090.

322. Zhou, G.; Yang, L.T.; Li, Y.R.; Zou, C.L.; Huang, L.P.; Qiu, L.H.; Huang, X.; Srivastava, M.K. Proteomic analysis of osmotic stress-responsive proteins in sugarcane leaves. Plant Mol. Biol. Rep 2012, 30, 349–359.

323. Shah, P.; Powell, A.L.T.; Orlando, R.; Bergmann, C.; Gutierrez-Sanchez, G. Pro-
     teomic analysis of ripening tomato fruit infected by Botrytis cinerea. J. Proteome
     Res 2012, 11, 2178–2192.
324. Ge, P.; Ma, C.; Wang, S.; Gao, L.; Li, X.; Guo, G.; Ma, W.; Yan, Y. Comparative pro-
     teomic analysis of grain development in two spring wheat varieties under drought
     stress. Anal. Bioanal. Chem 2012, 402, 1297–1313.
325. Kang, G.Z.; Li, G.Z.; Xu, W.; Peng, X.Q.; Han, Q.X.; Zhu, Y.J.; Guo, T.C. Pro-
     teomics reveals the effects of salicylic acid on growth and tolerance to subsequent
     drought stress in wheat. J. Proteome Res 2012, 11, 6066–6079.
326. Vitamvas, P.; Prasil, I.T.; Kosova, K.; Planchon, S.; Renaut, J. Analysis of proteome
     and frost tolerance in chromosome 5A and 5B reciprocal substitution lines between
     two winter wheats during long-term cold acclimation. Proteomics 2012, 12, 68–85.
327. Gunnaiah, R.; Kushalappa, A.C.; Duggavathi, R.; Fox, S.; Somers, D.J. Integrated
     metabolo-proteomic approach to decipher the mechanisms by which wheat QTL
     (Fhb1) contributes to resistance against Fusarium graminearum. PLoS One 2012,
     7, e40695.
328. Ravalason, H.; Grisel, S.; Chevret, D.; Favel, A.; Berrin, J.G.; Sigoillot, J.C.; Her-
     poel-Gimbert, I. Fusarium verticillioides secretome as a source of auxiliary enzymes
     to enhance saccharification of wheat straw. Bioresour. Technol 2012, 114, 589–596.
329. Kang, G.Z.; Li, G.Z.; Zheng, B.B.; Han, Q.X.; Wang, C.Y.; Zhu, Y.J.; Guo, T.C. Pro-
     teomic analysis on salicylic acid-induced salt tolerance in common wheat seedlings
     (Triticum aestivum L.). Biochim. Biophys. Acta 2012, 1824, 1324–1333.

*There are several supplemental files that are not available in this version
of the article. To view this additional information, please use the citation
on the first page of this chapter.*

CHAPTER 6

SNP Discovery Through Next-Generation Sequencing and Its Applications

SANTOSH KUMAR, TRAVIS W. BANKS, AND SYLVIE CLOUTIER

6.1 INTRODUCTION

Molecular markers are widely used in plant genetic research and breeding. Single Nucleotide Polymorphisms (SNPs) are currently the marker of choice due to their large numbers in virtually all populations of individuals. The applications of SNP markers have clearly been demonstrated in human genomics where complete sequencing of the human genome led to the discovery of several million SNPs [1] and technologies to analyze large sets of SNPs (up to 1 million) have been developed. SNPs have been applied in areas as diverse as human forensics [2] and diagnostics [3], aquaculture [4], marker assisted-breeding of dairy cattle [5], crop improvement [6], conservation [7], and resource management in fisheries [8]. Functional genomic studies have capitalized upon SNPs located within regulatory genes, transcripts, and Expressed Sequence Tags (ESTs)

SNP Discovery through Next-Generation Sequencing and Its Applications. © Kumar S, Banks TW, and Cloutier S. International Journal of Plant Genomics **2012** *(2012). http://dx.doi. org/10.1155/2012/831460. Used with the authors' permission.*

[9, 10]. Until recently large scale SNP discovery in plants was limited to maize, *Arabidopsis*, and rice [11–15]. Genetic applications such as linkage mapping, population structure, association studies, map-based cloning, marker-assisted plant breeding, and functional genomics continue to be enabled by access to large collections of SNPs. *Arabidopsis thaliana* was the first plant genome sequenced [16] followed soon after by rice [17, 18]. In the year 2011 alone, the number of plant genomes sequenced doubled as compared to the number sequenced in the previous decade, resulting in currently, 31 and counting, publicly released sequenced plant genomes (http://www.phytozome.net/). With the ever increasing throughput of next-generation sequencing (NGS), de novo and reference-based SNP discovery and application are now feasible for numerous plant species.

Sequencing refers to the identification of the nucleotides in a polymer of nucleic acids, whether DNA or RNA. Since its inception in 1977, sequencing has brought about the field of genomics and increased our understanding of the organization and composition of plant genomes. Tremendous improvements in sequencing have led to the generation of large amounts of DNA information in a very short period of time [19]. The analyses of large volumes of data generated through various NGS platforms require powerful computers and complex algorithms and have led to a recent expansion of the bioinformatics field of research. This book chapter focuses on the a priori discovery of SNPs through NGS, bioinformatics tools and resources, and the various downstream applications of SNPs.

6.2 HISTORY AND EVOLUTION OF SEQUENCING TECHNOLOGIES

6.2.1 INVENTION OF SEQUENCING

In 1977, two sequencing methods were developed and published. The Sanger method is a sequencing-by-synthesis (SBS) method that relies on a combination of deoxy- and dideoxy-labeled chain terminator nucleotides [20]. The first complete genome sequencing, that of bacteriophage *phi X174*, was achieved that same year using this pioneering method [21]. The chemical modification followed by cleavage at specific sites method also published in 1977 [22] quickly became the less favored of the two methods because of its technical complexities, use of hazardous chemicals, and inherent difficulty in scale-up. In contrast, the Sanger method, for which Frederick Sanger was awarded his second Nobel Prize in chemistry in 1980, was quickly adopted by the biotechnology industry which implemented it using a broad array of chemistries and detection methods [19].

6.2.2 SEQUENCING TECHNOLOGIES

In the last decade, new sequencing technologies have outperformed Sanger-based sequencing in throughput and overall cost, if not quite in sequence length and error rate [23]. This section will focus on the three main NGS platforms as well as the two main third-generation sequencing (TGS) platforms, their throughput and relative cost. We made every effort to ensure the accuracy of the data at the time of submission. However, the cost and throughput of these sequencing platforms change rapidly and, as such, our analysis only represents a snapshot in time. The flux of innovation in this field imposes a need for constant assessment of the technologies' potentials and realignment of research goals.

6.2.2.1 ROCHE (454) SEQUENCING

Pyrosequencing was the first of the new highly parallel sequencing technologies to reach the market [24]. It is commonly referred to as 454 sequencing after the name of the company that first commercialized it. It is an SBS method where single fragments of DNA are hybridized to a capture bead array and the beads are emulsified with regents necessary to PCR amplifying the individually bound template. Each bead in the emulsion acts as an independent PCR where millions of copies of the original template are produced and bound to the capture beads which then serve as the templates for the subsequent sequencing reaction. The individual beads are deposited into a picotiter plate along with DNA polymerase, primers, and the enzymes necessary to create fluorescence through the consumption of inorganic phosphate produced during sequencing. The instrument washes the picotiter plate with each of the DNA bases in turn. As template-specific incorporation of a base by DNA polymerase occurs, a pyrophosphate (PPi) is produced. This pyrophosphate is detected by an enzymatic luminometric inorganic pyrophosphate detection assay (ELIDA) through the generation of a light signal following the conversion of PPi into ATP [25]. Thus, the wells in which the current nucleotides are being incorporated by the sequencing reaction occurring on the bead emit a light signal proportional to the number of nucleotides incorporated, whereas wells in which the nucleotides are not being incorporated do not. The instrument repeats the sequential nucleotide wash cycle hundreds of times to lengthen the sequences. The 454 GS FLX Titanium XL$^+$ platform currently generates up to 700 MB of raw 750 bp reads in a 23 hour run. The technology has difficulty quantifying homopolymers resulting in insertions/deletions and has an overall error rate of approximately 1%. Reagent costs are approximately $6,200 per run [26].

6.2.2.2 ILLUMINA SEQUENCING

Illumina technology, acquired by Illumina from Solexa, followed the release of 454 sequencing. With this sequencing approach, fragments of DNA are hybridized to a solid substrate called a flow cell. In a process called bridge amplification, the bound DNA template fragments are amplified in an isothermal reaction where copies of the template are created in close proximity to the original. This results in clusters of DNA fragments on the flow cell creating a "lawn" of bound single strand DNA molecules. The molecules are sequenced by flooding the flow cell with a new class of cleavable fluorescent nucleotides and the reagents necessary for DNA polymerization [27]. A complementary strand of each template is synthesized one base at a time using fluorescently labeled nucleotides. The fluorescent molecule is excited by a laser and emits light, the colour of which is different for each of the four bases. The fluorescent label is then cleaved off and a new round of polymerization occurs. Unlike 454 sequencing, all four bases are present for the polymerization step and only a single molecule is incorporated per cycle. The flagship HiSeq2500 sequencing instrument from Illumina can generate up to 600 GB per run with a read length of 100 nt and 0.1% error rate. The Illumina technique can generate sequence from opposite ends of a DNA fragment, so called paired-end (PE) reads. Reagent costs are approximately $23,500 per run [26].

6.2.2.3 APPLIED BIOSYSTEMS (SOLID) SEQUENCING

The SOLiD system was jointly developed by the Harvard Medical School and the Howard Hughes Medical Institute [28]. The library preparation in SOLiD is very similar to Roche/454 in which clonal bead populations are prepared in microreactors containing DNA template, beads, primers, and PCR components. Beads that contain PCR products amplified by emulsion PCR are enriched by a proprietary process. The DNA templates on the beads are modified at their 3′ end to allow attachment to glass slides. A primer is annealed to an adapter on the DNA template and a mixture of fluorescently tagged oligonucleotides is pumped into the flow cell. When the oligonucleotide matches the template sequence, it is ligated onto the primer and the unincorporated nucleotides are washed away. A charged couple device (CCD) camera captures the different colours attached to the primer. Each fluorescence wavelength corresponds to a particular dinucleotide combination. After image capture, the fluorescent tag is removed and new set of oligonucleotides are injected into the flow cell to begin the next round of DNA ligation [19]. This sequencing-by-ligation method in SOLiD-5500x1 platform generates up to 1,410 million PE reads of 75 + 35 nt each with an error rate of 0.01% and reagent cost of approximately $10,500 per run [26].

   Although widely accepted and used, the NGS platforms suffer from amplification biases introduced by PCR and dephasing due to varying extension of templates. The TGS technologies use single molecule sequencing which eliminates the need for prior amplification of DNA thus overcoming the limitations imposed by NGS. The advantages offered by TGS technology are (i) lower cost, (ii) high throughput, (iii) faster turnaround, and (iv) longer reads [19, 29]. The TGS can broadly be classified into three different categories: (i) SBS where individual nucleotides are observed as they incorporate (Pacific Biosciences single molecule real time (SMART), Heliscope true single molecule sequencing (tSMS), and Life

Technologies/Starlight and Ion Torrent), (ii) nanopore sequencing where single nucleotides are detected as they pass through a nanopore (Oxford/Nanopore), and (iii) direct imaging of individual molecules (IBM).

6.2.2.4 HELICOS BIOSCIENCES CORPORATION (HELISCOPE) SEQUENCING

Heliscope sequencing involves DNA library preparation and DNA shearing followed by addition of a poly-A tail to the sheared DNA fragments. These poly-A tailed DNA fragments are attached to flow cells through poly-T anchors. The sequencing proceeds by DNA extension with one out of 4 fluorescent tagged nucleotides incorporated followed by detection by the Heliscope sequencer. The fluorescent tag on the incorporated nucleotide is then chemically cleaved to allow subsequent elongation of DNA [30]. Heliscope sequencers can generate up to 28 GB of sequence data per run (50 channels) with maximum read length of 55 bp at ~99% accuracy [31]. The cost per run per channel is approximately $360.

6.2.2.5 PACIFIC BIOSCIENCES SMART SEQUENCING

The Pacific Biosciences sequencer uses glass anchored DNA polymerases which are housed at the bottom of a zero-mode waveguide (ZMW). DNA fragments are added into the ZMW chamber with the anchored DNA polymerase and nucleotides, each labeled with a different colour fluorophore, and are diffused from above the ZMW. As the nucleotides circulate through the ZMW, only the incorporated nucleotides remain at the bottom of the ZMW while unincorporated nucleotides diffuse back above the ZMW. A laser placed below the ZMW excites only the fluorophores of the incorporated nucleotides as the ZMW entraps the light and does not allow it to reach the unincorporated nucleotides above [32]. The Pacific Biosciences sequencers can generate up to 140 MB of sequences per run (per smart cell) with reads of 2.5 Kbp at ~85% accuracy. The cost per run per smart cell is approximately $600.

Among the TGS technologies, Pacific Biosciences SMART and Heliscope tSMS have been used in characterizing bacterial genomes and in human-disease-related studies [31]; however, TGS has yet to be capitalized upon in plant genomes. The Heliscope generates short reads (55 bp) which may cause ambiguous read mapping due to the presence of paralogous sequences and repetitive elements in plant genomes. The Pacific Biosciences reads have high error rates which limit their direct use in SNP discovery. However, their long reads offer a definite advantage to fill gaps in genomic sequences and, at least in bacterial genomes, NGS reads have proven capable of "correcting" the base call errors of this TGS technology [33–36]. Hybrid assemblies incorporating short (Illumina, SOLiD), medium (454/Roche), and long reads (Pac-Bio) have the potential to yield better quality reference genomes and, as such, would provide an improved tool for SNP discovery.

The choice of a sequencing strategy must take into account the research goals, ability to store and analyze data, the ongoing changes in performance parameters, and the cost of NGS/TGS platforms. Some key considerations include cost per raw base, cost per consensus base, raw and consensus accuracy of bases, read length, cost per read, and availability of PE or single end reads. The pre- and postprocessing protocols such as library construction [37] and pipeline development and implementation for data analysis [38] are also important.

6.2.3 RNA AND CHIP SEQUENCING

Genome-wide analyses of RNA sequences and their qualitative and quantitative measurements provide insights into the complex nature of regulatory networks. RNA sequencing has been performed on a number of plant species including Arabidopsis [39], soybean [40], rice [41], and maize [42] for transcript profiling and detection of splice variants. RNA sequencing has been used in de novo assemblies followed by SNP discovery performed in nonmodel plants such as *Eucalyptus grandis* [43], *Brassica napus* [44], and *Medicago sativa* [45].

RNA deep-sequencing technologies such as digital gene expression [46] and Illumina RNASeq [47] are both qualitative and quantitative in nature and permit the identification of rare transcripts and splice variants [48]. RNA sequencing may be performed following its conversion into cDNA that can then be sequenced as such. This method is, however, prone to error due to (i) the inefficient nature of reverse transcriptases (RTs) [49], (ii) DNA-dependent DNA polymerase activity of RT causing spurious second strand DNA [50], and (iii) artifactual cDNA synthesis due to template switching [51]. Direct RNA sequencing (DRS) developed by Helicos Biosciences Corporation is a high throughput and cost-effective method which eliminates the need for cDNA synthesis and ligation/amplification leading to improved accuracy [52].

Chromatin immunoprecipitation (ChIP) is a specialized sequencing method that was specifically designed to identify DNA sequences involved in in vivo protein DNA interaction [53]. ChIP-sequencing (ChIP-Seq) is used to map the binding sites of transcription factors and other DNA binding sites for proteins such as histones. As such, ChIP-Seq does not aid SNP discovery, but the availability of SNP data along with ChIP-Seq allows the study of allele-specific states of chromatin organization. Deep sequence coverage leading to dense SNP maps permits the identification of transcription factor binding sites and histone-mediated epigenetic modifications [54]. ChIP-Seq can be performed on serial analysis of gene expression (SAGE) tags or PE using Sanger, 454, and Illumina platforms [55, 56].

The DNA, RNA, and ChIP-Seq data is analysed using a reference sequence if available or, in the absence of such reference, it requires de novo assembly, all of which is performed using specialized software, algorithms, pipelines, and hardware.

6.3 COMPUTING RESOURCES FOR SEQUENCE ASSEMBLY

The next-generation platforms generate a considerable amount of data and the impact of this with respect to data storage and processing time can be overlooked when designing an experiment. Bioinformatics research is constantly developing new software and algorithms, data storage approaches, and even new computer architectures to better meet the computation requirements for projects incorporating NGS. This chapter describes the state-of-the-art with respect to software for NGS alignment and analysis at the time of writing.

6.3.1 SOFTWARE FOR SEQUENCE ANALYSIS

Both commercial and noncommercial sequence analysis software are available for Windows, Macintosh, and Linux operating systems. NGS companies offer proprietary software such as consensus assessment of sequence and variation (Cassava) for Illumina data and Newbler for 454 data. Such software tend to be optimized for their respective platform but have limited cross applicability to the others. Web-based portals such as Galaxy [57] are tailored to a multitude of analyses, but the requirement to transfer multigigabyte sequence files across the internet can limit its usability to smaller datasets. Commercially available software such as CLC-Bio (http://www.clcbio.com/) and SeqMan NGen (http://www.dnastar.com/t-sub-products-genomics-seqman-ngen.aspx) provide a friendly user interface, are compatible with different operating systems, require minimal computing knowledge, and are capable of performing multiple downstream analyses. However, they tend to be relatively expensive, have narrow customizability, and require locally available high computing power. A recent review by Wang et al. [58] recommends Linux-based programs because they are often free, not specific to any sequencing platform, and less computing power hungry and, as a consequence, tend to perform faster. Flexibility in the parameter's choice for read assembly is another major advantage. However, most biologists are unfamiliar with Linux operating systems, its structure and command lines, thereby imposing a steep learning curve for adoption. Linux-based software such as Bowtie [59], BWA [60], and SOAP2/3 [61] have been used widely for the analysis of NGS data. Other software may not have gained broad acceptance but may have unique features worth noting. For reviews on NGS software, see Li and Homer [62], Wang et al. [58], and Treangen and Salzberg [63]. Characteristics of the most common NGS software and their attributes are listed in Table 1, and their download information can be found in Table 4.

TABLE 1: List of most cited/used software for sequence assembly of NGS data. Source locations for these software are compiled in Table 4.

Name (current version)	Assembly type (algorithm)	Supported parameters				Output format	Platform
		Color space	Read length	Gapped alignment	Paired-end		
CLC-Bio1	Reference2	Yes	Arbitrary	Yes	Yes	CLC-Bio	Linux/ Windows/ Mac OS X
SeqMan NGen1	Reference2	Yes	Arbitrary	Yes	Yes	ACE, BAM	Windows/ Mac OS X
NextGENe1 GENe	Reference2 Windows/ Mac OS X	Yes	Arbitrary	Yes	Yes	Next	
Bowtie (2)	Reference (FM-index)	Yes	Arbitrary	Yes	Yes	SAM	Linux/ Windows/ Mac OS X
BWA	Reference (FM-index)	Yes	Arbitrary	Yes	Yes	SAM	Linux
SOAP (3)	Reference (FM-index)	Yes	Arbitrary	No	Yes	SOAP2/3	Linux
MAQ (0.6.6)	Reference (Hashing reads)	Yes	≤ 127	Yes	Yes	MAQ	Linux/Solaris/Mac OS X
Novoalign (2.07.07)	Reference (Hashing reference)	Yes	Arbitrary	Yes	Yes	SAM	Linux/Mac OS X
Mosaik (1.1.0018)	Reference (Hashing reference)	Yes	Arbitrary	Yes	Yes	SAM	Linux/ Windows/ Mac OS X/Solaris
SHRiMP (2.2.2)	Reference (Hashing reference)	Yes	Arbitrary	Yes	Yes	SAM	Linux/Mac OS X
Mira (3.4)	Reference2	Yes	Arbitrary	Yes	Yes	FASTA, ACE	Linux

[1]Commercial software. [2]Option for de novo assembly and modules included for variant calling.

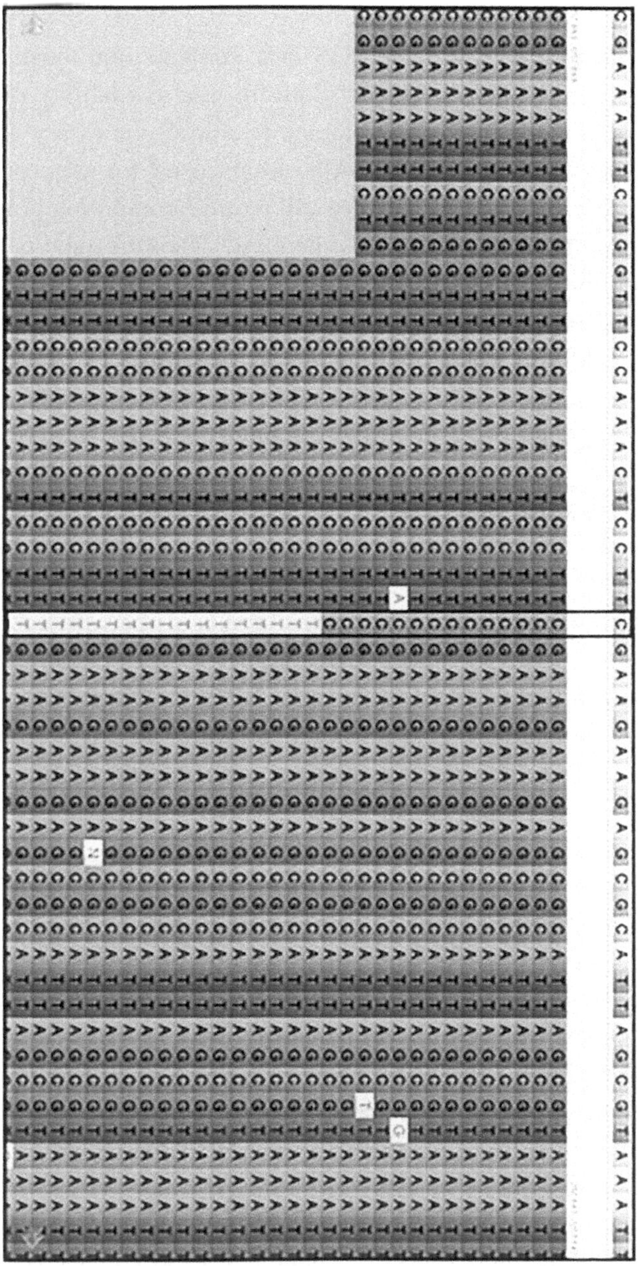

FIGURE 1: Graphical user interface of Tablet, an assembly visualization program, displays the reference genome on top and the mapped reads with color-coded SNPs on the bottom.

6.3.2 CONSIDERATION FOR SOFTWARE SELECTION

In selecting software for NGS data analysis one must consider, among other things, the sequencing platform, the availability of a reference genome, the computing and storage resources necessary, and the bioinformatics expertise available. Algorithms used for sequence analysis have matured significantly but may still require computing power beyond what is currently available in most genomics facilities and/or long processing time. For example, in aligning $2 \times 13,326,195$ paired-end reads (76 bp) from The Cancer Genome Atlas project (SRR018643) [64], SHRiMP [65] took 1,065 hrs with a peak memory footprint of 12 gigabytes to achieve the mapping of 81% of the reads to the human genome reference whereas Bowtie used 2.9 gigabytes of memory, a run time of 2.2 hrs but only achieved a 67% mapping rate [58]. Both time and memory become critical when dealing with a very large NGS dataset. Fast and memory efficient sequence mapping seems to be preferred over slower, memory demanding software even at the cost of a reduced mapping rate. It should be noted that a higher percentage of mapped reads is not a strict measure of quality because it may be indicative of a higher level of misaligned reads or reads aligned against repetitive elements, features that are not desirable [63].

In the absence of a reference genome, de novo assembly of a plant genome is achieved using sequence information obtained through a combination of Sanger and/or NGS of bacterial artificial chromosome (BAC) clones, or by whole genome shotgun (WGS) with NGS [66]. De novo assemblies are time consuming and require much greater computing power than read mapping onto a reference genome. The assembly accuracy depends in part on the read length and depth as well as the nature of the sequenced genome. The genomes of *Arabidopsis thaliana* [16], rice [67], and maize [68] were generated using a BAC-by-BAC approach while poplar [69], grape [70], and sorghum [71] genomic sequences were obtained through WGS. All genomes sequenced to date are fragmented to varying degrees because of the inability of sequencing technologies and bioinformatics algorithms to assemble through highly conserved repetitive elements. A list of current plant genome sequencing projects, their sequencing strategies, and status from standard draft to finished can be found in the review by Feuillet et al. [72].

Software programs such as Mira [73], SOAPdenovo [74], ABySS [75], and Velvet [76] have been used for de novo assembly. MIRA is well documented and can be readily customized, but it requires substantial computing memory and is not suited for large complex genomes. Of the freely available software, SOAPdenovo is one of the fastest read assembly programs and it uses a comparatively moderate amount of computing memory. The assembly generated by SOAPdenovo can be used for SNP discovery using SOAPsnp as implemented for the apple genome [77]. ABySS can be deployed on a computer cluster. It requires the least amount of memory and can be used for large genomes. Velvet requires the largest amount of memory. It can use mate-pair information to resolve and correct assembly errors.

6.4 SNP DISCOVERY

The most common application of NGS is SNP discovery, whose downstream usefulness in linkage map construction, genetic diversity analyses, association mapping, and marker-assisted selection has been demonstrated in several species [78]. NGS-derived SNPs have been reported in humans [79], *Drosophila* [80], wheat [81, 82], eggplant [83], rice [84–86], *Arabidopsis* [87, 88], barley [14, 89], sorghum [90], cotton [91], common beans [78], soybean [92], potato [93], flax [94], *Aegilops tauschii* [95], alfalfa [96], oat [97], and maize [98] to name a few.

SNP discovery using NGS is readily accomplished in small plant genomes for which good reference genomes are available such as rice and *Arabidopsis* [86, 99]. Although SNP discovery in complex genomes without a reference genome such as wheat [81, 82], barley [14, 89], oat [97], and beans [78] can be achieved through NGS, several challenges remain in other nonmodel but economically important crops. The presence of repeat elements, paralogs, and incomplete or inaccurate reference genome sequences can create ambiguities in SNP calling [63]. NGS read mapping can also suffer from sequencing error (erroneous base calling) and misaligned reads. The following section focuses on programs tailored for SNP discovery and emphasizes some of the precautions and considerations to minimize erroneous SNP calling.

6.4.1 SOFTWARE AND PIPELINES FOR SNP DISCOVERY

In theory, a SNP is identified when a nucleotide from an accession read differs from the reference genome at the same nucleotide position. In the absence of a reference genome, this is achieved by comparing reads from different genotypes using de novo assembly strategies [95]. Read assembly files generated by mapping programs are used to perform SNP calling. In practice, various empirical and statistical criteria are used to call SNPs, such as a minimum and maximum number of reads considering the read depth, the quality score and the consensus base ratio for examples [95]. Thresholds for these criteria are adjusted based on the read length and the genome coverage achieved by the NGS data. In assemblies generated allowing single nucleotide variants and insertions/ deletions (indels), a list of SNP and indel coordinates is generated and the read mapping results can be visualized using graphical user interface programs such as Tablet [100] (Figure 1), SNP-VISTA [101], or Savant [102] (refer to Table 4 for download information). Tablet has a user-friendly interface and is widely used because it supports a wide array of commonly used file formats such as SAM, BAM, SOAP, ACE, FASTQ, and FASTA generated by different read assemblers such as Bowtie, BWA, SOAP, MAQ, and SeqMan NGen. It displays contig overview, coverage information, read names and it allows searching for specific coordinates on scaffolds.

Broadly used SNP calling software include Samtools [103], SNVer [104], and SOAPsnp [74]. Samtools is popular because of its various modules for file conversion (SAM to BAM and vice-versa), mapping statistics, variant calling, and assembly visualization. Recently, SOAPsnp has gained popularity because of its tight integration with SOAP aligner and other SOAP modules which are constantly upgraded and provide a one stop shop for the sequencing analysis continuum. Variant calling algorithms such as Samtools and SNVer can be used as stand-alone programs or incorporated into pipelines for SNP calling. Reviews of SNP calling software have been published [63, 105]. Some of the main features of the current commonly used software are listed in Table 2 (refer to Table 4 for download information).

TABLE 2: Commonly used NGS variant calling software. Download information for these software is compiled in Table 4. A more comprehensive list of variant calling programs is available at http://seqanswers.com/wiki/Software/list.

Software	Multisample support	Reference	Features	Platform
Samtools	Yes	Aligned reads	Include computation of genotype likelihoods and variant calling	Linux
SOAPsnp	No	Variant database	Part of SOAP3 for variant calling	Linux
GATK	Yes	Aligned reads	Include variant caller, SNP filter, and SNP quality calibrator	Linux
SNVer	Yes	Aligned reads	Fast variant caller, assigning SNP significance based on read depth	Windows, Linux, Mac OS X
SHORE	Yes	Aligned reads	Variant calling based on reference sequence even from other species	Linux, Mac OS X
MaCH	Yes	Genotype likelihoods	Variant calling with or without LD information	Windows, Linux, Mac OSX
IMPUTE2	Yes	Candidate SNPs and genotype likelihoods	Variant calling and linkage map-based SNP imputation	Windows, Linux, Mac OSX

6.4.2 SNP DISCOVERY FROM MULTIPLE INDIVIDUALS AND COMPLEX GENOMES

SNP discovery is more robust when multiple and divergent genotypes are used simultaneously, creating the necessary basis to capture the genetic variability of a species. Large parts of plant genomes consist of repetitive elements [106] which can cause spurious SNP calling by erroneous read mapping to paralogous repeat element sequences. In polyploid genomes such as cotton (allotetraploid), homoeologous sequences can cause similar misalignment [91]. Improved read assembly and filtering of SNPs become even more important factors for accurate SNP calling in these cases because they can mitigate the effects of errors caused by paralogs and homoeologs.

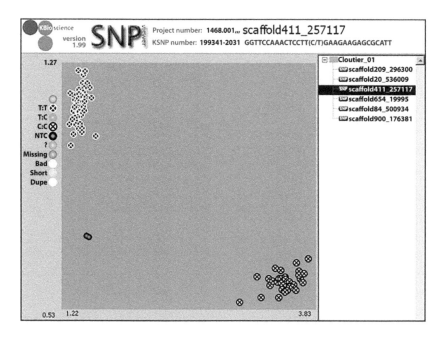

FIGURE 2: Validation of a T/C SNP by a KASPar assay (KBiosciences, Herts, England). Genotypes with a "T" are represented by black dots with a white cross clustered in the upper left and those with a "C" by white dots with a black cross in the bottom right cluster. The two black dots near the bottom left are negative controls. No heterozygous individuals were present in this population.

Read assembly algorithms such as Bowtie and SOAP as well as variant calling/genotyping softwares such as GATK [107] are rapidly evolving to accommodate an ever increasing number of reads, increased read length, nucleotide quality values, and mate-pair information of PE reads. Assembly programs such as Novoalign (http://www.novocraft.com/main/index.php) and STAMPY [108], although memory and time intensive, are highly sensitive for simultaneous mapping of short reads from multiple individuals [105].

SNP calls can be significantly improved using filtering criteria that are specific to the genome characteristics and the dataset. For instance, projects aimed at resequencing can compare different datasets from the same genotype and thus eliminate data with large discrepancies. This strategy identifies the most common sources of error and is applied in the 1000 genome project [109]. Reduced representation libraries (RRLs), that is, sequencing an enriched subset of a genome by eliminating a proportion of its repetitive fractions [79], reduce the probability of misalignments to repeats and thus potential downstream erroneous SNP calling. Filtering criteria that can improve SNP accuracy include (i) a minimum read depth (often ≥3 per genotype), (ii) >90% nucleotides within a genotype having identical call at a given position (~<10% sequencing error), (iii) a read depth ≤ mean of the sequence depth over the entire mapping assembly, (iv) the elimination of ribosomal DNA and other repetitive elements in the 50 nt flanking any SNP call, and (v) masking of homopolymer SNPs with a given base string length (often ≥2). Additionally, in polyploid species, separate assembly of homoeologs using stringent mapping parameters is often essential for genome-wide SNP identification to avoid spurious SNP calls caused by erroneous homoeologous read mapping [91].

6.4.3 SNP VALIDATION

Prior to any SNP applications, the discovered SNPs must be validated to identify the true SNPs and get an idea of the percentage of potentially false SNPs resulting from an SNP discovery exercise. The need for validation arises because a proportion of the discovered SNPs could have been wrongly called for various reasons including those outlined above. SNP validation can be accomplished using a variety of material such as a bi-parental segregating population or a diverse panel of genotypes. Usually a small subset of the SNPs is used for validation through assays such as the Illumina Goldengate [110], KBiosciences Competitive AlleleSpecific-PCR SNP genotyping system (KASPar) (http://www.lgcgenomics.com/) or the High Resolution Melting (HRM) curve analysis. Validation can serve as an iterative and informative process to modify and optimize the SNP filtering criteria to improve SNP calling. For example, a subset of 144 SNPs from a total of 2,113,120 SNPs were validated using the Goldengate assay on 160 accessions in apple [77]. Another example is illustrated in Figure 2 where a KASPar assay was performed on 92 genotypes from a segregating popula-tion illustrating the validation of a single "T/C" SNP in two distinct clus-ters. Other validation strategies used in nonmodel organisms are tabulated in Garvin et al. [111]. With the continuously competitive pricing of NGS, genotyping-by-sequencing (GBS) is becoming a viable SNP validation method. Either biparental segregating populations or a collection of diverse genotypes can be sequenced at a reasonable cost using indexing, that is, combining multiple independently tagged genotypes in a single NGS run to obtain genome-wide or reduced representation genome sequences at a lower coverage but potentially validating a much larger number of SNPs than the methods described above. Sequencing of segregating populations or diverse genotypes may also lead to the discovery of additional SNPs.

The two major factors affecting the SNP validation rate are sequencing and read mapping errors as discussed above. NGS platforms have differ-ent levels of sequencing accuracies, and this may be the most important factor determining the variation in the validation, from 88.2% for SOLiD followed by Illumina at 85.4% and Roche 454 at 71% [95]. The SNP vali-dation rates can be improved using RRL for SNP discovery and choosing SNPs within the nonrepetitive sequences including predicted single copy genes and single copy repeat junctions shown to have high validation rates [95].

6.5 SNP GENOTYPING

SNP genotyping is the downstream application of SNP discovery to iden-tify genetic variations. SNP applications include phylogenic analysis, marker-assisted selection, genetic mapping of quantitative trait loci (QTL), bulked segregant analysis, genome selection, and genome-wide association studies (GWAS). The number of SNPs and individuals to screen are of primary importance in choosing an SNP genotyping assay, though cost of the assay and/or equipment and the level of accuracy are also important considerations.

TABLE 3: Commonly used genotyping platforms.

Name	Assay type	Technology	Throughput (samples)	Multi-plexing	Relative scale (no. of SNP/no. of individuals)
Genechip	Hybridization	Oligo nucleotide array	96/5 days	Up to 18 × 10^6	Small/large
Infinium II	Hybridization	Bead array	Up to 128/5 days	Up to 13 × 10^6	Large/small-large
Goldengate	Primer extension-ligation	Bead array	172/3 days	Up to 3,072	Medium/large
iPlex	Primer extension	Mass spectrometry (MALDI-TOF)	3840/2.5 days	Up to 40	Medium/large
Taqman	PCR	Taqman probe	Up to 1536/day	Up to 256	Medium/medium
SNPlex	PCR	Capillary electrophoresis	Up to 1536/3 days	Up to 48	Medium/large
KASPar	PCR	FRET quenching oligos	Up to 96/day	—	Medium/large
Invader	Primer annealing/endonuclease digestion	FRET quenching oligos	Up to 384/day	Up to 200,000	Medium/large
HRM	PCR	Melting curve analysis	Up to 1536/day	—	Medium/large

Illumina Goldengate is a commonly used genotyping assay because of its flexibility in interrogating 96 to 3,072 SNP loci simultaneously (http://www.illumina.com/). HRM analysis is suitable for a few to an intermediate number of SNPs and can be performed within a typical laboratory setting. KASPar and SNPline genotyping systems (http://www.lgcgenomics.com/) can be used for genotyping a few to thousands of SNPs in a laboratory setting. The SNPline system is available in SNPlite or SNPline XL versions to allow flexibility in sample number and SNP assays. The iPLEX Gold technology developed by Sequenom (http://www.sequenom.com/) is based on the MassARRAY system which uses primer extension chemistry and matrix-assisted laser desorption/ionisation-time of flight (MALDI-TOF) mass spectrometry for genotyping.

TABLE 4: Download information of software used for NGS data.

Software	Source
Bowtie	http://bowtie-bio.sourceforge.net/bowtie2/index.shtml
BWA	http://bio-bwa.sourceforge.net/
SOAP	http://soap.genomics.org.cn/soap3.html#down2
MAQ	http://sourceforge.net/projects/maq/
Novoalign	http://www.novocraft.com/main/index.php
CLC-Bio Genomics	http://www.clcbio.com/index.php?id=1240
SeqManNGen	http://www.dnastar.com/t-products-seqman-ngen.aspx
NextGENe	http://softgenetics.com/NextGENe.html
Mosaik	http://bioinformatics.bc.edu/marthlab/Mosaik
SHRiMP	http://compbio.cs.toronto.edu/shrimp/
Mira	http://sourceforge.net/projects/mira-assembler/files/MIRA/stable/
Cassava	http://www.illumina.com/software/genome_analyzer_software.ilmn
Newbler	http://www.454.com/products/analysis-software/index.asp
Novoalign	http://www.novocraft.com/main/downloadpage.php
Tablet	http://bioinf.scri.ac.uk/tablet/
SNP-VISTA	http://genome.lbl.gov/vista/snpvista/
Samtools	http://sourceforge.net/projects/samtools/
Savant	http://genomesavant.com/savant/download.php
SOAPsnp	http://soap.genomics.org.cn/soapsnp.html
GATK	http://www.broadinstitute.org/gsa/wiki/index.php/ The_Genome_Analysis_Toolkit
SNver	http://snver.sourceforge.net/
MaCH	http://www.sph.umich.edu/csg/abecasis/MACH/
IMPUTE2	http://mathgen.stats.ox.ac.uk/impute/impute_v2.html# download_impute2
MEGA	http://www.megasoftware.net/
PHYLIP	http://evolution.genetics.washington.edu/phylip.html

The iPLEX Gold system has gained acceptance due to its high precision and cost-effective implementation. High throughput chip-based genotyping assays such as the Affymetrix GeneChip arrays (http://www.affymetrix.com/estore/) and the Illumina BeadChips (http://www.illumina.com/) are capable of validating up to a million SNPs per reaction across an entire genome. Detailed analyses of SNP genotyping assays and their features are reviewed in Tsuchihashi and Dracopoli [112], Sobrino and Carracedo [113], Giancola et al. [114], Kim and Misra [115], Gupta et al. [116], and Ragoussis [117]. A list of the most commonly used genotyping assays describing the assay type, technology, throughput, multiplexing ability, and relative scalability can be found in Table 3.

Array-based technologies such as Infinium and Goldengate substantially improved SNP genotyping efficiency, but they are species-specific, expensive to design and require specific equipment and chemistry. PCR and primer extension technologies like KASPar and Taqman (http://www.lifetechnologies.com/global/en/home.html) are limited by their low SNP throughput but can be useful to assay a large number of genotypes with few SNPs. NGS technologies have become viable for genotyping studies and may offer advantages over other genotyping methods in cost and efficiency.

6.5.1 GENOTYPING-BY-SEQUENCING (GBS)

There have been a number of approaches developed that use complexity reduction strategies to lower the cost and simplify the discovery of SNP markers using NGS, RNA-Seq, complexity reduction of polymorphic sequences (CRoPS), restriction-site-associated DNA sequencing (RAD-Seq), and GBS [118]. Of these methodologies GBS holds the greatest promise to serve the widest base of plant researchers because of its ability to allow simultaneous marker discovery and genotyping with low cost and a simple molecular biology workflow. Briefly, GBS involves digesting the genome of each individual in a population to be studied with a restriction enzyme [119]. One unique and one common adapter are ligated to the fragments and a PCR is carried out which is biased towards amplifying smaller DNA fragments. The resulting PCR products are then pooled and sequenced using an Illumina platform. The amplicons are not fragmented so only the ends of the PCR products are sequenced. The unique adapter acts as an ID tag so sequencing reads can be associated with an individual. The technique can be applied to species with or without a reference genome. The choice of enzyme has an effect on the number of markers identified and the amount of sequence coverage required. The more frequent the restriction recognition site, the higher the number of fragments and therefore more potential markers. Use of more frequent cutters may necessitate greater amounts of sequencing depending on the application. Poland et al. [120] recently demonstrated the use of two restriction enzymes to perform GBS in bread wheat, a hexaploid genome.

GBS has the potential to be a truly revolutionary technology in the arena of plant genomics. It brings high density genotyping to the vast majority of plant species that, until now, have had almost no investment in genomics resources. With little capital investment requirement and an affordable per sample cost, all plant researchers now have powerful genomic and genetic methodologies available to them. Uses of GBS include applications in marker discovery, phylogenetics, bulked segregant analysis, QTL mapping in biparental lines, GWAS, and genome selection. GBS can also be applied to fine mapping in candidate gene discovery and be used to generate high-density SNP genetic maps to assist in de novo genome assembly. We predict tremendous advances in functional genomics and plant breeding from the implementation of GBS because it is truly a democratizing application for NGS in nonmodel plant systems.

6.6 APPLICATIONS OF SNPS

NGS and SNP genotyping technologies have made SNPs the most widely used marker for genetic studies in plant species such as *Arabidopsis* [121] and rice [122]. SNPs can help to decipher breeding pedigree, to identify genomic divergence of species to elucidate speciation and evolution, and to associate genomic variations to phenotypic traits [85]. The ease of SNP development, reasonable genotyping costs, and the sheer number of SNPs present within a collection of individuals allow an assortment of applications that can have a tremendous impact on basic and applied research in plant species.

6.6.1 SNPS IN GENETIC MAPPING

A genetic map refers to the arrangement of traits, genes, and markers relative to each other as measured by their recombination frequency. Genetic maps are essential tools in molecular breeding for plant genetic improvement as they enable gene localization, map-based cloning, and the identification of QTL [123]. SNPs have greatly facilitated the production of much higher density maps than previous marker systems. SNPs discovered using RNA-Seq and expressed sequence tags (ESTs) have the added advantage of being gene specific [124]. Their high abundance and rapidly improving genotyping technologies make SNPs an ideal marker type for generating new genetic maps as well as saturating existing maps created with other markers. Most SNPs are biallelic thereby having a lower polymorphism information content (PIC) value as compared to most other marker types which are often multiallelic [125]. The limited information associated with their biallelic nature is greatly compensated by their high frequency, and a map of 700–900 SNPs has been found to be equivalent to a map of 300–400 simple sequence repeat (SSR) markers [125]. SNP-based linkage maps have been constructed in many economically important species such as rice [126], cotton [91] and *Brassica* [127]. The identification of candidate genes for flowering time in *Brassica* [127] and maize [128] are practical examples of gene discovery through SNP-based genetic maps.

6.6.2 GENOME-WIDE ASSOCIATION MAPPING

Association mapping (AM) panels provide a better resolution, consider numerous alleles, and may provide faster marker-trait association than biparental populations [129]. AM, often referred to as linkage disequilibrium (LD) mapping, relies on the nonrandom association between markers and traits [130]. LD can vary greatly across a genome. In low LD regions, high marker saturation is required to detect marker-trait association, hence the need for densely saturated maps. In general, GWASs require 10,000–100,000 markers applied to a collection of genotypes representing a broad genetic basis [130].

In the past few years, NGS technologies have led to the discovery of thousands, even millions of SNPs, and novel application platforms have made it possible to produce genome-wide haplotypes of large numbers of genotypes, making SNPs the ideal marker for GWASs. So far, 951 GWASs have been reported in humans (http://www.bing.com/search?q=www.genome.gov%2Fgwastudies%2F&src=ie9tr). In plants, such a study was first reported in *Arabidopsis* for flowering time and pathogen-resistance genes [131]. A GWAS performed in rice using ~3.6 million SNPs identified genomic regions associated with 14 agronomic traits [132]. The genetic structure of northern leaf blight, southern leaf blight, and leaf architecture was studied using ~1.6 million SNPs in maize [133–135]. SNP-based GWAS was also performed on species such as barley for which a reference genome sequence is not available [136]. Soto-Cerda and Cloutier [137] have reviewed the concepts, benefits, and limitations of AM in plants.

6.6.3 EVOLUTIONARY STUDIES

SSRs and mitochondrial DNA have been used in evolutionary studies since the early 1990s [138]. However, the biological inferences from results of these two marker types may be misinterpreted due to homoplasy, a phenomenon in which similarity in traits or markers occurs due to reasons other than ancestry, such as convergent evolution, evolutionary reversal, gene duplication, and horizontal gene transfer [139]. The advantage of SNPs over microsatellites and mitochondrial DNA resides in the fact that SNPs represent single base nucleotide substitutions and, as such, they are less affected by homoplasy because their origin can be explained by mutation models [140]. SNPs have been employed to quantify genetic variation, for individual identification, to determine parentage relatedness and population structure [138]. Seed shattering (or loss thereof) has been associated with an SNP through a GWAS aimed at unraveling the evolution of rice that led to its domestication [141]. SNPs have also been used to study the evolution of genes such as WAG-2 in wheat [142]. Algorithms such as neighbor-joining and maximum likelihood implemented in the PHYLIP [143] and MEGA [144] software are commonly used to generate phylogenetic trees.

The main advantage of SNPs is unquestionably their large numbers. As with all marker systems the researcher must be aware of ascertainment biases that exist in the panel of SNPs being used. These biases exist because SNPs are often developed from examining a small group of individuals and selecting the markers that maximize the amount of polymorphism that can be detected in the population used. This results in a collection of markers that sample only a fraction of the diversity that exists in the species but that are nevertheless used to infer relatedness and determine genetic distance for whole populations. Ideally, a set of SNP markers randomly distributed throughout the genome would be developed for each population studied. GBS moves us closer to this goal by incorporating simultaneous discovery of SNPs and genotyping of individuals. With this approach genome sample bias remains but can be mitigated by careful restriction enzyme selection.

6.7 FUTURE PERSPECTIVES

SNP discovery incontestably made a quantum leap forward with the advent of NGS technologies and large numbers of SNPs are now available from several genomes including large and complex ones (see Section 4). Unlike model systems such as humans and *Arabidopsis,* SNPs from crop plants remain limited for the time being, but broad access to reasonable cost NGS promises to rapidly increase the production of reference genome sequences as well as SNP discovery. Many issues remain to be addressed, such as the ascertainment bias of popular biparental populations and the low validation rate of some array-based genotyping platforms [145]. The area of epigenetic regulation of various genome components can be better understood as accurate and deeper sequencing is achieved. RNA and ChIP-sequencing projects, similar to RNA-Seq in the nonmodel plant sweet cherry to identify SNPs and haplotypes [146], can be undertaken to study functional genomics. A great deal of knowledge that is still elusive about the noncoding and repetitive elements can be determined with the next wave of modern and efficient sequencing technologies.

The first (Sanger) and the second (next) generation sequencing technologies have enabled researchers to characterize DNA sequence variation, sequence entire genomes, quantify transcript abundance, and understand mechanisms such as alternative splicing and epigenetic regulation [29].

Numerous plant genomes are now sequenced at various levels of completion and many more are underway [72]. The NGS technologies have made SNP discovery affordable even in complex genomes and the technologies themselves have improved tremendously in the past decade. Improvements in TGS promise synergies with NGS technologies to further assist our understanding of plant genetics and genomics. NGS has revolutionized genomics-related research, and it is our belief that the NGS-enabled discoveries will continue in the next decade.

REFERENCES

1. K. A. Frazer, D. G. Ballinger, D. R. Cox et al., "A second generation human haplotype map of over 3.1 million SNPs," Nature, vol. 449, no. 7164, pp. 851–861, 2007.
2. C. H. Brenner and B. S. Weir, "Issues and strategies in the DNA identification of World Trade Center victims," Theoretical Population Biology, vol. 63, no. 3, pp. 173–178, 2003.
3. M. I. McCarthy, G. R. Abecasis, L. R. Cardon et al., "Genome-wide association studies for complex traits: consensus, uncertainty and challenges," Nature Reviews Genetics, vol. 9, no. 5, pp. 356–369, 2008.
4. Z. J. Liu and J. F. Cordes, "DNA marker technologies and their applications in aquaculture genetics," Aquaculture, vol. 238, no. 1–4, pp. 1–37, 2004.
5. L. R. Schaeffer, "Strategy for applying genome-wide selection in dairy cattle," Journal of Animal Breeding and Genetics, vol. 123, no. 4, pp. 218–223, 2006.
6. H. Yu, W. Xie, J. Wang et al., "Gains in QTL detection using an ultra-high density SNP map based on population sequencing relative to traditional RFLP/SSR markers," PLoS ONE, vol. 6, no. 3, Article ID e17595, 2011.
7. J. M. Seddon, H. G. Parker, E. A. Ostrander, and H. Ellegren, "SNPs in ecological and conservation studies: a test in the Scandinavian wolf population," Molecular Ecology, vol. 14, no. 2, pp. 503–511, 2005.
8. C. T. Smith, C. M. Elfstrom, L. W. Seeb, and J. E. Seeb, "Use of sequence data from rainbow trout and Atlantic salmon for SNP detection in Pacific salmon," Molecular Ecology, vol. 14, no. 13, pp. 4193–4203, 2005.
9. B. N. Chorley, X. Wang, M. R. Campbell, G. S. Pittman, M. A. Noureddine, and D. A. Bell, "Discovery and verification of functional single nucleotide polymorphisms in regulatory genomic regions: current and developing technologies," Mutation Research, vol. 659, no. 1-2, pp. 147–157, 2008.
10. K. Faber, K. H. Glatting, P. J. Mueller, A. Risch, and A. Hotz-Wagenblatt, "Genome-wide prediction of splice-modifying SNPs in human genes using a new analysis

pipeline called AASsites," BMC Bioinformatics, vol. 12, supplement 4, article S2, 2011.

11. S. Atwell, Y. S. Huang, B. J. Vilhjálmsson et al., "Genome-wide association study of 107 phenotypes in Arabidopsis thaliana inbred lines," Nature, vol. 465, no. 7298, pp. 627–631, 2010.

12. W. B. Barbazuk, S. J. Emrich, H. D. Chen, L. Li, and P. S. Schnable, "SNP discovery via 454 transcriptome sequencing," Plant Journal, vol. 51, no. 5, pp. 910–918, 2007.

13. A. Ching, K. S. Caldwell, M. Jung et al., "SNP frequency, haplotype structure and linkage disequilibrium in elite maize inbred lines," BMC Genetics, vol. 3, article 19, 2002.

14. T. J. Close, P. R. Bhat, S. Lonardi et al., "Development and implementation of high-throughput SNP genotyping in barley," BMC Genomics, vol. 10, article 582, 2009.

15. X. Xu, X. Liu, S. Ge et al., "Resequencing 50 accessions of cultivated and wild rice yields markers for identifying agronomically important genes," Nature Biotechnology, vol. 30, no. 1, pp. 105–111, 2012.

16. S. Kaul, H. L. Koo, J. Jenkins et al., "Analysis of the genome sequence of the flowering plant Arabidopsis thaliana," Nature, vol. 408, no. 6814, pp. 796–815, 2000.

17. S. A. Goff, D. Ricke, T. H. Lan et al., "A draft sequence of the rice genome (Oryza sativa L. ssp. japonica)," Science, vol. 296, no. 5565, pp. 92–100, 2002.

18. J. Yu, S. Hu, J. Wang et al., "A draft sequence of the rice genome (Oryza sativa L. ssp. indica)," Science, vol. 296, no. 5565, pp. 79–92, 2002.

19. J. A. Shendure, G. J. Porreca, and G. M. Church, "Overview of DNA sequencing strategies," Current Protocols in Molecular Biology, chapter 7, no. 81, pp. 7.1.1–7.1.11, 2008.

20. F. Sanger, S. Nicklen, and A. R. Coulson, "DNA sequencing with chain-terminating inhibitors," Proceedings of the National Academy of Sciences of the United States of America, vol. 74, no. 12, pp. 5463–5467, 1977.

21. F. Sanger, G. M. Air, B. G. Barrell, et al., "Nucleotide sequence of bacteriophage phiX174 DNA," Nature, vol. 265, no. 5596, pp. 687–695, 1977.

22. A. M. Maxam and W. Gilbert, "A new method for sequencing DNA," Proceedings of the National Academy of Sciences of the United States of America, vol. 74, no. 2, pp. 560–564, 1977.

23. M. Kircher and J. Kelso, "High-throughput DNA sequencing—concepts and limitations," BioEssays, vol. 32, no. 6, pp. 524–536, 2010.

24. M. Ronaghi, M. Uhlén, and P. Nyrén, "A sequencing method based on real-time pyrophosphate," Science, vol. 281, no. 5375, pp. 363–365, 1998.

25. M. Ronaghi, S. Karamohamed, B. Pettersson, M. Uhlén, and P. Nyrén, "Real-time DNA sequencing using detection of pyrophosphate release," Analytical Biochemistry, vol. 242, no. 1, pp. 84–89, 1996.

26. T. C. Glenn, "Field guide to next-generation DNA sequencers," Molecular Ecology Resources, vol. 11, no. 5, pp. 759–769, 2011.

27. G. Turcatti, A. Romieu, M. Fedurco, and A. P. Tairi, "A new class of cleavable fluorescent nucleotides: synthesis and optimization as reversible terminators for DNA sequencing by synthesis," Nucleic Acids Research, vol. 36, no. 4, article e25, 2008.

28. J. Shendure, G. J. Porreca, N. B. Reppas et al., "Molecular biology: accurate multiplex polony sequencing of an evolved bacterial genome," Science, vol. 309, no. 5741, pp. 1728–1732, 2005.

29. E. E. Schadt, S. Turner, and A. Kasarskis, "A window into third-generation sequencing," Human Molecular Genetics, vol. 19, no. 2, pp. R227–R240, 2010.

30. T. D. Harris, P. R. Buzby, H. Babcock et al., "Single-molecule DNA sequencing of a viral genome," Science, vol. 320, no. 5872, pp. 106–109, 2008.

31. C. S. Pareek, R. Smoczynski, and A. Tretyn, "Sequencing technologies and genome sequencing," Journal of Applied Genetics, vol. 52, no. 4, pp. 413–435, 2011.

32. J. Eid, A. Fehr, J. Gray et al., "Real-time DNA sequencing from single polymerase molecules," Science, vol. 323, no. 5910, pp. 133–138, 2009.

33. S. Koren, M. C. Schatz, B. P. Walenz et al., "Hybrid error correction and de novo assembly of single-molecule sequencing reads," Nature Biotechnology, vol. 30, no. 7, pp. 693–700, 2012.

34. F. Ribeiro, D. Przybylski, S. Yin, et al., "Finished bacterial genomes from shotgun sequence data," Genome Research. In press.

35. A. Bashir, A. A. Klammer, W. P. Robins et al., "A hybrid approach for the automated finishing of bacterial genomes," Nature Biotechnology, vol. 30, no. 7, pp. 701–707, 2012.

36. X. Zhang, K. W. Davenport, W. Gu et al., "Improving genome assemblies by sequencing PCR products with PacBio," BioTechniques, vol. 53, no. 1, pp. 61–62, 2012.

37. P. Kothiyal, S. Cox, J. Ebert, B. J. Aronow, J. H. Greinwald, and H. L. Rehm, "An overview of custom array sequencing," Current Protocols in Human Genetics, no. 61, chapter 7, pp. 7.17.1–17.17.11, 2009.

38. J. D. McPherson, "Next-generation gap," Nature Methods, vol. 6, no. 11, supplement, pp. S2–S5, 2009.

39. A. P. M. Weber, K. L. Weber, K. Carr, C. Wilkerson, and J. B. Ohlrogge, "Sampling the arabidopsis transcriptome with massively parallel pyrosequencing," Plant Physiology, vol. 144, no. 1, pp. 32–42, 2007.

40. M. Libault, A. Farmer, T. Joshi et al., "An integrated transcriptome atlas of the crop model Glycine max, and its use in comparative analyses in plants," Plant Journal, vol. 63, no. 1, pp. 86–99, 2010.

41. T. Lu, G. Lu, D. Fan et al., "Function annotation of the rice transcriptome at single-nucleotide resolution by RNA-seq," Genome Research, vol. 20, no. 9, pp. 1238–1249, 2010.

42. W. B. Barbazuk, S. Emrich, and P. S. Schnable, "SNP mining from maize 454 EST sequences," Cold Spring Harbor Protocols. In press.

43. E. Novaes, D. R. Drost, W. G. Farmerie et al., "High-throughput gene and SNP discovery in Eucalyptus grandis, an uncharacterized genome," BMC Genomics, vol. 9, article 312, 2008.

44. M. Trick, Y. Long, J. Meng, and I. Bancroft, "Single nucleotide polymorphism (SNP) discovery in the polyploid Brassica napus using Solexa transcriptome sequencing," Plant Biotechnology Journal, vol. 7, no. 4, pp. 334–346, 2009.

45. S. S. Yang, Z. J. Tu, F. Cheung et al., "Using RNA-Seq for gene identification, polymorphism detection and transcript profiling in two alfalfa genotypes with divergent cell wall composition in stems," BMC Genomics, vol. 12, no. 1, article 199, 2011.

46. F. Ozsolak, D. T. Ting, B. S. Wittner et al., "Amplification-free digital gene expression profiling from minute cell quantities," Nature Methods, vol. 7, no. 8, pp. 619–621, 2010.

47. Z. Wang, M. Gerstein, and M. Snyder, "RNA-Seq: a revolutionary tool for transcriptomics," Nature Reviews Genetics, vol. 10, no. 1, pp. 57–63, 2009.

48. H. Xu, Y. Gao, and J. Wang, "Transcriptomic analysis of rice (Oryza sativa) developing embryos using the RNA-Seq technique," PLoS ONE, vol. 7, no. 2, Article ID e30646, 2012.

49. J. D. Roberts, B. D. Preston, L. A. Johnston, A. Soni, L. A. Loeb, and T. A. Kunkel, "Fidelity of two retroviral reverse transcriptases during DNA-dependent DNA synthesis in vitro," Molecular and Cellular Biology, vol. 9, no. 2, pp. 469–476, 1989.

50. U. Gubler, "Second-strand cDNA synthesis: mRNA fragments as primers," Methods in Enzymology, vol. 152, pp. 330–335, 1987.

51. J. Cocquet, A. Chong, G. Zhang, and R. A. Veitia, "Reverse transcriptase template switching and false alternative transcripts," Genomics, vol. 88, no. 1, pp. 127–131, 2006.

52. F. Ozsolak, A. R. Platt, D. R. Jones et al., "Direct RNA sequencing," Nature, vol. 461, no. 7265, pp. 814–818, 2009.

53. M. J. Solomon, P. L. Larsen, and A. Varshavsky, "Mapping protein-DNA interactions in vivo with formaldehyde: evidence that histone H4 is retained on a highly transcribed gene," Cell, vol. 53, no. 6, pp. 937–947, 1988.

54. T. S. Mikkelsen, M. Ku, D. B. Jaffe et al., "Genome-wide maps of chromatin state in pluripotent and lineage-committed cells," Nature, vol. 448, no. 7153, pp. 553–560, 2007.

55. P. Ng, J. J. Tan, H. S. Ooi et al., "Multiplex sequencing of paired-end ditags (MS-PET): a strategy for the ultra-high-throughput analysis of transcriptomes and genomes," Nucleic Acids Research, vol. 34, no. 12, p. e84, 2006.

56. G. Robertson, M. Hirst, M. Bainbridge et al., "Genome-wide profiles of STAT1 DNA association using chromatin immunoprecipitation and massively parallel sequencing," Nature Methods, vol. 4, no. 8, pp. 651–657, 2007.

57. B. Giardine, C. Riemer, R. C. Hardison et al., "Galaxy: a platform for interactive large-scale genome analysis," Genome Research, vol. 15, no. 10, pp. 1451–1455, 2005.

58. W. Wang, Z. Wei, T.-W. Lam, and J. Wang, "Next generation sequencing has lower sequence coverage and poorer SNP-detection capability in the regulatory regions," Scientific Reports, vol. 1, article 55, 2011.

59. B. Langmead, C. Trapnell, M. Pop, and S. L. Salzberg, "Ultrafast and memory-efficient alignment of short DNA sequences to the human genome," Genome Biology, vol. 10, no. 3, article R25, 2009.

60. H. Li and R. Durbin, "Fast and accurate short read alignment with Burrows-Wheeler transform," Bioinformatics, vol. 25, no. 14, pp. 1754–1760, 2009.

61. R. Li, C. Yu, Y. Li et al., "SOAP2: an improved ultrafast tool for short read alignment," Bioinformatics, vol. 25, no. 15, pp. 1966–1967, 2009.

62. H. Li and N. Homer, "A survey of sequence alignment algorithms for next-generation sequencing," Briefings in Bioinformatics, vol. 11, no. 5, Article ID bbq015, pp. 473–483, 2010.

63. T. J. Treangen and S. L. Salzberg, "Repetitive DNA and next-generation sequencing: computational challenges and solutions," Nature Reviews Genetics, vol. 13, no. 1, pp. 36–46, 2012.

64. R. McLendon, A. Friedman, D. Bigner et al., "Comprehensive genomic characterization defines human glioblastoma genes and core pathways," Nature, vol. 455, no. 7216, pp. 1061–1068, 2008.

65. S. M. Rumble, P. Lacroute, A. V. Dalca, M. Fiume, A. Sidow, and M. Brudno, "SHRiMP: accurate mapping of short color-space reads," PLoS Computational Biology, vol. 5, no. 5, Article ID e1000386, 2009.

66. S. Rounsley, P. R. Marri, Y. Yu et al., "De novo next generation sequencing of plant genomes," Rice, vol. 2, no. 1, pp. 35–43, 2009.

67. T. Sasaki, "The map-based sequence of the rice genome," Nature, vol. 436, no. 7052, pp. 793–800, 2005.

68. E. Pennisi, "Plant sciences: corn genomics pops wide open," Science, vol. 319, no. 5868, p. 1333, 2008.

69. G. A. Tuskan, S. DiFazio, S. Jansson et al., "The genome of black cottonwood, Populus trichocarpa (Torr. & Gray)," Science, vol. 313, no. 5793, pp. 1596–1604, 2006.

70. O. Jaillon, J. M. Aury, B. Noel et al., "The grapevine genome sequence suggests ancestral hexaploidization in major angiosperm phyla," Nature, vol. 449, no. 7161, pp. 463–467, 2007.

71. A. H. Paterson, J. E. Bowers, R. Bruggmann et al., "The Sorghum bicolor genome and the diversification of grasses," Nature, vol. 457, no. 7229, pp. 551–556, 2009.

72. C. Feuillet, J. E. Leach, J. Rogers, P. S. Schnable, and K. Eversole, "Crop genome sequencing: lessons and rationales," Trends in Plant Science, vol. 16, no. 2, pp. 77–88, 2011.

73. B. Chevreux, T. Pfisterer, B. Drescher et al., "Using the miraEST assembler for reliable and automated mRNA transcript assembly and SNP detection in sequenced ESTs," Genome Research, vol. 14, no. 6, pp. 1147–1159, 2004.

74. R. Li, Y. Li, X. Fang et al., "SNP detection for massively parallel whole-genome resequencing," Genome Research, vol. 19, no. 6, pp. 1124–1132, 2009.

75. J. T. Simpson, K. Wong, S. D. Jackman, J. E. Schein, S. J. M. Jones, and I. Birol, "ABySS: a parallel assembler for short read sequence data," Genome Research, vol. 19, no. 6, pp. 1117–1123, 2009.

76. D. R. Zerbino and E. Birney, "Velvet: algorithms for de novo short read assembly using de Bruijn graphs," Genome Research, vol. 18, no. 5, pp. 821–829, 2008.

77. D. Chagné, R. N. Crowhurst, M. Troggio et al., "Genome-wide SNP detection, validation, and development of an 8K SNP array for apple," PLoS ONE, vol. 7, no. 2, Article ID e31745, 2012.

78. A. J. Cortés, M. C. Chavarro, and M. W. Blair, "SNP marker diversity in common bean (Phaseolus vulgaris L.)," Theoretical and Applied Genetics, vol. 123, no. 5, pp. 827–845, 2011.

79. D. Altshuler, V. J. Pollara, C. R. Cowles et al., "An SNP map of the human genome generated by reduced representation shotgun sequencing," Nature, vol. 407, no. 6803, pp. 513–516, 2000.

80. J. Berger, T. Suzuki, K. A. Senti, J. Stubbs, G. Schaffner, and B. J. Dickson, "Genetic mapping with SNP markers in Drosophila," Nature Genetics, vol. 29, no. 4, pp. 475–481, 2001.

81. A. M. Allen, G. L. Barker, S. T. Berry et al., "Transcript-specific, single-nucleotide polymorphism discovery and linkage analysis in hexaploid bread wheat (Triticum aestivum L.)," Plant Biotechnology Journal, vol. 9, no. 9, pp. 1086–1099, 2011.

82. D. Trebbi, M. Maccaferri, P. de Heer et al., "High-throughput SNP discovery and genotyping in durum wheat (Triticum durum Desf.)," Theoretical and Applied Genetics, vol. 123, no. 4, pp. 555–569, 2011.

83. L. Barchi, S. Lanteri, E. Portis et al., "Identification of SNP and SSR markers in eggplant using RAD tag sequencing," BMC Genomics, vol. 12, article 304, 2011.

84. F. A. Feltus, J. Wan, S. R. Schulze, J. C. Estill, N. Jiang, and A. H. Paterson, "An SNP resource for rice genetics and breeding based on subspecies Indica and Japonica genome alignments," Genome Research, vol. 14, no. 9, pp. 1812–1819, 2004.

85. K. L. McNally, K. L. Childs, R. Bohnert et al., "Genomewide SNP variation reveals relationships among landraces and modern varieties of rice," Proceedings of the National Academy of Sciences of the United States of America, vol. 106, no. 30, pp. 12273–12278, 2009.

86. T. Yamamoto, H. Nagasaki, J. I. Yonemaru et al., "Fine definition of the pedigree haplotypes of closely related rice cultivars by means of genome-wide discovery of single-nucleotide polymorphisms," BMC Genomics, vol. 11, no. 1, article 267, 2010.

87. G. Jander, S. R. Norris, S. D. Rounsley, D. F. Bush, I. M. Levin, and R. L. Last, "Arabidopsis map-based cloning in the post-genome era," Plant Physiology, vol. 129, no. 2, pp. 440–450, 2002.

88. X. Zhang and J. O. Borevitz, "Global analysis of allele-specific expression in Arabidopsis thaliana," Genetics, vol. 182, no. 4, pp. 943–954, 2009.

89. R. Waugh, J. L. Jannink, G. J. Muehlbauer, and L. Ramsay, "The emergence of whole genome association scans in barley," Current Opinion in Plant Biology, vol. 12, no. 2, pp. 218–222, 2009.

90. J. C. Nelson, S. Wang, Y. Wu et al., "Single-nucleotide polymorphism discovery by high-throughput sequencing in sorghum," BMC Genomics, vol. 12, article 352, 2011.

91. R. L. Byers, D. B. Harker, S. M. Yourstone, P. J. Maughan, and J. A. Udall, "Development and mapping of SNP assays in allotetraploid cotton," Theoretical and Applied Genetics, vol. 124, no. 7, pp. 1201–1214, 2012.

92. D. L. Hyten, S. B. Cannon, Q. Song et al., "High-throughput SNP discovery through deep resequencing of a reduced representation library to anchor and orient scaffolds in the soybean whole genome sequence," BMC Genomics, vol. 11, no. 1, article 38, 2010.

93. J. P. Hamilton, C. N. Hansey, B. R. Whitty et al., "Single nucleotide polymorphism discovery in elite north American potato germplasm," BMC Genomics, vol. 12, article 302, 2011.

94. Y.-B. Fu and G. W. Peterson, "Developing genomic resources in two Linum species via 454 pyrosequencing and genomic reduction," Molecular Ecology Resources, vol. 12, no. 3, pp. 492–500, 2012.

95. F. M. You, N. Huo, K. R. Deal et al., "Annotation-based genome-wide SNP discovery in the large and complex Aegilops tauschii genome using next-generation sequencing without a reference genome sequence," BMC Genomics, vol. 12, article 59, 2011.

96. Y. Han, Y. Kang, I. Torres-Jerez et al., "Genome-wide SNP discovery in tetraploid alfalfa using 454 sequencing and high resolution melting analysis," BMC Genomics, vol. 12, p. 350, 2011.

97. R. E. Oliver, G. R. Lazo, J. D. Lutz et al., "Model SNP development for complex genomes based on hexaploid oat using high-throughput 454 sequencing technology," BMC Genomics, vol. 12, no. 1, article 77, 2011.

98. E. Jones, W. C. Chu, M. Ayele et al., "Development of single nucleotide polymorphism (SNP) markers for use in commercial maize (Zea mays L.) germplasm," Molecular Breeding, vol. 24, no. 2, pp. 165–176, 2009.

99. S. Ossowski, K. Schneeberger, R. M. Clark, C. Lanz, N. Warthmann, and D. Weigel, "Sequencing of natural strains of Arabidopsis thaliana with short reads," Genome Research, vol. 18, no. 12, pp. 2024–2033, 2008.

100. I. Milne, M. Bayer, L. Cardle et al., "Tablet-next generation sequence assembly visualization," Bioinformatics, vol. 26, no. 3, pp. 401–402, 2009.

101. N. Shah, M. V. Teplitsky, S. Minovitsky et al., "SNP-VISTA: an interactive SNP visualization tool," BMC Bioinformatics, vol. 6, no. 1, article 292, 2005.

102. M. Fiume, V. Williams, A. Brook, and M. Brudno, "Savant: genome browser for high-throughput sequencing data," Bioinformatics, vol. 26, no. 16, Article ID btq332, pp. 1938–1944, 2010.

103. H. Li, B. Handsaker, A. Wysoker et al., "The sequence alignment/map format and SAMtools," Bioinformatics, vol. 25, no. 16, pp. 2078–2079, 2009.

104. Z. Wei, W. Wang, P. Hu, G. J. Lyon, and H. Hakonarson, "SNVer: a statistical tool for variant calling in analysis of pooled or individual next-generation sequencing data," Nucleic acids research, vol. 39, no. 19, article e132, 2011.

105. R. Nielsen, J. S. Paul, A. Albrechtsen, and Y. S. Song, "Genotype and SNP calling from next-generation sequencing data," Nature Reviews Genetics, vol. 12, no. 6, pp. 443–451, 2011.

106. R. Ragupathy, R. Rathinavelu, and S. Cloutier, "Physical mapping and BAC-end sequence analysis provide initial insights into the flax (Linum usitatissimum L.) genome," BMC Genomics, vol. 12, article 217, 2011.

107. A. McKenna, M. Hanna, E. Banks et al., "The genome analysis toolkit: a MapReduce framework for analyzing next-generation DNA sequencing data," Genome Research, vol. 20, no. 9, pp. 1297–1303, 2010.

108. G. Lunter and M. Goodson, "Stampy: a statistical algorithm for sensitive and fast mapping of Illumina sequence reads," Genome Research, vol. 21, no. 6, pp. 936–939, 2011.

109. R. M. Durbin, "A map of human genome variation from population-scale sequencing," Nature, vol. 467, no. 7319, pp. 1061–1073, 2010.

110. J. B. Fan, M. S. Chee, and K. L. Gunderson, "Highly parallel genomic assays," Nature Reviews Genetics, vol. 7, no. 8, pp. 632–644, 2006.

111. M. R. Garvin, K. Saitoh, and A. J. Gharrett, "Application of single nucleotide polymorphisms to non-model species: a technical review," Molecular Ecology Resources, vol. 10, no. 6, pp. 915–934, 2010.

112. Z. Tsuchihashi and N. C. Dracopoli, "Progress in high throughput SNP genotyping methods," Pharmacogenomics Journal, vol. 2, no. 2, pp. 103–110, 2002.

113. B. Sobrino and A. Carracedo, "SNP typing in forensic genetics: a review," Methods in Molecular Biology, vol. 297, pp. 107–126, 2005.

114. S. Giancola, H. I. McKhann, A. Bérard et al., "Utilization of the three high-throughput SNP genotyping methods, the GOOD assay, Amplifluor and TaqMan, in diploid and polyploid plants," Theoretical and Applied Genetics, vol. 112, no. 6, pp. 1115–1124, 2006.

115. S. Kim and A. Misra, "SNP genotyping: technologies and biomedical applications," Annual Review of Biomedical Engineering, vol. 9, pp. 289–320, 2007.

116. P. K. Gupta, S. Rustgi, and R. R. Mir, "Array-based high-throughput DNA markers for crop improvement," Heredity, vol. 101, no. 1, pp. 5–18, 2008.

117. J. Ragoussis, "Genotyping technologies for genetic research," Annual Review of Genomics and Human Genetics, vol. 10, pp. 117–133, 2009.

118. J. W. Davey, P. A. Hohenlohe, P. D. Etter, J. Q. Boone, J. M. Catchen, and M. L. Blaxter, "Genome-wide genetic marker discovery and genotyping using next-generation sequencing," Nature Reviews Genetics, vol. 12, no. 7, pp. 499–510, 2011.

119. R. J. Elshire, J. C. Glaubitz, Q. Sun et al., "A robust, simple genotyping-by-sequencing (GBS) approach for high diversity species," PLoS ONE, vol. 6, no. 5, Article ID e19379, 2011.

120. J. A. Poland, P. J. Brown, M. E. Sorrells, and J.-L. Jannink, "Development of high-density genetic maps for barley and wheat using a novel two-enzyme genotyping-by-sequencing approach," PLoS ONE, vol. 7, no. 2, Article ID e32253, 2012.

121. M. W. Horton, A. M. Hancock, Y. S. Huang et al., "Genome-wide patterns of genetic variation in worldwide Arabidopsis thaliana accessions from the RegMap panel," Nature Genetics, vol. 44, no. 2, pp. 212–216, 2012.

122. G. K. Subbaiyan, D. L. E. Waters, S. K. Katiyar, A. R. Sadananda, S. Vaddadi, and R. J. Henry, "Genome-wide DNA polymorphisms in elite indica rice inbreds discovered by whole-genome sequencing," Plant Biotechnology Journal, vol. 10, no. 6, pp. 623–634, 2012.

123. J. C. Nelson, "Methods and software for genetic mapping," in The Handbook of Plant Genome Mapping, pp. 53–74, Wiley-VCH, Weinheim, Germany, 2005.

124. A. Rafalski, "Applications of single nucleotide polymorphisms in crop genetics," Current Opinion in Plant Biology, vol. 5, no. 2, pp. 94–100, 2002.

125. L. Kruglyak, "The use of a genetic map of biallelic markers in linkage studies," Nature Genetics, vol. 17, no. 1, pp. 21–24, 1997.

126. W. Xie, Q. Feng, H. Yu et al., "Parent-independent genotyping for constructing an ultrahigh-density linkage map based on population sequencing," Proceedings of the National Academy of Sciences of the United States of America, vol. 107, no. 23, pp. 10578–10583, 2010.

127. F. Li, H. Kitashiba, K. Inaba, and T. Nishio, "A Brassica rapa linkage map of EST-based SNP markers for identification of candidate genes controlling flowering time and leaf morphological traits," DNA Research, vol. 16, no. 6, pp. 311–323, 2009.

128. E. S. Buckler, J. B. Holland, P. J. Bradbury et al., "The genetic architecture of maize flowering time," Science, vol. 325, no. 5941, pp. 714–718, 2009.

129. S. A. Flint-Garcia, J. M. Thornsberry, and S. B. Edward, "Structure of linkage disequilibrium in plants," Annual Review of Plant Biology, vol. 54, pp. 357–374, 2003.

130. P. K. Gupta, S. Rustgi, and P. L. Kulwal, "Linkage disequilibrium and association studies in higher plants: present status and future prospects," Plant Molecular Biology, vol. 57, no. 4, pp. 461–485, 2005.

131. M. J. Aranzana, S. Kim, K. Zhao et al., "Genome-wide association mapping in Arabidopsis identifies previously known flowering time and pathogen resistance genes," PLoS Genetics, vol. 1, no. 5, p. e60, 2005.

132. X. Huang, X. Wei, T. Sang et al., "Genome-wide asociation studies of 14 agronomic traits in rice landraces," Nature Genetics, vol. 42, no. 11, pp. 961–967, 2010.

133. K. L. Kump, P. J. Bradbury, R. J. Wisser et al., "Genome-wide association study of quantitative resistance to southern leaf blight in the maize nested association mapping population," Nature Genetics, vol. 43, no. 2, pp. 163–168, 2011.

134. J. A. Poland, P. J. Bradbury, E. S. Buckler, and R. J. Nelson, "Genome-wide nested association mapping of quantitative resistance to northern leaf blight in maize," Proceedings of the National Academy of Sciences of the United States of America, vol. 108, no. 17, pp. 6893–6898, 2011.

135. F. Tian, P. J. Bradbury, P. J. Brown et al., "Genome-wide association study of leaf architecture in the maize nested association mapping population," Nature Genetics, vol. 43, no. 2, pp. 159–162, 2011.

136. R. K. Pasam, R. Sharma, M. Malosetti et al., "Genome-wide association studies for agronomical traits in a world wide spring barley collection," BMC Plant Biology, vol. 12, article 16, 2012.

137. B. J. Soto-Cerda and S. Cloutier, "Association mapping in plant genomes," in Genetic Diversity in Plants, M. Çalişkan, Ed., pp. 29–54, InTech, 2012.

138. P. A. Morin, G. Luikart, and R. K. Wayne, "SNPs in ecology, evolution and conservation," Trends in Ecology and Evolution, vol. 19, no. 4, pp. 208–216, 2004.

139. P. W. Hedrick, "Perspective: highly variable loci and their interpretation in evolution and conservation," Evolution, vol. 53, no. 2, pp. 313–318, 1999.

140. A. Vignal, D. Milan, M. SanCristobal, and A. Eggen, "A review on SNP and other types of molecular markers and their use in animal genetics," Genetics Selection Evolution, vol. 34, no. 3, pp. 275–305, 2002.

141. S. Konishi, T. Izawa, S. Y. Lin et al., "An SNP caused loss of seed shattering during rice domestication," Science, vol. 312, no. 5778, pp. 1392–1396, 2006.

142. O. Wei, Z. Peng, Y. Zhou, Z. Yang, K. Wu, and Z. Ouyang, "Nucleotide diversity and molecular evolution of the WAG-2 gene in common wheat (Triticum aestivum L.) and its relatives," Genetics and Molecular Biology, vol. 34, no. 4, pp. 606–615, 2011.

143. J. D. Retief, "Phylogenetic analysis using PHYLIP," Methods in Molecular Biology, vol. 132, pp. 243–258, 2000.

144. K. Tamura, J. Dudley, M. Nei, and S. Kumar, "MEGA4: Molecular Evolutionary Genetics Analysis (MEGA) software version 4.0," Molecular Biology and Evolution, vol. 24, no. 8, pp. 1596–1599, 2007.
145. M. W. Ganal, T. Altmann, and M. S. Röder, "SNP identification in crop plants," Current Opinion in Plant Biology, vol. 12, no. 2, pp. 211–217, 2009.
146. T. Koepke, S. Schaeffer, V. Krishnan et al., "Rapid gene-based SNP and haplotype marker development in non-model eukaryotes using 3'UTR sequencing," BMC Genomics, vol. 13, no. 1, article 18, 2012.

CHAPTER 7

SNP Discovery and Genetic Mapping Using Genotyping by Sequencing of Whole Genome Genomic DNA from a Pea RIL Population

GILLES BOUTET, SUSETE ALVES CARVALHO, MATTHIEU FALQUE, PIERRE PETERLONGO, EMELINE LHUILLIER, OLIVIER BOUCHEZ, CLÉMENT LAVAUD, MARIE-LAURE PILET-NAYEL, NATHALIE RIVIÈRE, AND ALAIN BARANGER

7.1 INTRODUCTION

SNPs (Single Nucleotide Polymorphisms) are genetic markers of choice for both linkage and association mapping and for population structure and evolution analysis. They are virtually unlimited, evenly distributed along the genome, bi-allelic and co-dominant. Massive SNP discovery was first limited to the few species with an available reference genome. Recently, with the advances in cheaper next generation sequencing (NGS) technologies, various accessions within species even with complex genomes could be sequenced [1]. The challenge of sequencing large genomes with high levels of repeated sequences first led to the development of novel approaches for reducing genome complexity [2]. cDNA sequencing, which specifically addresses the expressed genic fraction, was largely developed and reviewed in Duarte et al. [3]. Restriction site Associated DNA (RAD)

© *Boutet et al. 2016.* BMC Genomics, *2016, 17:121; DOI: 10.1186/s12864-016-2447-2. distributed under the terms of the Creative Commons Attribution 4.0 International License (http://creativecommons.org/licenses/by/4.0/).*

tags have been applied to a large range of organisms such as *Drosophila melanogaster* [4], fish and fungi [5]. In plants, RAD-Seq has been applied to a number of species for both large-scale SNP discovery and the mapping of SNP subsets in barley [6] and rye-grass [7]. In legume species, Deokar et al. [8] first reported the use of RAD-Seq in chickpea to discover 29,000 SNPs and subsequently map 604 recombination bins. Restriction enzyme digest to reduce genome complexity followed by direct Genotyping-by-Sequencing was reported for maize RILs and barley doubled haploid lines [9], where 2,382 markers were eventually mapped on the barley genetic map. In legume species, Sonah et al. [10] first used GBS in soybean to develop 10,120 high quality SNPs. Thus all these studies used genome reduction and various assembling tools.

Pea is the third production in the world among temperate grain legume crops after soybean and common bean and is a major source of protein for humans and livestock. Pea is particularly relevant in temperate cropping systems due to its capacity to fix nitrogen through symbiosis. Nevertheless, the species suffers from significant yield instability due to its high susceptibility to abiotic and biotic stresses, among which Aphanomyces root rot disease, due to the oomycete *Aphanomyces euteiches* Drechs. Resistance Quantitative Trait Loci (QTL) have been described, but the QTL confidence intervals are still large, especially due to the lack of markers and low resolution of existing genetic maps. It remains a challenge to reduce QTL confidence intervals, to discover underlying candidate genes and develop breeding programs using molecular markers strongly associated with phenotypes.

Although pea has actually entered the genomic era [11], it still suffers from limited genomic resources compared to other crops. The pea genome is 4.3 Gb, which is around 10 times larger than the genome of the model species *M.truncatula* [12]. This includes repeats mostly derived from transposon-based sequences [13]. Recent reports indicated that large new sequencing resources are under development [14] and that a consortium for pea genome sequencing is at work (http://www.coolseasonfoodlegume.org/pea_genome), however no full genome sequence is available yet. Large numbers of new molecular markers are still needed to saturate pea maps and significantly improve QTL mapping both for research and breeding objectives. Although transcriptome sequencing has recently been used in pea for SNP discovery [3, 15, 16] and mapping [3, 17, 18], available genetic maps remain at low to medium density, and are mainly based on a few hundred SSRs [19] and on a few hundred [20, 21] up to a few thousand [3, 18, 22] SNPs, usually developed through dedicated genotyping facilities. The development of larger resources is therefore required for mapping and genetic improvement purposes.

To complement the existing resources, our objective was to develop a comprehensive SNP resource in pea using genotyping by HiSeq sequencing of whole genome DNA and then to apply it for substantial genetic mapping. To our knowledge, this is the first report, in a species lacking a sequenced reference genome, of a whole genome genomic DNA sequencing strategy for high-throughput SNP discovery, genotyping (below called WGGBS) and genetic mapping. This novel approach was carried out at low sequence coverage, without prior genome complexity reduction and without sequence read assembly, on a RIL population segregating for *A.euteiches* resistance. The quality of the SNPs was then validated through genotyping using a benchmark technology [23]. This was made possible by optimizing SNP discovery tools that can work without data assembly or a reference genome [24], and developing appropriate tools to map very large numbers of SNPs from mapping populations comprising a few individuals [25].

7.2 RESULTS

7.2.1 SNP DISCOVERY AND SELECTION OF A SUBSET OF HIGHLY DESIGNABLE MARKERS

To maximize the identification of relevant polymorphic SNPs, four genetically distant *P. sativum* genotypes were selected for genomic DNA preparation and HiSeq sequencing. These were the parental lines of the 'Champagne' x 'Terese' and 'Baccara' x 'PI180693' RIL mapping populations. Raw and pre-processed sequencing data analysis across the four samples showed low levels of contamination, and unexpectedly low levels of sequence repeats. In addition, considering that the nuclei were not isolated prior to DNA extraction, there were also rather low levels of organelle contaminants (with a higher level for the 'Baccara' sample, which was not etiolated). The final clean sequences represented 69 to 80 % of the raw sequence data depending on the genotype (Additional file 1: Table S1).

From a total of 1.32 billion cleaned reads from four *P. sativum* lines, the discoSnp tool identified 419,024 SNPs. A "post-discoSnp" filtering step, based on the availability of sequence data for all the four lines, homozygosity of each pea line, global sequence coverage and minor allele coverage, was used to remove putative "false heterozygous" and multi-locus SNPs. Finally, 213,030 SNPs considered as robust for genotyping were selected. Most of them showed coverage between 6X and 14X (Additional file 2: Figure S1), which was consistent with the corresponding sequencing coverage for each line at approximately 7-fold the pea estimated genome size (Additional file 1: Table S1). As expected discoSnp filtering excluded SNPs with less than 5X coverage (Additional file 2: Figure S1). To optimize potential future GoldenGate® or KASP™ genotyping assay designs, only 131,850 SNPs were retained in two subsets of (i) 88,864 SNPs with a context sequence showing no other polymorphism at least 50 bp on either side of the SNP (considered as very highly designable) (ii) 42,986 SNPs with a context sequence showing no other polymorphism at least 50 bp on one side of the SNP and no other polymorphism at least 27 bp on the other side (considered as highly designable). Within the resulting 131,850 SNPs, polymorphic SNPs between pairs of parental lines ranged from 23,760 between the two spring sown field pea lines 'Baccara' and 'Terese' to over 97,000 between each of these two lines and the fodder pea genotype 'Champagne' (Table 1).

7.2.2 SNP INFERRING ON THE 'BACCARA' X 'PI180693' RIL MAPPING POPULATION

Forty-eight *P. sativum* RILs from the 'Baccara' x 'PI180693' mapping population were selected for genomic DNA extraction, Hiseq sequencing and genotyping.

Raw and pre-processed sequencing data analysis (Additional file 1: Table S1) across the 48 RILs was consistent with previous sequence analysis on the four parental lines, even with the twofold lower sequencing effort

TABLE 1: Number of SNPs that were polymorphic between sequenced pairs of pea parental lines, from a subset of 131,850 highly or very highly designable SNPs.

	Baccara	PI180693	Terese	Champagne
Baccara		88,851	23,760	97,705
PI180693			88,799	59,428
Terese				97,285
Champagne				

on the RILs (two lines per lane) than on the parental lines (one line per lane).

A total of 88,851 SNPs (out of the 131,850 selected SNPs) were polymorphic between the 'Baccara' and PI180693' parental lines. The kiss-reads module [24] of the discoSnp tool was then used to infer which of these SNPs were present in the 48 RILs. Most of these 88,851 SNPs showed coverage ranging between 3X and 7X (Additional file 2: Figure S1). This is consistent with the sequencing coverage for each of the 48 RILs of approximately 3.5-fold the estimated pea genome size (Additional file 1: Table S1). 13,187 SNPs were genotyped on all of the 48 sequenced RILs (Fig. 1). A total of 64,754 SNPs, which showed less than ten missing data points and less than 10 % of heterozygous data points among the 48 sequenced RILs, was retained.

A strict alignment of SNP context sequences with the Glint tool (http://lipm-bioinfo.toulouse.inra.fr/download/glint/) showed that only 482 SNPs are 100 % identical between the 35,455 SNP set developed by Duarte et al. [3], and the 88,851 'Baccara' x 'PI180693' polymorphic SNP set generated in this study. Only 45 SNPs were found to be identical between the 604 'Baccara' x 'PI180693' polymorphic SNP set genotyped in a GoldenGate® assay and mapped to the Duarte et al. [3] reference consensus map, and the 64,754 'Baccara' x 'PI180693' polymorphic SNP set retained in this study from SNP inferring on the 48 RILs. Genotyping data obtained with both methods (GoldenGate® vs WGGBS) for the 45 common SNPs were identical for 39 SNPs, apart from a few missing data with each method and a few heterozygous loci identified with WGGBS but not GoldenGate®. The remaining six SNPs totally failed in the GoldenGate®

assay but were successfully genotyped by direct sequencing (Additional file 3: Table S2).

7.2.3 HIGH DENSITY GENETIC MAPS

A first genetic 'Baccara' x 'PI180693' Duarte-derived map (below called BP-Duarte map), was constructed from the 'Baccara' x 'PI180693' polymorphic markers used in Hamon et al. [26] and Duarte et al. [3]. Positions of the resulting 914 mapped markers, covering 1073 cM, were generally colinear with those of the reference consensus map [3]. Linkage group (LG) lengths in cM were either similar or smaller in the BP-Duarte map than in the consensus map (Table 2).

A second high-density 'Baccara' x 'PI180693' WGGBS-derived map (below referred to as the BP-WGGBS map), including 64,263 markers and covering 1027 cM, was constructed from genotyping data used for the BP-Duarte genetic map by adding the data for the selection of 64,754 SNPs "genotyped-by-sequencing" on the 48 'Baccara' x 'PI180693' RILs (Table 2) to the previous matrix. This genetic map included 910 previously mapped markers and 63,353 (98.5 %) newly mapped genomic SNPs. SNP context sequences, genotyping data on parental lines and positions on the new BP-WGGBS of the 63,353 newly mapped genomic SNPs are described in Additional file 4: Table S3. Detailed mapping and polymorphism data in the 48 'Baccara'x'PI180693' RILs are included in Additional file 5: Table S4.

This new BP-WGGBS map showed an average density of 62.6 markers per cM. Marker density was very high for all *P. sativum* LGs (PsLGs), and ranged from 52 markers/cM (PsLGI) to 74 markers/cM (PsLGIV) (Table 2). Overall, new SNP markers were usually densely and homogeneously distributed along the seven pea LGs, with a few notable exceptions: (i) two large areas remained without markers, i.e. two gaps larger than 10 cM between two contiguous markers (Table 2), located on PsLGI and PsLGII (Fig. 2), (ii) several spots showed 400 to 800 markers at the same genetic position, on PsLGI, PsLGII, PsLGIII (two very close spots), and PsLGVII. PsLGVI showed a "staircase" curve alternating marker dense and marker poor areas (Fig. 2).

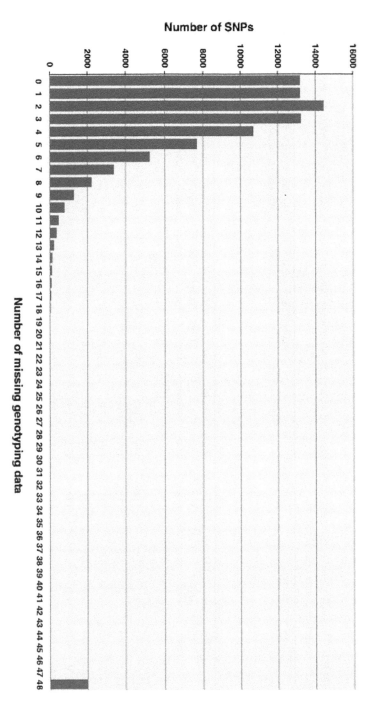

FIGURE 1: Frequency histogram of the number of missing data points in WGGBS of 48 RILs with the 88,851 reliable SNPs that are polymorphic between the 'Baccara' and 'PI180693' parentals (For example, 13,187, 13,186,and 14,452, were genotyped with 0, 1, or 2 missing data points, respectively among the 48 sequenced RILs).

TABLE 2: Comparative marker numbers, maps lengths and marker distributions per linkage group between the BP-WGGBS, the BP-Duarte and the consensus Duarte et al. [3] maps.

	Number of markers			Length (cM)			Number of markers/cM			Number of gaps > 10 cM between two contiguous markers			Number of WGGBS developed SNPs newly mapped
	BP-WGGBS map	BP-Duarte map	Duarte et al. [3] reference consensus map	BP-WGGBS map	BP-Duarte map	Duarte et al. [3] reference consensus map	BP-WGGBS map	BP-Duarte map	Duarte et al. [3] reference consensus map	BP-WGGBS map	BP-Duarte map	Duarte et al. [3] reference consensus map	BP-WGGBS map
PsLGI	6163	93	235	118	140	147	52.2	0.7	1.6	1	1	0	6071
PsLGII	8995	102	260	171	173	218	52.7	0.6	1.2	1	1	1	8898
PsLGIII	12,868	162	339	181	189	203	70.9	0.9	1.7	0	0	0	12,706
PsLGIV	9785	133	270	133	146	169	73.6	0.9	1.6	0	0	0	9652
PsLGV	7634	120	265	139	134	156	55.3	0.9	1.7	0	0	0	7514
PsLGVI	8490	116	298	119	111	142	71.5	1	2.1	0	0	0	8373
PsLG-VII	10,328	188	404	166	179	220	62.3	1	1.8	0	0	0	10,139
Whole	64,263	914	2071	1027	1073	1255	62.6	0.9	1.7	2	2	1	63,353

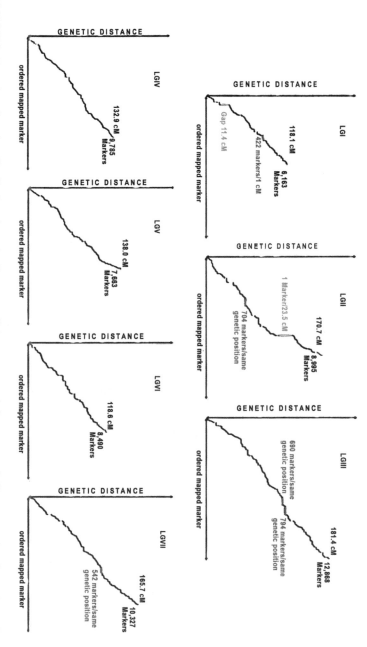

FIGURE 2: Dot-plot of marker distribution along the *P. sativum* linkage groups. A flatter curve indicates a region denser in markers. The red vertical ellipses indicate gaps without markers. The blue horizontal ellipses indicate hot-spots of markers at the same genetic position.

Overall, 3.2 % of the newly developed SNPs showed segregation distortion (P < 0.01) among the 48 'Baccara' x 'PI180693' derived RILs (Additional file 6: Figure S3). Most of the distorted markers clustered into genomic regions, mainly on PsLGs II (10.9 % of the mapped markers), III (5.4 %), IV (0.6 %), V (1.6 %) and VII (1.7 %). Markers in a defined cluster always distorted towards one of the parental lines ('Baccara' for clusters on PsLGs IV, V and VII, 'PI180693' for the cluster on PsLG III, and both parental lines on PsLGII for clusters on either side of the above described gap). PsLGI and PsLGVI showed negligible proportions of distorted markers.

Positions of the 914 markers common to the BP-Duarte and BP-WGGBS maps were colinear between the two maps, as well as with their published positions on the Duarte et al. consensus map [3] (Additional file 7: Figure S2). Except for a few local inversions, colinearity of these markers was maintained along the three maps. Map sizes were similar between the BP-Duarte and BP-WGGBS maps (respectively 1073 and 1027 cM), and significantly lower than the size of the Duarte et al. [3] reference consensus map from four populations (1255 cM). The number of mapped markers in the BP-WGGBS map was increased 31-fold and 70-fold compared to the consensus reference (comprising 2071 markers [3]) and to the BP-Duarte maps (Table 2).

7.2.4 SNP INFERRING ON OTHER PEA PARENTAL LINES

SNP inferring on four supplementary sequenced parental lines showed that coverage of most of the 88,851 SNPs as expected ranged between 6X and 13X (Additional file 2: Figure S1). Among these the 63,353 newly mapped SNPs indicated a fair to high level of polymorphism between four other pairs of mapping population parents, ranging from 33 % to 64 % (Table 3, Additional file 4: Table S3). The SNP context sequence and polymorphism data for the eight *P. sativum* parents of the mapping populations are described in Additional file 4: Table S3.

7.2.5 SELECTION AND VALIDATION OF A SNP SUB-SET IN A KASP™ GENOTYPING ASSAY

Based on mapping positions within QTL confidence intervals for biotic stress resistance and polymorphism data, we selected 1000 SNPs out of

TABLE 3: Percentage of polymorphic SNPs among the 63,353 newly developed SNPs mapped to the BP-WGGBS map, between five pairs of parental lines of pea mapping populations.

P. Sativum LG	Nb of 'WG-GBS' SNPs	% of polymorphic SNPs for 5 couples of mapping populations parents				
		Baccara & PI180693	Baccara & 552	Champagne & Terese	JI296 & DP	JI296 & FP
PsLGI	6071	100	35	64	34	37
PsLGII	8898	100	29	61	42	30
PsLGIII	12,706	100	40	67	40	29
PsLGIV	9652	100	38	66	36	34
PsLGV	7514	100	35	60	40	45
PsLGVI	8373	100	40	64	34	35
PsLGVII	10,139	100	43	63	32	26
Whole	63,353	100	38	64	37	33

the 63,353 mapped on the BP-WGGBS map. These were used for genotyping 1511 samples corresponding to 1438 different pea accessions (including the eight parental lines and the 48 RILs sequenced in this study), in a KASP™ 1,536-well plate format assay (see Methods).

Among these 1000 SNPs selected, 47 (5 %) failed due to missing or non-readable signals. Successful genotyping results were obtained with the other 953 SNPs (95 %), among which 949 (99.6 %) revealed the biallelic codominant polymorphism expected in the 'Baccara' x 'PI180693' RIL population. The remaining four markers (0.4 %) revealed a dominant polymorphism probably due to a failed probe design.

The 1,439,983 genotyping data points consisted of 47.3 % homozygous loci for alleles from 'Baccara', 48.1 % homozygous loci for alleles from 'PI180693', 1.3 % heterozygous loci, and 3.3 % unassigned data (failed, outside clusters, or null alleles). The entire polymorphism data set generated by the KASP™ genotyping assay in the eight sequenced pea parental lines and 4 F1s from crosses between pairs of these lines is described in Additional file 8: Table S5.

Forty-five thousand seven hundred forty-four comparisons were made between KASP™ and WGGBS genotyping data obtained with 953 SNPs on 48 RILs. Although there were higher levels of unassigned data generated with WGGBS than KASP™ genotyping, both methods gave very similar results for the 48 RILs genotypes (Additional file 9: Table S6). Conflicting data between the two methods remained below 1.5 % and included data: (i) where one method assigned a homozygote but the other a heterozygote (1 %), and which were mostly loci lying within regions showing residual heterozygosity and could therefore be re-assigned as heterozygotes; (ii) showing two different homozygotes assigned by either method (0.5 %), most of which (2/3) clustered in specific areas of a single RIL (BAP8_172) and may therefore correspond to a seed divergence and the remaining (1/3) that could be true genotyping errors from either method (Additional file 9: Table S6).

7.3 DISCUSSION

In this study, we discovered 419,024 genomic SNPs in whole genome sequences of four pea lines, but without prior genome reduction or sequence assembly. Among them, 213,030 appear robust for genotyping. For the first time in pea a WGGBS-derived map was produced. It contains 64,263 markers including 63,353 new genomic SNPs added to 910 other markers that allowed this new high-density genetic map to be aligned to a previously published reference consensus map (Duarte et al., [3]).

7.3.1 WHOLE GENOME DNA SEQUENCING: A STRATEGY WITHOUT GENOME REDUCTION FOR HIGH DENSITY DEVELOPMENT OF MARKERS IN AN ORPHAN SPECIES

SNP development strategies usually call for a genome reduction step, either focusing on the expressed fraction of the genome, using tags or methylation at restriction sites, or capture, in order to produce sequencing data with a sufficient coverage to avoid false polymorphisms. Although the pea genome is large (4.3 Gb) and thought to contain a large proportion of repeated sequences, we chose not to go through this genome reduction step and to work on direct whole genome DNA sequencing, associated with reduced coverage thresholds to validate the SNPs. This strategy generated more than ten billion 100 bp high quality reads from eight pea lines and 48

RILs, with a high level of polymorphism validation in other sequenced parental lines or in derived RILs. Results of the cleaning process were consistent for all lines, except 'Baccara' which was not etiolated prior to tissue sampling. The unexpectedly low level of repeated sequences (10 to 15 %) could be because the mapping parameters in the cleaning method were very stringent or some pea genome repeated sequences are under-represented in the REPbase and Genbank databases. Regardless, the remaining repeated sequences were taken into consideration during SNP discovery with the discoSnp tool and post discoSnp filtering processes. Thus we did not observe high SNP coverage that could have corresponded to organelle or repeated sequence regions (Additional file 2: Figure S1). The upcoming availability of the whole pea genome sequence will give further insights into the quantity and nature of its repeated sequences [14].

7.3.2 SNP CALLING WITHOUT DATA ASSEMBLY AND WITHOUT A REFERENCE GENOME

For a genomic SNP discovery approach, a tool appropriate for a large genome without reference sequence and a large amount of data was a prerequisite. We therefore used discoSnp software [24] that filled the specifications for SNP discovery from non-assembled reads. This tool, designed for calling SNPs from any number of input read sets, avoids the assembly and mapping processes, and therefore needed very limited time and memory to process ten billion reads. SNP discovery on the four parental lines only needed 1 CPU, 4 GB RAM and 48 h. SNP inferring with the kissreads module of discoSnp on the 48 RILs used 24 CPU, 4 GB RAM and 20 h. discoSnp outputs pairs of sequences in a fasta file. Sequences in the pair differ by a unique isolated nucleotide polymorphism, ie without any other polymorphisms in the context sequence on either side of the SNP. Although this is a strong limitation to the comprehensive detection of all possible SNPs, it is perfect for specifically selecting SNPs that can be directly used to design SNP-based assays for genotyping such as Illumina GoldenGate®, LGC Kasp™ or Affymetrix Axiom®.

The header of the discoSnp output sequences contains information about the read quality and coverage of each of the two alleles in each of the input read sets. Thus we could apply filters specifically adjusted to our biological parameters. One filter removed false heterozygotes (previously undetected) due to low coverage and low depth of whole genome sequencing data. Another filter, based on the coverage of the minor allele for SNPs detected on four pea lines, was used to remove questionable SNPs for complexity, multilocus status, copy number variations, or potential location in genome repeats. After this stringent filtering step, around 50 % of the 419,024 SNPs for all four pea parental lines were retained. This is a much higher figure than currently reported for GBS strategies which usually result in high levels of missing sequence data and a large percentage of uncalled genotypes [27, 28]. Indeed, whole genome sequencing (WGS) inherently provides very good quality sequence data that are more comprehensive than GBS. However the combined use of a tool such as discoSnp and a "post-discoSnp" specific filtering strategy were needed for efficient SNP discovery process. Furthermore, the level of missing data after sequencing 48 RILs was also very low, with 64,754 (among 88,851) SNPs showing less than 20 % missing sequence data (Fig. 1). To our knowledge this is the first study conducted using discoSnp on a real data set of reads from HiSeq sequencing. It had been previously used on a much smaller 454 read set after reduction of the *Ixodes ricinus* genome, to detect "ready to genotype" SNPs [29].

Only 482 of the 88,851 newly discovered polymorphic SNPs (between 'Baccara' and 'PI180693') were in common with the 35,455 gene-based SNPs developed by Duarte et al. [3]. This may be due to (i) relatively lower coverage and lower sequencing depth provided in this study, (ii) the specific detection of only isolated SNPs by discoSnp. Recently, Alves Carvalho et al. compared three SNP calling methods, including discoSnp, on the same set of sequence data. Each detected the same numbers of robust SNPs, but only a quarter of the SNPs were in common in the three sets and only half between two sets [30].

7.3.3 VALIDATION OF THE WGGBS GENOTYPING STRATEGY USING A SNP KASP™ GENOTYPING ASSAY

Despite the rather low average sequencing coverage in SNP discovery (around 7x) and in further SNP inferring (around 3.5x), several biological and technical observations confirmed the robustness of our strategy and showed its full efficiency in revealing reliable and designable SNPs. The polymorphism levels revealed between pairs of lines were consistent with previously described genetic distances between these lines [3]. Indeed, polymorphism levels were lower for 'Baccara'/'Terese' than for 'Baccara'/'Champagne', 'Champagne'/'Terese' or 'Baccara'/'PI180693'. 'Baccara' and 'Terese' belong to the spring field pea group, 'Champagne' to the winter fodder pea group, and 'PI180693' to the garden pea group. Similarly, genotyping data for the 45 SNPs in common with the 604 SNP set genotyped in a GoldenGate® assay and mapped by Duarte et al. were entirely consistent with the 64,754 genomic SNPs generated and genotyped-by-sequencing in this study (Additional file 3: Table S2). Furthermore, almost all of the 88,851 SNPs polymorphic between 'Baccara' and 'PI180693' were successfully genotyped on the RIL population (Fig. 1). Genotyping of 1438 different pea accessions with a subset of 1000 SNPs (1.4 % of the overall newly generated SNP resource) using the KASP™ technology also revealed a very high rate of 953 true and reliable SNPs (95 %). Thus, the chosen strategy of low average sequencing coverage (around 7x) followed by discoSnp-based SNPs discovery was fully validated.

Finally, genotyping data obtained with both KASP™ and WGGBS methods for the 953 SNPs on the 48 RILs were very similar. The level of missing data (almost twofold higher with WGGBS than with KASP™), as well as the 1.5 % inconsistent data between methods, suggests that WGGBS is less robust in heterozygote discrimination, probably due to the low sequencing depth. However, WGGBS was highly efficient and robust in revealing homozygous genotypes. Thus we confirmed that the low average sequencing coverage (around 3.5X) coupled with the use of the kiss-reads module of discoSnp inferred robust and reliable genotyping data.

7.3.4 HIGH-DENSITY GENETIC MAP CONSTRUCTION FROM A P. SATIVUM RIL POPULATION

Most developments in genotyping techniques in the last ten years have been clearly associated with multiplexing large numbers of markers (up to millions for example in Illumina or Affymetrix genotyping arrays) for each plant analyzed. WGGBS approaches tend to also provide huge quantities of SNP allele information for each plant. This leads to new challenges when building linkage maps. The main problem is that the number of plants in the mapping population is always far too small for a robust ordering of all the polymorphic markers. Thus, two distinct objectives have to be considered to produce: (i) a robust map with the largest possible number of markers (but still moderate due to limited population size) for which the order is statistically supported for a given LOD threshold; and (ii) a "complete map" in which all polymorphic markers are given a position, but without the order between markers being statistically supported (this is the principle of "bin-mapping" [31]). Another technical problem is the need for most mapping software to load the whole set of marker segregation data in memory, which makes it impossible to work with data sets beyond a certain size. The pipeline used in this study, developed in response to these different challenges, makes it possible to build both framework maps with orders supported by a LOD threshold and complete maps saturating the genomic space with all polymorphic markers [25]. Thus the map is suitable for QTL detection, GWAS and genomic selection approaches. In addition, marker data are loaded sequentially in small batches, so there is no limit to the number of markers, regardless of computer memory size. This sequential approach needs a much shorter computation time compared to the direct use of the annealing command of the CAR_H ^{T}A GENE tool: about 10 h were needed to map 64,000 SNPs on the seven PsLGs, whereas annealing commands in CAR_H ^{T}A GENE would probably have taken months or years. We are currently working on parallelizing the code for multicore computers to further reduce this computation time.

This pipeline allowed us to generate the first high-density pea genetic map mainly based on SNPs developed and genotyped by sequencing whole genome genomic DNA from a RIL population. The map size obtained was significantly smaller than those observed in previous reference composite maps [3, 18, 32], and similar to the sizes of recently reported individual maps [22] and the BP-Duarte individual map (this study), but with a more than 70-fold increase in marker density. A very high level of colinearity was observed for the markers that were in common with other *P. sativum* composite [3, 18, 32] or individual ([22], BP-Duarte-derived) maps. Thus the newly developed BP-WGGBS map is a useful tool for future studies focusing on variation in a given genomic region in pea.

A 3.2 % segregation distortion (P < 0.01) was observed for the newly developed SNPs among the 48 'Baccara'x'PI180693' derived RILs. This figure is very close to the 3 % identified for 224 markers in the 'Baccara'/'PI180693'/'552' consensus map [33] and the 4.5 % identified for SNPs in the 'Baccara' x 'PI180693' individual population data [3]. However it is much lower than the segregation distortion observed in other individual pea populations where large numbers of SNPs were genotyped and mapped [3, 17, 18]. The much higher marker number involved and the high resolution of the BP-WGGBS map allowed regions of segregation distortion to be identified in this population towards one or the other parental line, probably carrying detrimental lethal or sublethal alleles. Although the two parental lines are actually genetically distant [3], sterility barriers were not noted in the RIL production and fixation process. It is possible however that gametic or zygotic selection did take place and the issue of whether floral development or fertility genes lie in these regions still needs to be addressed. Genotypic data for distorted markers was not specifically considered during the mapping process. It is possible therefore that some marker order inversions occur in the distorted regions. However, considering that all the markers are not distorted in a specific region and the WGGBS-derived map aligns well with other maps (Additional file 7: Figure S2), such inversions are probably localized and infrequent. Information on marker distortion combined with future knowledge of localized recombination hot and cold spots, is of vital interest for some of the genomic regions that control traits of interest in this population. These are currently transferred through back-cross assisted breeding from 'Baccara' x 'PI180693' RILs to field pea spring and winter sown elite genetic backgrounds [34]. Interestingly a 23 cM gap on PsLGII separated two regions where markers showed distortions towards either parental line. Although a single marker ensures the link between these two regions, it is validated by other individual 'Baccara' x 'PI180693' maps (Duarte derived map from this study) or consensus maps where the gap was also observed [19, 33] or not [3, 18]. It is likely that in this region showing no recombination a chromosomal rearrangement occurred in the 'Baccara' x 'PI180693' population.

By choosing to sequence whole genomic DNA, we developed a large original resource of genomic SNP markers corresponding to genic and extra-genic sequences distributed all along the pea genome. Like genomic SSRs, but unlike cDNA-derived SNPs, it can be considered as neutral regarding evolutionary selection pressure. This was confirmed by the dense and homogeneous distribution of newly developed SNP markers along the seven pea LGs. A remaining 23 cM gap on PsLGII (Fig. 2) corresponds to a misassembly of two large blocks more than 30 cM apart, or to a 25 cM region without polymorphic markers between 'Baccara' and 'PI180693' comprising three gaps of 13.1, 4.6, 4.4 cM on two previous reference consensus maps [3, 19]. Interestingly no such gap was observed on the Sindhu et al. consensus map [18] that did not include the 'Baccara' x 'PI180693' RIL population. Furthermore, the atypical distribution of markers along the PsLGVI with a "staircase" curve alternating marker dense and marker poor areas (Fig. 2), is consistent with what we know of the complex synteny with *M.truncatula* pseudo chromosomes 2 & 6 [3].

7.3.5 A COMPREHENSIVE TOOL FOR ACADEMIC RESEARCH AND BREEDING IN PEA: THE CASE OF AN A. EUTEICHES RESISTANCE QTL

Sequencing and mapping a RIL population segregating for partial resistance to *A. euteiches* should lead to breakthroughs in the study of QTLs involved in *A.euteiches* resistance [26, 33]. There is now the potential for densification, fine mapping and candidate gene identification within the QTL confidence intervals. In addition, alignment of the BP-WGGBS map with the Duarte et al. [3] consensus map [3], anchors some of the markers to the genome of the model species *M.truncatula*, and opens the way for large-scale syntenic studies. The "metaQTL" area described by Hamon et al. [26] on a PsLGVII region corresponding to flowering MetaQTL Morpho8, and *A. euteiches* resistance MetaQTLs Ae26 and Ae27 is for instance defined between the SSR markers AA505 and AB101. This region originally covered 52.6 cM and included eight markers. Now it covers a much shorter genetic distance (23.6 cM) and includes 2,477 markers on the 'BP-WGGBS' map. The upper section of this region (the distal part has no clearly defined syntenic block) can also be linked to a 4.2 Mb region of *M.truncatula* chromosome 4 using around 30 gene-derived bridge markers (Fig. 3). Mapping, alignment and redetection of *A.euteiches* resistance QTL from Hamon et al. [26] are currently in progress [26].

The alignment of this new BP-WGGBS map to the consensus reference map [3] from four RIL populations including 'Baccara' x 'PI180693', will also facilitate the use of the SNP resource to fine map genomic regions and QTLs associated with traits of interest identified from other RIL populations where maps with common markers are available [22, 32, 35]. As shown by the level of polymorphism observed in five pairs of mapping population parents (Table 3) from sequencing only four pea lines (Table 3), any pea line can be rapidly and cheaply sequenced to obtain polymorphism information for the newly developed and mapped SNPs. For example, the PsLGVI region bordered by Ps001502 and FVE markers including a major QTL for winter frost damage [22] and syntenic with a 6.97 Mb region on the MtrChr8 [3] potentially includes 1,669 markers from the BP-WGGBS map.

7.4 CONCLUSION

In this study, a major genomic SNP marker resource was generated. It will be extremely valuable in future research of the genetic control of traits of interest, and breeding for the introgression and management of these traits into cultivated gene pools. The results show that it is now possible to generate and map reliable SNP markers in a plant species with a large unsequenced genome. The novelty of our approach included using whole sequencing without genome reduction, direct SNP discovery excluding sequence assembly, and a sequential mapping approach to handle tens of thousands of markers. The entire set can be used in single gene, association, or linkage mapping studies to capture new QTLs and refine QTL localizations. This can be done in any pea genetic background since mapping alignments are provided and the markers show high levels of polymorphism in alien backgrounds. The resource can also be screened in linkage studies (fine mapping, introgression) aimed at specific regions when generating large segregating populations. Finally, we generated a

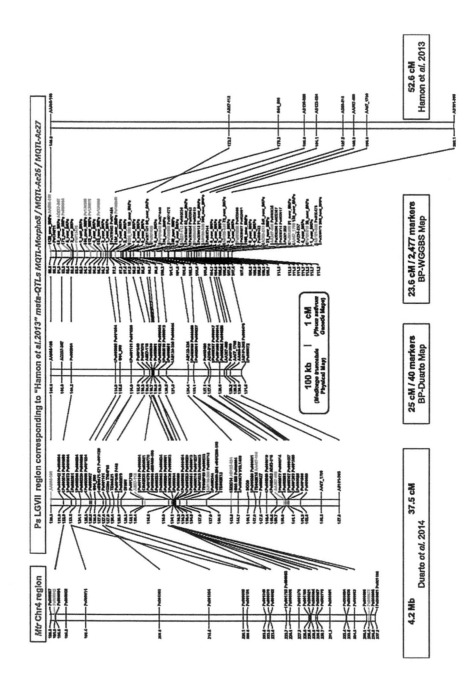

FIGURE 3: Marker densification in a MetaQTL region controlling partial resistance to A.euteiches between the SSR AA505 and AB101 reference markers on PsLGVII. The left hand side shows this region on the Duarte et al. [3] consensus map and its projection on *M. truncatula* pseudochromosome 4 (from Duarte et al. [3] - Additional file 5: Table S4). The center shows the same region on the two individual BP-Duarte and BP-WGGBS maps, covering respectively 25 cM and 23.6 cM. The right hand side shows the same region, detailed in Hamon et al. [26], covering 52.6 cM and corresponding to three MetaQTLs Morpho8, Ae26 and Ae27 (from Hamon et al. [26] - Additional files 9: Table S6 and Additional file 10: Table S7).

large choice of polymorphic markers internal to and bordering QTL confidence intervals. These provide an unprecedented tool for Marker Assisted Selection.

7.5 METHODS

7.5.1 PLANT MATERIAL AND TISSUE SAMPLING

A first set of plant material consisting of eight *P. sativum* genotypes was selected for (i) being parental lines of mapping populations (ii) including sources of partial resistance or tolerance to biotic and abiotic stresses and (iii) examining polymorphisms within or between *P. sativum* cultivated types, i.e. fodder, field (winter sown and spring sown), and garden pea types (Additional file 10: Table S7).

A second set consisted of 48 F8 Recombinant Inbred Lines (RILs) selected from a 178 RIL mapping population developed by Single Seed Descent from the cross between the cultivar 'Baccara' and the ecotype 'PI180693' [26, 33]. The 48 RILs were part of a 90 RIL set of the population that was used to establish a previous reference composite map [3]. The 48 RILs were sampled using MapPop 1.0 software [36] which was designed to select informative individuals optimizing the distribution of recombination points all over the genome. The "sampleexp" command was used to select the RIL set minimizing the expected Average Bin Length (eABL), i.e. the distance between two recombination points, using 'Baccara' x 'PI180693' genotyping and genetic map data from Hamon et al. [33] as the input file.

The 56 *P. sativum* genotypes were grown in a climate-controlled chamber (16 h photoperiod, temperature 15 °C night/20 °C day, 60 % minimal hygrometry). At least five plants per genotype were collected 15 days after sowing including a final etiolation period of five days (except for the Baccara sample). Tissues were flash frozen in liquid nitrogen and stored at −80 °C until further use.

7.5.2 DNA EXTRACTION

Genomic DNA was extracted from leaf tissue using a CTAB method as described by Rogers and Bendich [37]. The quality and quantity of extracted DNA was evaluated using a NanoDrop 8000 spectrophotometer (Thermo Fisher Scientific) and the Quant-iTTM dsDNA Assay Kit (Invitrogen). An estimated quantity of 3 µg of total genomic DNA was used to prepare each library, in a volume of 130 µL of ultra-purified water.

7.5.3 LIBRARY PREPARATION AND SEQUENCING

DNA libraries were prepared using the Illumina TruSeq DNA protocol following manufacturer's guidelines. Briefly, DNA was fragmented using a Covaris M220 sonicator. DNA was then end repaired and A-tailed, followed by the ligation of adapters and 12 cycles of PCR. Library profiles were controlled on a BioAnalyzer High Sensitivity chip. Quantities of usable material for each of the libraries were estimated by qPCR (KAPA Library Quantification Kit–Illumina Genome Analyzer-SYBR Fast Universal) and then normalized and pooled. The quality of the pools was then checked using qPCR and immediately followed by sequencing on the HiSeq2000 platform (Plateforme Genomique - Genopole Toulouse Midi-Pyrenees, France), using TruSeq PE Cluster Kit v3 (2 x 100 pb) and TruSeq SBS Kit v3. Eight lanes, each generating on average 36 Gb of sequences (360 M reads x 100 pb length) were necessary to sequence the set of eight lines. Twenty-four lanes, each generating on average 38 Gb of sequence (190 M reads x 100 pb length), were necessary to sequence the 48 RILs. Raw data were produced as sff files.

7.5.4 SEQUENCE CLEANING

Raw sequence data were processed with a dedicated pipeline enchaining different cleaning steps: after excluding contaminants, Flexbar [38] was used to remove sequence adapters. Strict read mapping was then performed with BWA [39] at first against RepBase [40], Medicago repeat library from TIGR Plant Repeats (ftp://ftp.plantbiology.msu.edu/pub/data/TIGR_Plant_Repeats/TIGR_Medicago_Repeats.v2), and available pea repeat sequences from Genbank. BWA was used subsequently to remove mitochondrial (from *M.truncatula*, *A.thaliana* and *Glycine max* NCBI) and chloroplastic (from pea NCBI RefSeq NC_014057.1) contamination. Low complexity sequences were finally removed using a custom perl script.

7.5.5 SNP DISCOVERY ON FOUR PEA LINES WITHOUT DATA ASSEMBLY

The discoSnp tool [24] calls SNPs from one or several read sets without using a reference genome or any other source of information. DiscoSnp aims to predict isolated SNPs (well suited for being easily amplified by PCR), k nucleotides apart from any other polymorphism source, with k being the main parameter. A micro assembly approach generates a fasta file containing each identified SNP with the contig it belongs to, represented by a pair of sequences which differ only at the polymorphic site. Each sequence comment provides information on average read coverage and average read quality per input read set. We applied the discoSnp tool to cleaned reads from four pea lines ('Baccara', 'PI180693', 'Champagne' and 'Terese'), with "k-mer = 27" as the input parameter, which was empirically shown to maximize the number of predictions in our conditions [24]. The kissreads module of the discoSnp tool provides the coverage for each SNP on the reads for each RIL. A dedicated perl script then enchained several filters to retain only the most reliable SNPs based on available sequence data for the four pea lines, strict homozygozity of each pea accession, a minimum 5x coverage on at least one pea accession, and a coverage of the minor allele at least half that of the major allele.

7.5.6 SNP INFERENCE AND WHOLE GENOME GENOTYPING BY SEQUENCING

Robust SNPs (highly or very highly designable for genotyping assay designs) polymorphic between the 'Baccara' and 'PI180693' parental lines were inferred using the kissreads module of the discoSnp tool [24] on cleaned reads generated by sequencing the 48 'Baccara' x 'PI180693' RILs. Kissreads provides the coverage for each SNP on reads for each RIL. Considering that less sequence data was generated for RILs than the parental lines, and that a SNP which is inferred on already identified SNPs is robust, a 3x coverage threshold was applied to each RIL read set to select reliable data.

To identify potential polymorphisms in a larger set of parental lines, SNPs were inferred in the same SNP set as used for the 48 RILs using the kissreads module. It was applied to the cleaned reads generated by sequencing four additional parental lines of mapping populations ('552', 'JI296', 'DP' and 'FP'). A 5x coverage threshold was set in order to select reliable data.

For each SNP, the allele was coded "A" when identical to the 'Baccara' parent, or "B" when identical to the 'PI180693' parent.

7.5.7 CONSTRUCTION OF BP-DUARTE AND BP-WGGBS GENETIC MAPS

To first build a robust individual 'Baccara' x 'PI180693' map, we used genotyping data from the Duarte et al. [3] reference consensus map. All the markers that were polymorphic for the 90 'Baccara' x 'PI180693' RILs were used for the consensus map. We also used genotyping data obtained from mainly SSR [26] and SNP [3] markers on the entire 'Baccara' x 'PI180693' 178 RIL population. The final genotyping data matrix comprised a total of 928 markers, including 295 markers with data on 178 RILs and 633 markers with data on a sub-set of 90 RILs. The 1:1 allelic segregation ratio for each marker within the RIL population was verified using a Chi-square test (P > 0.01 and P > 0.001). Genetic linkage analyses were performed using the "group" command of $CAR_H{}^TA$ GENE software [41], with a minimum LOD score threshold of 3.0 and a recombination frequency <0.3. Marker order was refined using the "annealing 100 100 0.1 0.9" command of CAR_H ^TA GENE. The resulting map was called BP-Duarte.

To build the new map including the WGGBS data, we added to the previous matrix the "genotyping by sequencing" data obtained in this study on 48 'Baccara' x 'PI180693' RILs. The 1:1 allelic segregation ratio at each marker was verified and the LGs established as for the previous genetic map. Then each LG was constructed individually as described earlier [25] using CAR_H ^TA GENE called from custom R scripts (http://www.r-project.org/foundation). In the first step, statistically robust scaffold maps were constructed by elongating the map from one seed marker in both directions with the most strongly linked marker in the data set, located at a distance greater than 5 cM. This procedure ensured an extremely robust order of regularly-spaced markers. To avoid potential detrimental consequences due to the choice of the seed marker, for instance on incomplete map coverage, 10 independent replicates were performed using ten randomly drawn seed markers, and a consensus scaffold was built by merging all the ten scaffolds. In a second step, marker density of the consensus scaffolds was increased to produce framework maps containing as many markers as possible, while keeping a LOD score >3.0 for the robustness of marker orders. Finally, the complete map was obtained by individual placement of additional markers on the framework map using bin mapping [31]. Markers with a minimum allele frequency less than 6 % and markers with more than 78 % of missing data were not included on the scaffold and framework maps, but they were included in the final placement step, since this could not influence the placement of other markers. The resulting map was called BP-WGGBS.

For both the BP-Duarte and BP-WGGBS genetic maps, the Haldane function was used to calculate cM distances between markers [42] and MapChart 2.2 was used to draw the maps [43].

7.5.8 SELECTION AND VALIDATION OF A 1000 SNP SUBSET IN A 1536-WELL PLATE KASP™ GENOTYPING ASSAY

A subset of 1000 SNPs was chosen among the mapped SNPs on the BP-WGGBS genetic map to design a 1536 well plate KASP™ [23] assay.

SNP selection was based on: (i) an even distribution of markers in genomic regions containing QTL of interest for resistance to various biotic stresses; (ii) one single SNP per genetic position whatever the number of fully linked markers at this position; (iii) highest polymorphism levels between pairs of parental lines of mapping populations among the eight sequenced lines. The final selected SNP subset covered around 500 cM of the 1027 cM BP-WGGBS genetic map and can be considered as a quality-unbiased sampling of the whole SNP resource generated. SNP designability on KASP™ technology showed a 99 % success rate (LGC genomics service lab, UK). For each assay, at least 15 µg of genomic DNA, i.e. 16 96-well plates each containing 100 µL of samples were provided to LGC Genomics service lab, UK (http://www.lgcgenomics.com) for sample normalization and genotyping using KASP™ technology [23]. A total of 1511 DNA samples (corresponding to 1438 different pea accessions, including the eight parental lines and the 48 RILs sequenced in this study, and a large number of other RILs and accessions), were genotyped. Automatic allele calling for each locus was carried out using Klustercaller software [23]. The homozygous and heterozygous clusters were checked visually and were manually edited when necessary. Technical replicates and signal intensities were verified and only the most reliable calls were retained.

REFERENCES

1. Poland JA, Rife TW. Genotyping-by-sequencing for plant breeding and genetics. Plant Genome. 2012;5(3):92–102.
2. He J, Zhao X, Laroche A, Lu Z-X, Liu H, Li Z. Genotyping-by-sequencing (GBS), an ultimate marker-assisted selection (MAS) tool to accelerate plant breeding. Front Plant Sci. 2014;5:484.
3. Duarte J, Riviere N, Baranger A, Aubert G, Burstin J, Cornet L, et al. Transcriptome sequencing for high throughput SNP development and genetic mapping in Pea. BMC Genomics. 2014;15(1):126.
4. Miller M, Dunham J, Amores A, Cresko W, Johnson E. Rapid and cost effective polymorphism identification and genotyping using restriction site associated DNA (RAD) markers. Genome Res. 2007;17:240–8.
5. Baird NA, Etter PD, Atwood TS, Currey MC, Shiver AL, Lewis ZA, et al. Rapid SNP Discovery and Genetic Mapping Using Sequenced RAD Markers. PLoS One. 2008;3(10):e3376.

6. Chutimanitsakun Y, Nipper R, Cuesta-Marcos A, Cistue L, Corey A, Filichkina T, et al. Construction and application for QTL analysis of a Restriction Site Associated DNA (RAD) linkage map in barley. BMC Genomics. 2011;12(1):4.

7. Pfender WF, Saha MC, Johnson EA, Slabaugh MB. Mapping with RAD (restriction-site associated DNA) markers to rapidly identify QTL for stem rust resistance in Lolium perenne. Theor Appl Genet. 2011;122(8):1467–80.

8. Deokar A, Ramsay L, Sharpe A, Diapari M, Sindhu A, Bett K, et al. Genome wide SNP identification in chickpea for use in development of a high density genetic map and improvement of chickpea reference genome assembly. BMC Genomics. 2014;15(1):708.

9. Elshire RJ, Glaubitz JC, Sun Q, Poland JA, Kawamoto K, Buckler ES, et al. A Robust, Simple Genotyping-by-Sequencing (GBS) Approach for High Diversity Species. PLoS One. 2011;6(5), e19379.

10. Sonah H, Bastien M, Iquira E, Tardivel A, Légaré G, Boyle B, et al. An Improved Genotyping by Sequencing (GBS) Approach Offering Increased Versatility and Efficiency of SNP Discovery and Genotyping. PLoS One. 2013;8(1), e54603.

11. Smýkal P, Aubert G, Burstin J, Coyne CJ, Ellis NT, Flavell AJ, et al. Pea (Pisum sativum L.) in the genomic era. Agronomy. 2012;2(2):74–115.

12. Kalo P, Seres A, Taylor SA, Jakab J, Kevei Z, Kereszt A, et al. Comparative mapping between Medicago sativa and Pisum sativum. Mol Gen Genomics. 2004;272(3):235–46.

13. Macas J, Neumann P, Navrátilová A. Repetitive DNA in the pea (Pisum sativum L.) genome: comprehensive characterization using 454 sequencing and comparison to soybean and Medicago truncatula. BMC Genomics. 2007;8(1):1–16.

14. Burstin J, Alves-Carvalho S, Tayeh N, Aluome C, Bourion V, Klein A, Carrere S, Brochot A-L, Salloignon P, Siol M et al. Recent pea genomic resources will enhance complementary improvement strategies in this crop. In: VIIth Int Conf on Legumes Genetics and Genomics, 2014, July 6–11. Saskatoon, Canada, http://knowpulse2.usask.ca/iflrc-iclgg/?q=node/447; 2014. [Accessed 12 Feb 2016]

15. Franssen S, Shrestha R, Brautigam A, Bornberg-Bauer E, Weber A. Comprehensive transcriptome analysis of the highly complex Pisum sativum genome using next generation sequencing. BMC Genomics. 2011;12(1):227.

16. Kaur S, Pembleton LW, Cogan NO, Savin KW, Leonforte T, Paull J, et al. Transcriptome sequencing of field pea and faba bean for discovery and validation of SSR genetic markers. BMC Genomics. 2012;13:104.

17. Leonforte A, Sudheesh S, Cogan NO, Salisbury PA, Nicolas ME, Materne M, et al. SNP marker discovery, linkage map construction and identification of QTLs for enhanced salinity tolerance in field pea (Pisum sativum L.). BMC Plant Biol. 2013;13(1):161.

18. Sindhu A, Ramsay L, Sanderson L-A, Stonehouse R, Li R, Condie J, et al. Gene-based SNP discovery and genetic mapping in pea. Theor Appl Genet. 2014;1–17.

19. Loridon K, McPhee K, Morin J, Dubreuil P, Pilet-Nayel ML, Aubert G, et al. Microsatellite marker polymorphism and mapping in pea (Pisum sativum L.). Theor Appl Genet. 2005;111(6):1022–31.

20. Deulvot C, Charrel H, Marty A, Jacquin F, Donnadieu C, Lejeune-Henaut I, et al. Highly-multiplexed SNP genotyping for genetic mapping and germplasm diversity studies in pea. BMC Genomics. 2010;11(1):468.

21. Bordat A, Savois V, Nicolas M, Salse J, Chauveau A, Bourgeois M, et al. Translational Genomics in Legumes Allowed Placing In Silico 5460 Unigenes on the Pea Functional Map and Identified Candidate Genes in Pisum sativum L. G3. 2011;1(2):93–103.

22. Klein A, Houtin H, Rond C, Marget P, Jacquin F, Boucherot K, et al. QTL analysis of frost damage in pea suggests different mechanisms involved in frost tolerance. Theor Appl Genet. 2014;127(6):1319–30.

23. Semagn K, Babu R, Hearne S, Olsen M. Single nucleotide polymorphism genotyping using Kompetitive Allele Specific PCR (KASP): overview of the technology and its application in crop improvement. Mol Breeding. 2014;33(1):1–14.

24. Uricaru R, Rizk G, Lacroix V, Quillery E, Plantard O, Chikhi R, et al. Reference-free detection of isolated SNPs. Nucleic Acids Res. 2014;43(2):e11.

25. Bauer E, Falque M, Walter H, Bauland C, Camisan C, Campo L, et al. Intraspecific variation of recombination rate in maize. Genome Biol. 2013;14(9):R103.

26. Hamon C, Coyne C, McGee R, Lesne A, Esnault R, Mangin P, et al. QTL meta-analysis provides a comprehensive view of loci controlling partial resistance to Aphanomyces euteiches in four sources of resistance in pea. BMC Plant Biol. 2013;13(1):45.

27. Crossa J, Beyene Y, Kassa S, Pérez P, Hickey JM, Chen C, et al. Genomic prediction in maize breeding populations with genotyping-by-sequencing. G3: Genes| Genomes| Genetics. 2013;3(11):1903–26.

28. Swarts K, Li H, Romero Navarro JA, An D, Romay MC, Hearne S, Acharya C, Glaubitz JC, Mitchell S, Elshire RJ, Buckler ES, Bradbury PJ. Novel Methods to Optimize Genotypic Imputation for Low-Coverage, Next-Generation Sequence Data in Crop Plants. Madison, WI: Crop Science Society of America. Plant Genome. 2014;7(3). doi:10.3835/plantgenome2014.05.0023.

29. Quillery E, Quenez O, Peterlongo P, Plantard O. Development of genomic resources for the tick Ixodes ricinus: isolation and characterization of single nucleotide polymorphisms. Mol Ecol Resour. 2014;14(2):393–400.

30. Alves Carvalho S, Uricaru R, Duarte J, Lemaitre C, Rivière N, Boutet G, Baranger A, Peterlongo P. Reference-free high-throughput SNP detection in pea: an example of discoSnp usage for a non-model complex genome. In: European Conference on Computational Biology. vol. 5 F1000Posters; 2014

31. Falque M, Décousset L, Devrins D, Jacob A-M, Joets J, Martinant J-P, et al. Linkage mapping of 1454 new maize candidate gene loci. Genetics. 2005;170(4):1957–66.

32. Sudheesh S, Lombardi M, Leonforte A, Cogan NOI, Materne M, Forster JW, Kaur S. Consensus Genetic Map Construction for Field Pea (Pisum sativum L.), Trait Dissection of Biotic and Abiotic Stress Tolerance and Development of a Diagnostic Marker for the er1 Powdery Mildew Resistance Gene. Plant Mol Biol Rep. 2015;33(5):1391-1403. doi:10.1007/s11105-014-0837-7.

33. Hamon C, Baranger A, Coyne C, McGee R, Goff I, L'Anthoëne V, et al. New consistent QTL in pea associated with partial resistance to Aphanomyces euteiches in multiple French and American environments. Theor Appl Genet. 2011;123(2):261–81.

34. Lavaud C, Lesné A, Piriou C, Le Roy G, Boutet G, Moussart A, et al. Validation of QTL for resistance to Aphanomyces euteiches in different pea genetic backgrounds using near-isogenic lines. Theor Appl Genet. 2015;1–16.

35. Tayeh N, Bahrman N, Devaux R, Bluteau A, Prosperi J-M, Delbreil B, et al. A high-density genetic map of the Medicago truncatula major freezing tolerance QTL on chromosome 6 reveals colinearity with a QTL related to freezing damage on Pisum sativum linkage group VI. Mol Breeding. 2013;32(2):279–89.

36. Brown D, Vision T. MapPop 1.0: software for selective mapping and bin mapping. Computer program available from http://visionlab.web.unc.edu/software-and-data-bases/mappop/. [Accessed 12 Feb 2016] 2000.

37. Rogers S, Bendich A. Extraction of total cellular DNA from plants, algae and fungi. In: Gelvin S, Schilperoort R, editors. Plant Molecular Biology Manual. Springer Netherlands; 1994. p. 183–90.

38. Dodt M, Roehr JT, Ahmed R, Dieterich C. FLEXBAR—flexible barcode and adapter processing for next-generation sequencing platforms. Biology. 2012;1(3):895–905.

39. Li H, Durbin R. Fast and accurate short read alignment with Burrows–Wheeler transform. Bioinformatics. 2009;25(14):1754–60.

40. Jurka J, Kapitonov VV, Pavlicek A, Klonowski P, Kohany O, Walichiewicz J. Repbase Update, a database of eukaryotic repetitive elements. Cytogenet Genome Res. 2005;110(1–4):462–7.

41. de Givry S, Bouchez M, Chabrier P, Milan D, Schiex T. CARH TA GENE: multipopulation integrated genetic and radiation hybrid mapping. Bioinformatics. 2005;21(8):1703–4.

42. Haldane J. The combination of linkage values, and the calculation of distance between the loci of linked factors. J Genet. 1919;8(4):299–309.

43. Voorrips RE. MapChart: Software for the Graphical Presentation of Linkage Maps and QTLs. J Hered. 2002;93(1):77–8.

There are several supplemental files that are not available in this version of the article. To view this additional information, please use the citation on the first page of this chapter.

PART III

AVENUES OF CROP IMPROVEMENT UNDER CHANGING CLIMATE USING GENOMICS

CHAPTER 8

Improved Evidence-Based Genome-Scale Metabolic Models for Maize Leaf, Embryo, and Endosperm

SAMUEL M. D. SEAVER, LOUIS M. T. BRADBURY,
OCÉANE FRELIN, RAPHY ZARECKI, EYTAN RUPPIN,
ANDREW D. HANSON, AND CHRISTOPHER S. HENRY

8.1 INTRODUCTION

The ability of a plant to grow and survive is linked to its metabolic network (Stitt et al., 2010), which indicates that a capacity to predict and understand plant metabolism will improve our understanding of plant response to changing environments and genetic perturbations (Mo et al., 2009; Chang et al., 2011; Saha et al., 2011). Furthermore, the yield of a wide range of plant products is crucial to human society, particularly when inputs such as water are limited (Skirycz and Inze, 2010). Many classical biochemical and genetic experiments involve the elucidation of biological functions for individual gene products. However, many external and internal perturbations lead to systemic responses, and a systems-level understanding of plant metabolism is required to fully explain these system responses.

To build this systems-level understanding, several genome-scale metabolic reconstructions have recently been published for plant species (Poolman et al., 2009; de Oliveira Dal'molin et al., 2010a,b; Saha et al., 2011; Poolman et al., 2013). Each reconstruction consists of all reactions known to be catalyzed by one or more of the gene products in the plant genome. The methods employed to study these metabolic models, such as flux balance analysis (FBA), consider all reactions in the model when attempting to predict a biological phenotype, such as plant growth. Metabolic reconstructions are built from many data sources, notably public databases and individual publications. Reconstructions are validated by comparing the activity of well-characterized pathways in silico with biochemical evidence in the literature. Poolman et al. (2009) built the first genome-scale plant metabolic reconstruction, which could respire on heterotrophic media in silico and produce biomass components in proportions that matched in vivo observations. de Oliveira Dal'molin et al. (2010a) investigated autotrophic biosynthesis of plant biomass, showing that the model correctly predicted the reactions used for both photosynthesis and photorespiration. de Oliveira Dal'Molin et al. also developed a metabolic reconstruction of a C4 plant (de Oliveira Dal'molin et al., 2010b) containing plastidial reactions for photosynthesis. This reconstruction was shown to be capable of performing three known subtypes of C4 photosynthesis. In other work, Saha et al. (2011) show that genetic perturbations in the phenylpropanoid biosynthesis pathway could be simulated in silico, producing an impact on cell wall composition that compared favorably with experimental data from known maize mutants.

The validation approaches described above are based on a few well-known biochemical pathways, and involve large genome-scale metabolic reconstructions, built to provide a systems-level understanding of how a metabolic network behaves under certain conditions. For example, Schwender and Hay (2012) investigated how a metabolic reconstruction exhibited variation in reaction activity in response to variation in the biosynthetic demands of oil and protein as storage products in the plant embryo and were able to identify the utilization of a pathway within the network of reactions that was not yet characterized in the literature. Similarly,

Töpfer et al. (2013) explored the means with which a set of pathways in a metabolic reconstruction responded to various conditions of light and temperature, showing, in one case, the preference for methylerythritol 4-phosphate pathway over the mevalonate pathway in isoprenoid biosynthesis, and also generating a new hypothesis for the role of homocysteine–cysteine conversion.

Genome-scale metabolic reconstructions are generated based on the annotation of all gene products in the full genome, and, thus, they include every reaction that can be catalyzed by the plant. However, a multi-cellular organism will activate different subsets of their genes in different organs, tissues, developmental stages, and environmental conditions. To be accurate, genome-scale metabolic reconstructions must represent the reduced metabolism that truly exists in cells of a specific type and in a specific condition. Most reconstructions mentioned previously were either intended to represent a leaf cell or the primary metabolism of a generic plant cell. Other metabolic reconstructions have been built to target specific tissues and organs, such as the seeds of barley (*Hordeum vulgare*; Grafahrend-Belau et al., 2009), and the embryos of oilseed rape (*Brassica napus*; Hay and Schwender, 2011a,b; Pilalis et al., 2011). Grafahrend-Belau et al. followed up their study of barley seeds by building manually curated metabolic reconstructions of barley stem and leaf, and integrating the three reconstructions into a single model (Grafahrend-Belau et al., 2013). Recently, several new approaches have emerged to integrate large-scale data (Baerenfaller et al., 2008) in an automated manner to either generate new condition-specific models (Mintz-Oron et al., 2012), or to constrain the behavior of individual reactions in a full genome-scale model to better reflect the behavior of specific organs or tissues (Töpfer et al., 2013).

The ongoing explosion in plant transcriptome sequencing, driven by advances in next-generation sequencing (NGS) and by the relative ease of sequencing a collection of cDNAs as opposed to predicting gene models in plant chromosomes (Ozsolak and Milos, 2011), means that many transcript profiles are now publicly available, and individual laboratories can afford to generate new transcript profiles for individual experiments. Indeed, Töpfer et al. used their own transcript profiles, which they generated from Arabidopsis rosettes (Töpfer et al., 2013). Several computational methods have been developed that are able to integrate transcript profiles with a metabolic reconstruction to produce improved predictions of reaction utilization and flux.

Töpfer et al. used E-flux (Colijn et al., 2009), which fits flux predictions based on gene expression data, but does not attempt to reduce a full genome model to a tissue or organ specific version. The Töpfer et al. work was focused on several primary and secondary metabolic pathways that are known to be active with the rosettes of Arabidopsis. Mintz-Oron et al. used the iMAT approach (Jerby et al., 2010; Zur et al., 2010), which generates aggregate models based on random sampling of fluxes to fit gene expression data. While this approach provides a more comprehensive account of the metabolic network, the extensive sampling can be cumbersome. An updated method eliminates the need for random sampling and thereby runs faster (Wang et al., 2012). This method searches for an optimal solution by iteratively activating each reaction whose associated genes have high expression, which means that the method still performed many optimizations. We have developed a new approach that requires far fewer optimization steps, allowing for transcriptome-based metabolic reconstructions to be formed from transcript profiles at a greater speed and with less complexity. We note here the introduction of the term transcriptome-based to reflect this class of model, which is based on fitting a genome-scale model to a select subset of gene expression data. The term tissue-specific is often used for models of this type. However, expression data often does not capture the entire behavior or a tissue, nor does a single tissue necessarily reflect a single biological behavior (e.g., leaf tissue consists of several sub-cell types).

We demonstrate our approach for reconstruction of transcript-specific models with a new genome-scale metabolic reconstruction of maize. Our new genome-scale maize model includes three important enhancements over previously published models: (i) an expanded and improved biomass composition; (ii) improved gene-protein-reaction associations where low confidence gene-reaction mappings based on poor evidence or purely computational predictions have been removed; and (iii) improved compartmentalization of reactions to subcellular organelles based on a combination of literature evidence, curation, and gapfilling algorithms. The improved gene-reaction associations in our new model were critical to our use of maize transcript profiles (Davidson et al., 2011) to produce new transcriptome-based models of the leaf, embryo, and endosperm in maize. We applied our novel model reconstruction method to maximize the activity of reactions associated with high expression genes while removing as many reactions associated with low expression genes as possible. We also adjusted the biomass composition of our embryo and endosperm models to better fit the actual composition data for these tissues by curating data for individual components from a variety of literature sources. To test the accuracy of our models, we explored how well they replicate the flux profiles measured for central carbon metabolism in embryo and endosperm tissues (Alonso et al., 2010, 2011). This analysis demonstrates that our models have an improved fit between the fluxes generated in silico and the fluxes measured in vivo. All models produced from this work are available for download from the DOE Systems Biology Knowledgebase (http://kbase. us) and the PlantSEED resource (Seaver et al., 2014).

8.2 MATERIALS AND METHODS

8.2.1 BIOCHEMISTRY

We used the plant biochemistry database built for the PlantSEED project (Seaver et al., 2014). This database is notably built on KEGG (Kanehisa and Goto, 2000; Kanehisa et al., 2012) and MetaCyc (Caspi et al., 2012), which had been integrated using InChI (Heller et al., 2013) strings generated from mol files provided by both databases. The integration includes several plant biochemistry databases such as the BioCyc databases for *Arabidopsis thaliana* (*Arabidopsis*; AraCyc v11.5 Mueller et al., 2003; Zhang et al., 2010), and maize (MaizeCyc v2.2.2 Monaco et al., 2013 and CornCyc v4.0 Zhang et al., 2010), and several published metabolic models for *A. thaliana* (de Oliveira Dal'molin et al., 2010a,b, 2011; Saha et al., 2011; Mintz-Oron et al., 2012) and maize (de Oliveira Dal'molin et al., 2010b; Saha et al., 2011). The metabolic reconstructions we built depend on this integration, and the reactions for the respective *Arabidopsis* and maize metabolic reconstructions are thus drawn from this database.

8.2.2 COMPARTMENTS

An important aspect of plant metabolic models is the compartmentalization of reactions into plastids, mitochondria, and other organelles. To accurately capture this compartmentalization, we downloaded localization data for proteins from PPDB (Sun et al., 2009), SUBA (Tanz et al., 2013), AraCyc, MaizeCyc, and CornCyc. We systematically avoided any protein localizations generated solely via computational predictions. From PPDB, we only used data that the PPDB team had curated. From SUBA, we only used data from GFP experiments, which are more reliable than the data from mass spectrometry experiments. Finally, from AraCyc, MaizeCyc, and CornCyc, many reactions are localized according to biochemical support such as the histidine pathway in plastids (Ingle, 2011). Even if the genes associated with these pathways do not have localization data, we considered them to be localized if there was experimental evidence for the gene-reaction associations. Much of the localization data could only be applied directly to either of the two different species, and therefore we propagated the associations between Arabidopsis and maize by using the same conservative approach we applied to EnsemblCompara protein families in the PlantSEED project (Vilella et al., 2009; Kersey et al., 2014; Seaver et al., 2014).

8.2.3 MODEL PATHWAY-GAPFILLING

A new gapfilling algorithm was applied during the reconstruction of all our plant genome-scale models. This algorithm provides a means of identifying the minimal set of reactions that must be made reversible or added to the model in order to activate as many gene-associated reactions in the model as possible. The constraints of the optimization problem resemble the constraints for existing classical gapfilling approaches (Satish Kumar et al., 2007; Kumar and Maranas, 2009).

$$N_{super} \cdot v = 0 \qquad (1)$$

$$0 \leq v_i \leq 100z_i \qquad i = 1,\ldots,r_{gapfill} \tag{2}$$

$$z_{for,i} + z_{rev,i} \leq 1 \qquad i=1,\ldots,r_{gapfill} \tag{3}$$

$$-100 \leq v_{ex,i} \leq 100\gamma_i \quad i=1,\ldots,m_{transported} \tag{4}$$

Equation (1) represents the mass balance constraints, where N_{super} is the matrix of stoichiometric coefficients through all reactions in our model plus all candidate reactions added from our biochemistry database, while v is the vector of fluxes through all model and database reactions represented in the N_{super} matrix. In these and all other constraints, reversible reactions have been decomposed into separate forward and backward component reactions to ensure that all fluxes are always positive. Equation (2) sets the bounds on the flux through reaction i, where v_i is the flux and z_i is a binary use variable equal to zero when the flux is zero and equal to one otherwise. Equation (3) ensures that the forward and backward components of the same reaction may not both be active at the same time; in our formulation, this constraint is the sole reason for using binary variables. Equation (4) establishes the growth conditions for the gapfilling analysis; metabolites present in the growth media (e.g., heterotrophic media or autotrophic media) have a γ_i of 1 in Equation (4). Otherwise γ_i is zero.

In addition to these standard constraints, we applied a new constraint that introduces a slack flux for all reactions found in the original un-gap-filled model:

$$v_{for,i} + v_{rev,i} + \delta_i \geq 0.01 \qquad i=1,\ldots,r_{model} \tag{5}$$

Equation (5) states that the sum of the net flux through reaction i ($v_{for,i} + v_{rev,i}$) and the slack flux for reaction i (δ_i) must be greater than or equal to 0.01. As a result of this constraint, a reaction can only have a net flux of zero if the corresponding slack flux is 0.01. Thus, the slack flux is a variable used to identify reactions that carry no flux in the model. We utilize this new slack flux for this purpose in the objective function for our gapfilling.

8.2.3.1 OBJECTIVE

$$\text{Minimize} \quad \sum_{i=1}^{r_{annotated}} a\left(\gamma_{activate,i}\delta_i\right) + \sum_{i=1}^{r_{gapfilling}} a\left(\gamma_{gapfill,i}v_i\right) \tag{6}$$

This new objective function minimizes the sum of the slack fluxes associated with the reactions included in our original model while simultaneously minimizing the flux through all gapfilled reactions added to the model from our database. The purpose of this objective function is to maximize the number of gene-associated reactions that carry flux while minimizing the number of gapfilled reactions added to the model. This effectively gives precedence to the gene-associated reactions in our model. The activation coefficient, $\gamma_{activate,\,i}$, dictates the cost of leaving a gene-associated reaction inactive, while the gapfilling coefficient, $\gamma_{gapfill,\,i}$, dictates the cost of adding a gapfilled reaction to the model. In our gapfilling studies, we set $\gamma_{activate,\,i}$ equal to one for all gene-associated reactions, while we computed $\gamma_{gapfill,\,i}$ as described in our previous work (Henry et al., 2009, 2010).

We also used a scaling factor a in our objective function, which scales the cost of leaving some model reactions inactive against the cost of adding new reactions to the model from the database. We explored values for a ranging from 0.01 to 0.25, but we found only a small effect on the solutions produced. Generally, an a of 0.1 generated the most well-balanced gapfilling solutions.

In this gapfilling formulation, we utilize continuous linear flux variables in our objective function rather than the more typical binary variables (e.g., $z_{for,\,i}$ and $z_{rev,\,i}$) (Kumar et al., 2007). This adjustment reduced the compute time required to obtain a globally optimal solution by over

90% while having no appreciable impact on solutions obtained. This use of linear variables has been previously proposed in other published gapfilling algorithms, with detailed sensitivity analyses performed and similar results obtained (Latendresse, 2014). Thus, we do not repeat the sensitivity analysis here.

8.2.4 TRANSCRIPTOME-BASED PATHWAY-GAPFILLING

Our method for producing transcriptome-based models builds on the pathway-gapfilling approach (see previous used during the reconstruction of our models. Our pathway-gapfilling approach attempts to maximize the number of number of active gene-associated reactions. This approach further refines the model toward a specific transcriptome by maximizing the activity of reactions associated with highly expressed genes while minimizing active reactions associated with minimally expressed genes. This formulation includes flexibility permitting high-expression reactions to remain "off" if activating them requires the function of too many low expression reactions, and vice versa.

The first step of this algorithm is to categorize every reaction in the model as either high expression or low expression. This is done by assigning an expression score, $E_{exp,\,i}$, to every gene-associated reaction i as follows:

$$E_{exp,i} = \mathrm{Max}(C_{exp,i,j}) \qquad i = 1,\ldots,r \quad j = 1,\ldots,c_i \qquad (7)$$

$$C_{exp,j} = \mathrm{Min}(P_{exp,j,k}) \qquad j = 1,\ldots,c_i \quad k = 1,\ldots,p_j \qquad (8)$$

$$P_{exp,k} = \mathrm{Max}(G_{exp,k,l}) \qquad k = 1,\ldots,p_j \quad l=1,\ldots,g_k \qquad (9)$$

In Equations (7)–(9), the reaction expression score, $E_{exp,\,i}$, is equal to the maximum of the complex expression scores, $C_{exp,\,i,\,j}$ for all c_i protein complexes catalyzing reaction i; the complex expression scores are equal to the minimum of the protein expression scores, $P_{exp, j, k}$, for all p_j protein subunits of each complex j; and the protein expression scores, are equal to the maximum of all gene expression scores, $G_{exp, k, l}$, associated with the gk genes encoding each protein subunit. The gene expression score is equal to the normalized expression value of gene in the transcriptome being used as the basis to construct the model. In our analysis, the expression value of each gene was normalized by the median expression value for the same gene across all 37 conditions included in our data set, which included data from numerous organs, tissues, and growth conditions.

Reactions with an expression score falling below 0.2 were categorized as being "low expression." Biologically, a score of 0.2 means that the critical genes associated with the reaction are expressed at 20% of their average expression across all 37 conditions included in our transcriptomics data. This represents a conservative calling of "low expression" genes. We then applied the gapfilling algorithm as described in Equations (1)–(6) with two modifications: (i) the mass-balance constraints encoded by Equation (1) only included the stoichiometry of the reactions in the gapfilled full genome model (stoichiometry was not expanded to include the entire biochemistry database as done in full gapfilling); and (ii) the objective function was altered to maximize the high expression reaction activity while minimizing flux through low-expression reactions (Equation 10).

8.2.4.1 OBJECTIVE

$$\text{Minimize} \sum_{i=1}^{r_{high}} a\left(E_{exp-high,i}\delta_{high,i}\right) + \sum_{i=1}^{r_{low}} a\left(E_{low,i}v_{low,i}\right) \qquad (10)$$

Similar to our gapfilling formulation, this objective function minimizes the flux through the low expression reactions while also minimizing the slack fluxes associated with all high expression reactions. This maximizes the number of high expression reactions with a non-zero flux while setting the flux through as many low expression reactions as possible to zero.

Again, we use a scaling factor a in our objective function, which scales the cost of leaving some high expression reactions inactive against the cost of activating some low expression reactions. We explored values for a ranging from 0.01 to 0.25, with only minimal effect on the solutions produced. We found an a of 0.1 generated the most well-balanced solutions.

8.2.5 COMPARISON WITH ESTIMATED FLUXOMICS DATA FOR EMBRYO AND ENDOSPERM

In order to calculate how well the metabolic model can match experimentally measured flux data for a list of specific reactions, we applied a QP where we minimized the distance between the predicted fluxes and the experimentally measured fluxes. The QP utilized the standard FBA constraints:

$$N_{model} \bullet v = 0 \tag{11}$$

$$v_{min,i} \leq v_i \leq 1000 \qquad i = 1,\ldots,r_{model} \tag{12}$$

$$-50 \leq v_{ex,i} \leq 50\gamma_i \qquad i = 1,\ldots,m_{transported} \tag{13}$$

Equation (11) represents our mass balance constraints, where N_{model} is the stoichiometry matrix for all model reactions and v is the vector of fluxes through all model reactions. Unlike our gapfilling formulation, in this study, reversible reactions were not decomposed. Equation (12) represents the bounds on the flux through each reaction, with the lower bound vmin, i being zero if a reaction is irreversible and −1000 is a reaction is reversible. As in our gapfilling formulation, Equation (13) sets the bounds of uptake of nutrients from the environment.

In the quadratic objective function of our QP, we minimize the deviation of our predicted fluxes (v_i) from the experimentally measured fluxes ($v_{exp, i}$):

$$\text{Minimize} \sum_{i=1}^{r_{measured}} \left(v_{exp,i} - v_i\right)^2 \tag{14}$$

This approach is similar to that adopted by Lee et al. (2012), but by using QP, we find a single solution and avoid the iterative approach they describe. The calculations were done when the model was grown on heterotrophic media. After the minimal distance between experimental and model predicted fluxes was found via the QP problem as described above, we performed a Spearman correlation between the experimental flux values and the actual predicted flux values found by the solution when the model reached the minimal distance. The results in the form of the Spearman value and the p-value of the Spearman correlation are shown in Table 2.

8.3 RESULTS

8.3.1 A HIGH-QUALITY EVIDENCE-BASED GENOME-SCALE METABOLIC RECONSTRUCTION OF MAIZE

In order to generate a metabolic reconstruction based on available evidence, as described in the Materials and Methods Section, we started by building a full genome-scale metabolic reconstruction that integrated every reaction and gene-reaction association from all available resources. We then refined this model by removing the reactions and gene-reaction associations that did not have available support such as literature citation, human curation, or notation of presence in a specific compartment. We call this refined model an Evidence-Based Model. Here we described the process applied to complete this model refinement.

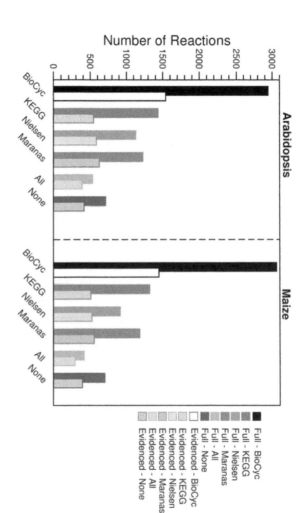

FIGURE 1: Number of reactions in the Full and Evidenced metabolic reconstructions for Arabidopsis and maize. The bars represent the number of reactions shared with each of the four primary biochemical sources used to build the Full metabolic reconstruction. Reactions are counted multiple times if they are present in multiple compartments. The "All" category corresponds with reactions that are shared between all four sources, and the "None" category corresponds with reactions present in compartments that are not otherwise found in the primary sources due to protein localization evidence. The dominant source of reactions was the BioCyc databases, ~50% more reactions originated from AraCyc and MaizeCyc/CornCyc than from KEGG. In addition, the dominant source of evidence for gene-reaction evidence came from AraCyc, and as a result, far fewer reactions are shared between the Evidenced metabolic reconstructions and the published counterparts, which were originally derived from KEGG.

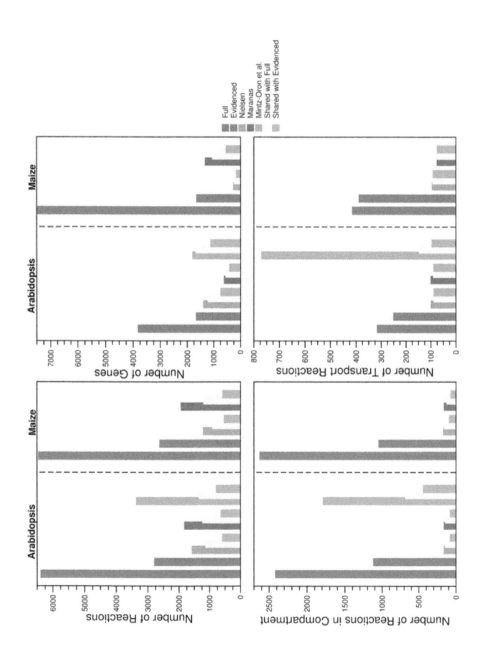

FIGURE 2: Comparison of the number of reactions, genes, compartmentalized reactions, and transport reactions found in the Full and Evidenced metabolic reconstructions for Arabidopsis and Maize and the published metabolic models. For each of the published metabolic models, we also show the number of reactions, genes, compartmentalized reactions, and transport reactions that are shared with both the Full and Evidenced metabolic reconstructions. The number of reactions and genes in the Full metabolic reconstructions dwarf the numbers in the other models. The model generated by Mintz-Oron et al. was the first plant metabolic model to be published for which integration from more than biochemical source was performed, and as such, it has more reactions than the other published models. However, AraCyc has since gone through several expansions, which explains why so many more reactions are in the Full metabolic reconstructions (Figure 1). The high number of genes in the Full maize model is indicative of the number of paralogs for which computational predictions are made by multiple sources. Only 40% of the genes in iRS1563 are found in the Evidenced maize metabolic reconstruction. The Evidenced metabolic reconstructions contain over 1000 reactions that are found in other compartments (notably in the plastid, see Figure 4), which is approximately 10 times more than the number of compartmentalized reactions found in the models from the Nielsen and Maranas labs. The process of creating the metabolic model of Mintz-Oron et al. predicted many more compartmentalized and transport reactions than those found in the Evidenced metabolic reconstruction for Arabidopsis, but only 25% of the compartmentalized reactions and 13% of the transport reactions are found in the Evidenced metabolic reconstruction.

8.3.2 INITIAL RECONSTRUCTION OF FULL GENOME-SCALE METABOLIC MODELS

We built our initial genome-scale metabolic reconstructions for *Arabidopsis* and maize using all reactions and genes obtained from all available resources. The resources included KEGG, the respective BioCyc databases, and the respective published metabolic models for Arabidopsis and maize (de Oliveira Dal'molin et al., 2010a,b; Saha et al., 2011). The two initial reconstructions are named "Full" and were composed of 6399 total reactions for Arabidopsis and 6458 for maize (Table 1).

Although we used multiple sources, we note that every published metabolic model available was in turn derived from KEGG and the respective BioCyc database. These databases are dynamic and improved over time, and, as a consequence, the published models are considered outdated. We therefore did not fully integrate the published metabolic models with two important exceptions: transport reactions and organellar reactions. These two sets of reactions, with the exception of those present in the model generated by Mintz-Oron et al. (2012), were manually reviewed in order to ensure that intra-organellar metabolic networks were active. We therefore ensure that these reactions are included.

TABLE 1: A list of metabolic models generated in our work and their statistics.

Species	Type/Organ/ Tissue	Reactions	Compounds	Gene-reaction associations	Gapfilled reactions
Arabidopsis thaliana	Full	6399	6236	16,577	1073
Arabidopsis thaliana	Evidenced	2801	2864	4262	697
Zea mays	Full	6458	6250	35,226	979
Zea mays	Evidenced	2629	2634	5540	667
Zea mays	Evidenced/ Leaf	2322	2635	4656	925
Zea mays	Evidenced/ Embryo	2304	2636	4680	885
Zea mays	Evidenced/ Endosperm	2280	2636	4602	920

The most telling statistic in comparing the Full metabolic reconstructions for both species is that maize has many more gene-reaction associations. This is partly because maize has undergone a recent whole-genome duplication event (Schnable et al., 2009), thus creating many paralogs, and partly because, for the MaizeCyc and CornCyc databases, many gene-reaction associations were predicted, and thereby included many similar homologs.

For each metabolic reconstruction, we showed the number of reactions that came from each source in Figure 1. In the Evidenced models, most of the reactions originated from BioCyc databases because KEGG provides comparatively little literature evidence for gene-reaction associations. In contrast, there is significant overlap between the KEGG database and the metabolic models published by the Nielsen/Maranas groups. This is because those metabolic models were generated from KEGG alone. We also highlight the variation in the number of reactions, compartmentalized reactions, transport reactions and genes between our models and those in the literature in Figure 2. In the case of the number of reactions, compartmentalized reactions and transport reactions in the Full and Evidenced models for both species, we show that the models created in this work are larger than the published models, with the exception of the model published by Mintz-Oron et al. Our models are larger than other published models primarily due to the more comprehensive database of biochemistry and plant annotations from which we generate our models, as well as the inclusion of recent database updates in our new model. The model published by Mintz-Oron et al. is larger still generally because it was expanded to include many computationally predicted compartmentalized reactions and transporters.

8.3.3 COMPARTMENTS

By using the protein localization data collected from various sources, we were able to confirm the presence of ~2000 reactions in eight compartments (plastid, mitochondrion, peroxisome, endoplasmic reticulum, nucleus, cell wall, vacuole, and Golgi body). We collected gene localization data for 12,398 Arabidopsis genes and 8737 maize genes for eight compartments in the metabolic reconstructions (see Materials and Methods), and we added reactions to the appropriate compartment whenever they were associated with a localized gene. We find that the gene localization data led to more than 700 reactions being placed in new locations that are not otherwise designated in the databases and published models used as sources; the "None" column in Figure 1 indicates this. In the next Section, we highlight two reactions as an example of this. We show a breakdown of the number of reactions found in each compartment (Figure 3), and this highlights that the majority of the reactions are found in the plastid. Furthermore, we qualitatively examined the contribution of each database to the localization of reactions (Figure 4). The total number of reactions assigned to any compartment in the Full maize metabolic reconstruction by PPDB data is 1675, by GFP data is 1077, and by AraCyc data is 429. The PPDB data accounts for more reactions in the plastid, mitochondrion, and peroxisome, and the GFP data accounts for more reactions in the remaining compartments. Whilst there is some agreement between the sources, the number of reactions assigned to a compartment by PPDB or GFP alone is a validation of our decision to use multiple sources of evidence-based localization data.

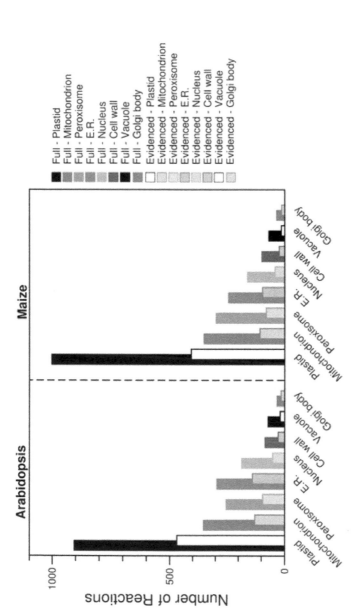

FIGURE 3: Comparison of the number of reactions found in each of the compartments in the Full and Evidenced metabolic reconstructions for Arabidopsis and maize. Evidence from AraCyc, PPDB, and SUBA was used to assign each reaction to different compartments. The substantial difference in the number of reactions in each compartment for the Full and Evidenced models is a result of the large number of gene-reaction associations in the Full model, which in turn is a result of the many computational predictions used to make the associations. As such, the use of protein localization evidence to assign reactions to compartments is far more reliable with the use of evidence for the gene-reaction associations in the Evidenced model. ER, Endoplasmic reticulum.

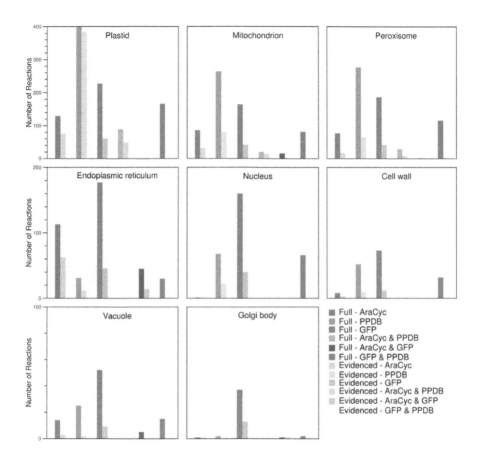

FIGURE 4: Comparison of the sources responsible for the number of reactions found in each of the compartments in the Full and Evidenced metabolic reconstructions for maize. The assignment of reactions to compartments is done by using evidence from AraCyc, PPDB, and SUBA. In general, the number of reactions in different compartments appears to be mostly influenced by a single source. More reactions are assigned by AraCyc to the endoplasmic reticulum; by PPDB to the plastid, mitochondrion, and peroxisome; and by GFP experiments listed in SUBA to nucleus, cell wall, vacuole, and Golgi body. The pairing of PPDB and SUBA shares evidence more frequently in all compartments with the exception of the endoplasmic reticulum where there is more agreement between AraCyc and SUBA.

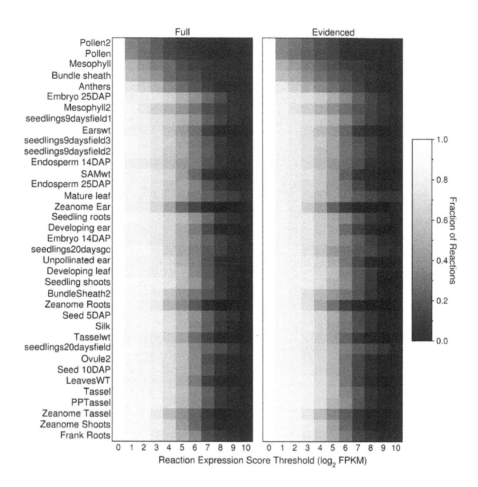

FIGURE 5: Fraction of active reactions in the Full and Evidenced metabolic reconstructions for maize at different thresholds. An expression score is computed for each reaction (Equations 7–9) using maize transcript profiles from qTeller (http://qteller.com). The transcript profiles are ordered by the sizes of the resulting metabolic reconstruction when the threshold applied is one, thus the smaller reconstructions with fewer active reactions are positioned at the top of the figure. Although the Evidenced metabolic reconstruction has half the number of reactions found in the Full metabolic reconstruction, both models appear to shrink at similar rates when increasing the threshold. Two sets of tissues, in general, have more inactive reactions at lower thresholds: (1) reproductive tissues, such as pollen and anthers, as well as tissues consisting of single cell types such as mesophyll and bundle sheath, and (2) tissues which originated from the Zeanome project.

8.3.4 EVIDENCE FOR GENE-REACTION ASSOCIATIONS

As stated above, we wish to refine our Full metabolic reconstructions to only contain reactions with reliable evidence for gene-reaction associations. Almost every gene-reaction association found in KEGG, and in any plant BioCyc databases that is not AraCyc, are computationally predicted (Zhang et al., 2010; Nakaya et al., 2013; Kanehisa et al., 2014; Seaver et al., 2014). Additionally, in many of the cases, and this problem is particularly acute in plants, the set of computationally predicted genes associated with reactions may be homologous, but do not perform the same catalytic function (i.e., they are out-paralogs). The large number of gene-reaction associations in the Full metabolic reconstruction for maize highlights this problem because maize, as a species, had a recent whole-genome duplication leading to additional paralogs (Schnable et al., 2009). It is important to identify the correct gene-reaction associations, because the genes duplicated by whole-genome duplication in maize appear to be down-regulated (Schnable and Freeling, 2011; Schnable et al., 2011).

We tackled this problem of over-annotation in two steps. First we included the gene-reaction associations for which there is evidence from two primary sources, AraCyc and PlantSEED (Mueller et al., 2003; Zhang et al., 2010; Seaver et al., 2014). The PathwayTools software enables users to assign evidence codes for gene-reaction associations, and in particular we were able to weed out all the gene-reaction associations where the evidence codes indicated that only a computational prediction was made. The PlantSEED project manually reviewed many of the gene-reaction associations found in AraCyc and elsewhere (Seaver et al., 2014), but also included many carefully reviewed in-paralogs (Sonnhammer and Koonin, 2002; Seaver et al., 2014), thus allowing us to include a greater number of gene-reaction associations in our metabolic reconstructions. By using these sources of evidence, we produced an evidence-based metabolic reconstruction for Arabidopsis that contained only those reactions for which there was gene-reaction association evidence from AraCyc and Plant-SEED, which we denote as "Evidenced." The Evidenced metabolic reconstruction for Arabidopsis is smaller, with 2801 reactions, and a smaller number of gene-reaction associations (Table 1). The number of reactions in the Evidenced metabolic reconstruction is 44% that of the Full metabolic reconstruction, but the number of gene-reaction associations is 26%, which is an indication of how many computational predictions are made for genes associated with reactions which otherwise have evidence for their associations with other genes.

In the second step of our model refinement, we considered the lack of evidence for any other species, given that much biochemical research in plants has been on *Arabidopsis* as a model organism. As a result, there exist only a tiny number of gene-reaction associations with evidence in MaizeCyc and CornCyc combined, and to create an Evidenced model for maize, one must consider propagating the gene-reaction associations from Arabidopsis. In order to avoid the pitfall of over-annotation, and yet create a reliable set of gene-reaction associations for maize, we used the same very conservative approach we applied to EnsemblCompara protein families in the PlantSEED project, described below (Vilella et al., 2009; Seaver et al., 2014). This approach greatly reduced the number of maize orthologs found in the same protein family as the *Arabidopsis* genes found in the Evidenced *Arabidopsis* metabolic reconstruction. In doing so, we are able to create an Evidenced metabolic reconstruction for maize by adding to the model only the reactions for which the associated genes have orthologs in the Evidenced metabolic reconstruction for *Arabidopsis*. The Evidenced metabolic reconstruction for maize has 2631 reactions and, ~30,000 fewer gene-reaction associations than found in the Full metabolic reconstruction (~84%; Table 1).

We highlight the utility of our approach with an example involving two reactions from the mevalonate pathway. Simkin et al. report, using YFP-fused constructs, that Phosphomevalonate kinase (PMK) and Mevalonate diphosphate decarboxylase (MVD) localize to the peroxisomes (Simkin et al., 2011). The complementary reactions for these two enzymes are found in AraCyc, MaizeCyc, and CornCyc, albeit without any localization data attached, and with experimental evidence only available for one enzyme in AraCyc. Thus, only one reaction (MVD) would be included in the Arabidopsis model and would only be cytosolic. The evidence for the gene-reaction associations is found in PlantSEED in the form of manual curation, and leads to both reactions being included in the Arabidopsis model. The results for the enzyme localization from Simkin et al. are found in SUBA, and the two reactions were therefore correctly added to the peroxisome in the Arabidopsis model. Finally, the use of EnsemblCompara protein families as described above leads to the correct maize genes being associated with the same reactions, and the reactions being thus added to the peroxisome in the maize model.

We generated a corresponding metabolic model for all four of our metabolic reconstructions by adding a biomass equation matching that used by the PlantSEED and containing more than 90 compounds. We also utilized a new pathway gapfilling method (see Materials and Methods) that attempts to generate biomass and simultaneously activate all reactions with associated genes. The pathway gapfilling recommended reactions to add to our models to produce biomass and improve the function of all the pathways included in the model. We tested our gapfilled models by simulating growth on heterotrophic media in the KBase environment before applying the transcript profiles.

8.3.5 TRANSCRIPTOME-BASED METABOLIC RECONSTRUCTIONS OF MAIZE

8.3.5.1 MAIZE TRANSCRIPTOMICS

We built transcriptome-based metabolic reconstructions of maize, derived directly from the gapfilled genome-scale Evidenced metabolic model, such that each transcriptome-based model will be a subset of the Evidenced metabolic model. To generate these transcriptome-based metabolic reconstructions, we used RNA-Seq data collated at qTeller (http://qteller.com/, downloaded on 02/04/2014). The data consists of 37 experiments from nine sources, covering a range of cells, tissues, organs, and conditions. As an initial exploration of how the transcript profiles may affect a transcriptome-based model, we computed, for each of the datasets, and at 10 different thresholds, the number of reactions in the genome-scale Full and Evidenced metabolic models for maize that would be active in the organ or tissue, and conditions from which the transcript profiles were retrieved (Figure 5). The threshold was applied to the reaction expression scores (Equations 7–9), and as the threshold increases, the number of reactions that would be active in the resulting metabolic reconstruction decreases. The results show that the smallest metabolic reconstructions are derived either from data from specific cell types (mesophyll and bundle sheath) or highly reproductive tissues (pollen and anthers); the other tissues and organs with larger reconstructions encompassed multiple cell types and in general, up to a threshold of four, show little difference in the sizes of the resulting metabolic network. Furthermore, qualitatively, it appears that the relative change in the network sizes is similar across organs and tissues in both the Full and Evidenced metabolic models. Finally, using several of the transcript profiles from the same source appears to consistently result in metabolic networks that are relatively smaller, notably those from the Zeanome dataset (http://www.ncbi.nlm.nih.gov/Traces/sra/?study=SRP011480), which is an important reminder that, when performing in silico experiments using transcript profiles, one must ensure they come from the same source.

To investigate further, we explored how the threshold creates gaps in the primary metabolism of transcriptome-based models. We aggregated the various pathways under nine different categories of primary metabolism as defined by the PlantSEED project (Seaver et al., 2014) and we explored how these pathways shrink in size as the threshold is increased (Figures 6, 7). Overall, within each pathway category, a similar pattern is observed where the sex organs and single-cell transcript profiles result in the smaller metabolic model, and for all transcript profiles, there appears to be a similar decrease in the sizes of the pathways. However, it is notable that this pattern varies from category to category and from organ to organ or tissue to tissue. Many essential reactions that may be necessary for a derived metabolic model to operate may be inactivated by the use of a simple expression threshold. For instance, within almost every category, there are reactions for which the computed expression score is zero or constitutively low (Figure 7), but the reactions are essential. Reactions in the "Fatty acids" category would appear to be the most impervious to the use of a low threshold as many, if not all, of the reactions appear to exhibit a medium to high expression score across most organs and tissues. A notable example is the set of reaction expression scores computed from the transcriptome labeled Embryo_25DAP (25 days after pollination), which matches our understanding of the embryo typically being rich in lipids. As it is therefore not reasonable to use a simplistic approach to generate transcriptome-based metabolic models, we thus develop a novel method for applying the gene expression levels in transcript profiles directly to the genome-scale metabolic model (see Materials and Methods). The method attempts to activate every reaction that is associated with a highly expressed gene whilst minimizing activity of reactions associated with minimally expressed genes. The results of the generation of these models from transcript profiles using this method are found in Section Generating the Transcriptome-Based Metabolic Models. However, first we address the derivation of new biomass compositions to represent the leaf, endosperm and embryo tissues.

8.3.5.2 HIGH-QUALITY MAIZE BIOMASS EQUATION FOR LEAF, ENDOSPERM AND EMBRYO TISSUE

One use for the metabolic models we build is to predict the biosynthesis of plant biomass components. This is done by creating a specialized biomass composition reaction that contains each of the biomass components in relative proportions, and by "maximizing" biomass production when simulating growth in the metabolic model. All of the prior published metabolic models for plants have assumed a basic biomass composition that contained mostly primary metabolites. Little emphasis was placed on the diversity of compounds that a plant biosynthesizes. For our transcriptome-based metabolic models, we aim to distinguish between the functions of the models by providing a high-quality biomass composition reaction representing the organ or tissue from which the modeled transcriptomes were collected. We constructed these reaction based on an extensive literature search. Here we describe a biomass that contains more cofactors and fatty acids, supported by almost 30 literature references, including detailed quantifications. The following paragraphs briefly described the biomass composition along with the relevant references.

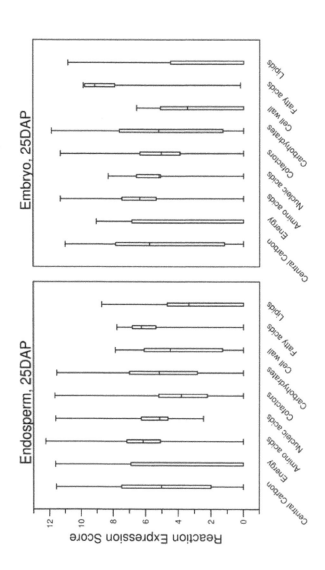

FIGURE 7: Boxplots describing the distribution of computed reaction expression scores (Equations 7–9) from the transcript profiles of two tissues and within the different categories of plant primary metabolism. Almost every category contained at least one reaction with a reaction expression score of zero. Furthermore, for the "Energy" and "Lipids" categories, more than half of the reaction expression scores are zero. It can be seen that the median reaction expression scores for "Embryo, 25DAP" are higher, which supports the observation made for this tissue in the previous figure (Figure 6). Additionally, the lower quartiles of the reaction expression scores in the "Carbohydrates" and "Cell wall" categories are higher for "Endosperm, 25DAP." Both of these categories include pathways involved in sugar metabolism, and this supports the known biological function of the endosperm as a storage of starch.

8.3.5.3 AMINO ACIDS

The biomass fraction attributable to protein is estimated to be 8 and 11.6% of dry weight in endosperm and embryo, respectively (Ingle et al., 1965). To quantify the relative contribution of each amino acid in the endosperm, the total amino acid context determined experimentally by Misra et al. (1972) was used with two exceptions. Firstly, the cysteine content was doubled as the reported value concerned cystine. Secondly, the glutamate:glutamine and aspartate:asparagine ratios were deduced from the composition of mature Zein proteins (Wu et al., 2012) to estimate their individual contribution. For composition of amino acids in embryo, the sequences of two globulins were used, which account for 20% of total embryo protein (Belanger and Kriz, 1989; Wallace and Kriz, 1991). Water loss due to formation of the peptide bond was taken into account.

8.3.5.4 NUCLEIC ACIDS

The biomass fraction attributable to DNA was reported to be 0.038 and 0.015% in endosperm and embryo, respectively, while that attributable to RNA was reported to be 0.3 and 0.1% in endosperm and embryo, respectively (Ingle et al., 1965). The biomass fraction attributed to each nucleotide was estimated using published GC content (Haberer et al., 2005).

8.3.5.5 CARBOHYDRATES

The endosperm biomass fraction attributable to carbohydrates was calculated to be about 90% of dry mass (Ingle et al., 1965; Alonso et al., 2011). Of this carbohydrate fraction, 77.6% is starch, 16.6% is cell walls (Alonso et al., 2011) and the remaining 5.8% is free sucrose, fructose, and glucose (Ingle et al., 1965). The reported composition of endosperm cell walls (Dewitt et al., 1999) was used to calculate the quantities of the majority of the monosaccharides. The embryo biomass fraction attributable to carbohydrates is calculated to be 58.5% (Rolletschek et al., 2005; Alonso et al.,

2010). Of this carbohydrate fraction, 49.6% is starch, 42.7% is cell walls (Alonso et al., 2010) and the remaining 7.7% is free sucrose, fructose, and glucose (Rolletschek et al., 2005). The reported composition of cell walls (McCann et al., 2007) was used to calculate the quantities of the majority of the monosaccharides and the ratio of monosaccharides found in the leaf (Penning de Vries et al., 1974) was used to calculate ribose, glucuronate, and galacturonate content.

For both endosperm and embryo, the galactose, glycerol, and sulfo-quinovose biomass fraction was estimated using values for galactolipids, glycerolipids, and sulfolipids, respectively (see the Section Lipids and Sterols). Finally, further evidence was used to deduce the biomass fraction of inositol (Teas, 1954).

8.3.5.6 PHENOLIC COMPOUNDS

The cell wall of maize is considered to contain two main types of phenolic derivatives: p-coumaric acid and ferulic acid (Assabgui et al., 1993; Saulnier et al., 1995).

8.3.5.7 VITAMINS AND COFACTORS

As key components of metabolism, we emphasized biosynthetic pathways of cofactors more than the published metabolic models, and specified the biomass fraction assigned to each of the B vitamins and other cofactors with greater accuracy. The list of vitamins and cofactors included biotin, thiamin diphosphate, NAD and derivatives, FAD and FMN, coenzyme A, 4-phosphopantetheine, tetrahydrofolate and its derivatives, α-tocopherol, ascorbate, ubiquinone-9, lipoic acid, heme, and pyridoxal-5′-phosphate (Cameron and Teas, 1948; Teas, 1954; Giri et al., 1960; Ingle et al., 1965; Metz et al., 1970; Weber, 1987; Battey and Ohlrogge, 1990; Shannon et al., 1996; Szal et al., 2003; Tumaney et al., 2004; Shi et al., 2005; Drozak and Romanowska, 2006; Hu et al., 2006; Naqvi et al., 2009; Perez-Lopez et al., 2010; Richter et al., 2010; Enami et al., 2011; Spielbauer et al., 2013; Seaver et al., 2014).

8.3.5.8 PIGMENTS

Two pigments were included in our endosperm and embryo biomass: β-carotene, and lutein (Weber, 1987).

8.3.5.9 LIPIDS AND STEROLS

Lipids represent 1.5 and 32.6% of the biomass of endosperm and embryo, respectively (Weber, 1979). The biomass composition of fatty acids and sitosterol, campesterol, stigmasterol, and phytosphingosine in this study were based on those reported by Weber (1979). Galactose, glycerol, and sulfoquinovose content were also calculated based on the lipid composition.

8.3.5.10 CARBOXYLIC ACIDS AND OTHER COMPOUNDS

Many other compounds compose plant biomass, and we included here a list of a subset of these for which a value is reported in the literature: cis-aconitate, citrate, malate, oxaloacetate lactate (Skogerson et al., 2010; Rolletschek et al., 2011), and S-adenosylmethionine (Apelbaum and Yang, 1981). Choline and ethanolamine were estimated from the values for phosphatidylcholine and phosphatidylethanolamine, respectively. Finally, the mineral content of the biomass was set at 5%, split evenly between potassium and chloride (Penning de Vries et al., 1974).

8.3.5.11 GENERATING THE TRANSCRIPTOME-BASED METABOLIC MODELS

We used the novel transcriptome-based gapfilling approach (see Materials and Methods) along with three separate transcript profiles to generate metabolic models that are specific to the leaf, endosperm and embryo, which are named "Leaves 20-day old seedling – field," "Endosperm 25 days after pollination," and "Embryo 25 days after pollination" (Davidson et al., 2011) http://www.ncbi.nlm.nih.gov/bioproject/80041). We used these three transcript profiles in particular because they came from the same experiments and, therefore, were processed in a similar manner.

We applied the three transcriptome profiles separately to the Evidenced metabolic model of maize (see Materials and methods Section) to generate three separate metabolic models that can grow in heterotrophic media. All three metabolic models contained an average of 2302 reactions (Table 1), which is 88% of the number of reactions in the Evidenced model, and there are 2153 reactions that are found in all three of them. By comparison, the final compartmentalized model created by Mintz-Oron et al. (2012) for *Arabidopsis* has 3508 reactions and the resulting tissue-specific models generated from their work has on average 2848 reactions, which is 81% of the reactions in their full model. This result indicates that our approach, while using a full set of gene expression data for the maize transcript profiles to generate smaller models, results in models whose sizes are similar to other work on generating organ and tissue-specific models for a plant.

8.3.6 COMPARISON OF FLUXES IN TISSUE-SPECIFIC METABOLIC RECONSTRUCTIONS TO FLUXOMICS DATA

We have described a process that generates and refines metabolic models in three steps, generating metabolic models at each step. We can now show how these metabolic models not only compare with fluxomics data, but how that comparison improves at each step, resulting in transcriptome-based models with the closest fit to the original fluxomics data.

The experimental data we used were fluxes for central carbon metabolism estimated using ^{14}C labeling in two different tissues, the embryo and endosperm (Alonso et al., 2010, 2011). The reactions from these two studies were matched to the reactions in the models, and we used the approach described in Section Comparison with Estimated Fluxomics Data for Embryo and Endosperm to fit the fluxes within the models to the experimentally determined fluxes. We report the Spearman correlation and its p-value in Table 2, showing that the correlation is high for both transcriptome-based models. This result indicates that the central carbon metabolism of the models generated in this work is able to perform as observed in the original tissues. The reactions used here have a median expression score of 6.60 and 7.17 in the embryo and endosperm transcriptomics dataset, respectively, but the lowest expression score is ~1.2 for both tissues. This last statement in turn exemplifies the importance of our approach, in ensuring that reactions with a low expression score are still included in model generated from a transcript profile if considered to be essential for the metabolic functioning of the organ or tissue.

TABLE 2: Comparison of prior published model of maize with the models generated by this work using the percentage of blocked reactions and the spearman rank correlation coefficient when using fluxomics data (p-value in parentheses).

Type/Tissue	Blocked reactions (%)	Endosperm	Embryo
iRS1563*	53	$0.69 \ (2.3 \times 10^{-3})$	$0.46 \ (7.5 \times 10^{-2})$
Full	30	$0.99 \ (9.4 \times 10^{-51})$	$0.99 \ (7.1 \times 10^{-54})$
Evidenced	21	$0.99 \ (4.7 \times 10^{-34})$	$0.83 \ (1.7 \times 10^{-10})$
Evidenced/Endosperm	16	$0.99 \ (1.3 \times 10^{-29})$	n/a
Evidenced/Embryo	16	n/a	$0.83 \ (8.6 \times 10^{-10})$

Saha et al. (2011)

8.4 DISCUSSION

In this manuscript, we created a total of seven metabolic reconstructions for two species (see Supplementary Material). In succession, we created two Full metabolic reconstructions for *Arabidopsis* and maize, comprised of many possible sources of plant biochemistry reconciled into single large networks. These Full models also included many predicted gene-reaction associations, a subset for which we found evidence either in the literature or via human inference, and we used these to create a more reliable metabolic reconstruction for the two species. Finally, via use of a novel, simple and fast organ and tissue-specific pathway gapfilling method, along with well-curated biomass for the leaf, endosperm and embryo, we generated three metabolic models specific for these organ and tissues. The evidence that we used, for both the genes whose products catalyze the reactions and the localization of gene products in different compartments, is comprehensive and reliable.

Our approach allows us to create relatively large metabolic reconstructions that compare favorably to the prior published metabolic models, albeit with a smaller set of gene-reaction associations. This enables us to apply transcriptome data with a high degree of confidence. The approach is validated by the fact that the embryo and endosperm models retained nearly every reaction of central carbon metabolism. This was done both by the body of evidence available for the gene-reaction associations, and the pathway gapfilling method which included reactions with a low expression score, but were essential to the models. Finally, it was shown that the same models can be active and able to replicate the activity observed in published experimental fluxomics datasets. To date, we believe we are the first to apply such wide-ranging body of evidence to the generation of large-scale metabolic reconstructions.

All of our work was carried out through the DOE Systems Biology Knowledgebase (KBase; http://kbase.us/), an open software and data platform that aim to enable researchers to predict and ultimately design biological function. The data is publicly available within KBase workspaces named "Maize_Tissue_Models" (https://narrative.kbase.us/functional-site/#/ws/objects/Maize_Tissue_Models) and also via the PlantSEED website (http://plantseed.theseed.org). The KBase software environment allows researchers to copy the individual metabolic models and to explore the models using the suite of modeling tools available.

REFERENCES

1.   Alonso, A. P., Dale, V. L., and Shachar-Hill, Y. (2010). Understanding fatty acid synthesis in developing maize embryos using metabolic flux analysis. Metab. Eng. 12, 488–497. doi: 10.1016/j.ymben.2010.04.002
2.   Alonso, A. P., Val, D. L., and Shachar-Hill, Y. (2011). Central metabolic fluxes in the endosperm of developing maize seeds and their implications for metabolic engineering. Metab. Eng. 13, 96–107. doi: 10.1016/j.ymben.2010.10.002
3.   Apelbaum, A., and Yang, S. F. (1981). Biosynthesis of stress ethylene induced by water deficit. Plant Physiol. 68, 594–596. doi: 10.1104/pp.68.3.594

4. Assabgui, R. A., Reid, L. M., Hamilton, R. I., and Arnason, J. T. (1993). Correlation of kernel (E)-ferulic acid content of maize with resistance to Fusarium graminearum. Phytopathology 83, 949–953. doi: 10.1094/Phyto-83-949

5. Baerenfaller, K., Grossmann, J., Grobei, M. A., Hull, R., Hirsch-Hoffmann, M., Yalovsky, S., et al. (2008). Genome-scale proteomics reveals Arabidopsis thaliana gene models and proteome dynamics. Science 320, 938–941. doi: 10.1126/science.1157956

6. Battey, J. F., and Ohlrogge, J. B. (1990). Evolutionary and tissue-specific control of expression of multiple acyl-carrier protein isoforms in plants and bacteria. Planta 180, 352–360. doi: 10.1007/BF01160390

7. Belanger, F. C., and Kriz, A. L. (1989). Molecular characterization of the major maize embryo globulin encoded by the glb1 gene. Plant Physiol. 91, 636–643. doi: 10.1104/pp.91.2.636ameron, J. W., and Teas, H. J. (1948). The relation between nicotinic acid and carbohydrates in a series of maize endosperm genotypes. Proc. Natl. Acad. Sci. U.S.A. 34, 390–398. doi: 10.1073/pnas.34.8.390aspi, R., Altman, T., Dreher, K., Fulcher, C. A., Subhraveti, P., Keseler, I. M., et al. (2012). The MetaCyc database of metabolic pathways and enzymes and the BioCyc collection of pathway/genome databases. Nucleic Acids Res. 40, D742–D753. doi: 10.1093/nar/gkr1014hang, R. L., Ghamsari, L., Manichaikul, A., Hom, E. F., Balaji, S., Fu, W., et al. (2011). Metabolic network reconstruction of Chlamydomonas offers insight into light-driven algal metabolism. Mol. Syst. Biol. 7:518. doi: 10.1038/msb.2011.52olijn, C., Brandes, A., Zucker, J., Lun, D. S., Weiner, B., Farhat, M. R., et al. (2009). Interpreting expression data with metabolic flux models: predicting Mycobacterium tuberculosis mycolic acid production. PLoS Comput. Biol. 5:e1000489. doi: 10.1371/journal.pcbi.1000489

8. Davidson, R. M., Hansey, C. N., Gowda, M., Childs, K. L., Lin, H., Vaillancourt, B., et al. (2011). Utility of RNA sequencing for analysis of maize reproductive transcriptomes. Plant Genome 4, 191–203. doi: 10.3835/plantgenome2011.05.0015

9. de Oliveira Dal'molin, C. G., Quek, L. E., Palfreyman, R. W., Brumbley, S. M., and Nielsen, L. K. (2010a). AraGEM, a genome-scale reconstruction of the primary metabolic network in Arabidopsis. Plant Physiol. 152, 579–589. doi: 10.1104/pp.109.148817

10. de Oliveira Dal'molin, C. G., Quek, L. E., Palfreyman, R. W., Brumbley, S. M., and Nielsen, L. K. (2010b). C4GEM, a genome-scale metabolic model to study C-4 plant metabolism. Plant Physiol. 154, 1871–1885. doi: 10.1104/pp.110.166488

11. de Oliveira Dal'molin, C. G., Quek, L. E., Palfreyman, R. W., and Nielsen, L. K. (2011). AlgaGEM–a genome-scale metabolic reconstruction of algae based on the Chlamydomonas reinhardtii genome. BMC Genomics 12(Suppl. 4):S5. doi: 10.1186/1471-2164-12-S4-S5

12. Dewitt, G., Richards, J., Mohnen, D., and Jones, A. M. (1999). Comparative compositional analysis of walls with two different morphologies: archetypical versus transfer-cell-like. Protoplasma 209, 238–245. doi: 10.1007/BF01453452

13. Drozak, A., and Romanowska, E. (2006). Acclimation of mesophyll and bundle sheath chloroplasts of maize to different irradiances during growth. Biochim. Biophys. Acta 1757, 1539–1546. doi: 10.1016/j.bbabio.2006.09.001

14. Enami, K., Ozawa, T., Motohashi, N., Nakamura, M., Tanaka, K., and Hanaoka, M. (2011). Plastid-to-nucleus retrograde signals are essential for the expression of nuclear starch biosynthesis genes during amyloplast differentiation in tobacco BY-2 cultured cells. Plant Physiol. 157, 518–530. doi: 10.1104/pp.111.178897

15. Giri, K. V., Rao, N. A., Cama, H. R., and Kumar, S. A. (1960). Studies on flavinadenine dinucleotide-synthesizing enzyme in plants. Biochem. J. 75, 381–386.

16. Grafahrend-Belau, E., Junker, A., Eschenroder, A., Muller, J., Schreiber, F., and Junker, B. H. (2013). Multiscale metabolic modeling: dynamic flux balance analysis on a whole-plant scale. Plant Physiol. 163, 637–647. doi: 10.1104/pp.113.224006

17. Grafahrend-Belau, E., Schreiber, F., Koschutzki, D., and Junker, B. H. (2009). Flux balance analysis of barley seeds: a computational approach to study systemic properties of central metabolism. Plant Physiol. 149, 585–598. doi: 10.1104/pp.108.129635

18. Haberer, G., Young, S., Bharti, A. K., Gundlach, H., Raymond, C., Fuks, G., et al. (2005). Structure and architecture of the maize genome. Plant Physiol. 139, 1612–1624. doi: 10.1104/pp.105.068718

19. Hay, J., and Schwender, J. (2011a). Computational analysis of storage synthesis in developing Brassica napus L. (oilseed rape) embryos: flux variability analysis in relation to (1)(3)C metabolic flux analysis. Plant J. 67, 513–525. doi: 10.1111/j.1365-313X.2011.04611.x

20. Hay, J., and Schwender, J. (2011b). Metabolic network reconstruction and flux variability analysis of storage synthesis in developing oilseed rape (Brassica napus L.) embryos. Plant J. 67, 526–541. doi: 10.1111/j.1365-313X.2011.04613.x

21. Heller, S., McNaught, A., Stein, S., Tchekhovskoi, D., and Pletnev, I. (2013). InChI–the worldwide chemical structure identifier standard. J. Cheminform. 5:7. doi: 10.1186/1758-2946-5-7

22. Henry, C. S., Dejongh, M., Best, A. A., Frybarger, P. M., Linsay, B., and Stevens, R. L. (2010). High-throughput generation, optimization, and analysis of genome-scale metabolic models. Nat. Biotechnol. 1672, 1–6. doi: 10.1038/nbt.1672

23. Henry, C. S., Zinner, J., Cohoon, M., and Stevens, R. (2009). iBsu1103: a new genome scale metabolic model of B. subtilis based on SEED annotations. Genome Biol. 10:R69. doi: 10.1186/gb-2009-10-6-r69

24. Hu, W. H., Shi, K., Song, X. S., Xia, X. J., Zhou, Y. H., and Yu, J. Q. (2006). Different effects of chilling on respiration in leaves and roots of cucumber (Cucumis sativus). Plant Physiol. Biochem. 44, 837–843. doi: 10.1016/j.plaphy.2006.10.016

25. Ingle, J., Beitz, D., and Hageman, R. H. (1965). Changes in composition during development and maturation of maize seeds. Plant Physiol. 40, 835–839. doi: 10.1104/pp.40.5.835

26. Ingle, R. A. (2011). Histidine biosynthesis. Arabidopsis Book 9:e0141. doi: 10.1199/tab.0141

27. Jerby, L., Shlomi, T., and Ruppin, E. (2010). Computational reconstruction of tissue-specific metabolic models: application to human liver metabolism. Mol. Syst. Biol. 6, 401. doi: 10.1038/msb.2010.56

28. Kanehisa, M., and Goto, S. (2000). KEGG: kyoto encyclopedia of genes and genomes. Nucleic Acids Res. 28, 27–30. doi: 10.1093/nar/28.1.27

29. Kanehisa, M., Goto, S., Sato, Y., Furumichi, M., and Tanabe, M. (2012). KEGG for integration and interpretation of large-scale molecular data sets. Nucleic Acids Res. 40, D109–D114. doi: 10.1093/nar/gkr988

30. Kanehisa, M., Goto, S., Sato, Y., Kawashima, M., Furumichi, M., and Tanabe, M. (2014). Data, information, knowledge and principle: back to metabolism in KEGG. Nucleic Acids Res. 42, D199–D205. doi: 10.1093/nar/gkt1076

31. Kersey, P. J., Allen, J. E., Christensen, M., Davis, P., Falin, L. J., Grabmueller, C., et al. (2014). Ensembl Genomes 2013: scaling up access to genome-wide data. Nucleic Acids Res. 42, D546–D552. doi: 10.1093/nar/gkt979

32. Kumar, V. S., Dasika, M. S., and Maranas, C. D. (2007). Optimization based automated curation of metabolic reconstructions. BMC Bioinform. 8:212. doi: 10.1186/1471-2105-8-212

33. Kumar, V. S., and Maranas, C. D. (2009). GrowMatch: an automated method for reconciling in silico/in vivo growth predictions. PLoS Comput. Biol. 5:e1000308. doi: 10.1371/journal.pcbi.1000308

34. Latendresse, M. (2014). Efficiently gap-filling reaction networks. BMC Bioinform. 15:225. doi: 10.1186/1471-2105-15-225

35. Lee, D., Smallbone, K., Dunn, W. B., Murabito, E., Winder, C. L., Kell, D. B., et al. (2012). Improving metabolic flux predictions using absolute gene expression data. BMC Syst. Biol. 6:73. doi: 10.1186/1752-0509-6-73

36. McCann, M. C., Defernez, M., Urbanowicz, B. R., Tewari, J. C., Langewisch, T., Olek, A., et al. (2007). Neural network analyses of infrared spectra for classifying cell wall architectures. Plant Physiol. 143, 1314–1326. doi: 10.1104/pp.106.093054

37. Metz, J., Lurie, A., and Konidaris, M. (1970). A note on the folate content of uncooked maize. S. Afr. Med. J. 44, 539–541.

38. Mintz-Oron, S., Meir, S., Malitsky, S., Ruppin, E., Aharoni, A., and Shlomi, T. (2012). Reconstruction of Arabidopsis metabolic network models accounting for subcellular compartmentalization and tissue-specificity. Proc. Natl. Acad. Sci. U.S.A. 109, 339–344. doi: 10.1073/pnas.1100358109

39. Misra, P. S., Jambunathan, R., Mertz, E. T., Glover, D. V., Barbosa, H. M., and McWhirter, K. S. (1972). Endosperm protein synthesis in maize mutants with increased lysine content. Science 176, 1425–1427. doi: 10.1126/science.176.4042.1425

40. Mo, M. L., Palsson, B. O., and Herrgard, M. J. (2009). Connecting extracellular metabolomic measurements to intracellular flux states in yeast. BMC Syst. Biol. 3:37. doi: 10.1186/1752-0509-3-37

41. Monaco, M. K., Sen, T. Z., Dharmawardhana, P. D., Ren, L., Schaeffer, M., Naithani, S., et al. (2013). Maize metabolic network construction and transcriptome analysis. Plant Gen. 6. doi: 10.3835/plantgenome2012.09.0025. Available online at: https://www.crops.org/publications/citation-manager/tpg/6/1/plantgenome2012.09.0025

42. Mueller, L. A., Zhang, P., and Rhee, S. Y. (2003). AraCyc: a biochemical pathway database for Arabidopsis. Plant Physiol. 132, 453–460. doi: 10.1104/pp.102.017236

43. Nakaya, A., Katayama, T., Itoh, M., Hiranuka, K., Kawashima, S., Moriya, Y., et al. (2013). KEGG OC: a large-scale automatic construction of taxonomy-based ortholog clusters. Nucleic Acids Res. 41, D353–D357. doi: 10.1093/nar/gks1239

44. Naqvi, S., Zhu, C., Farre, G., Ramessar, K., Bassie, L., Breitenbach, J., et al. (2009). Transgenic multivitamin corn through biofortification of endosperm with three vita-

mins representing three distinct metabolic pathways. Proc. Natl. Acad. Sci. U.S.A. 106, 7762–7767. doi: 10.1073/pnas.0901412106

45. Ozsolak, F., and Milos, P. M. (2011). RNA sequencing: advances, challenges and opportunities. Nat. Rev. Genet. 12, 87–98. doi: 10.1038/nrg2934

46. Penning de Vries, F. W., Brunsting, A. H., and van Laar, H. H. (1974). Products, requirements and efficiency of biosynthesis: a quantitative approach. J. Theor. Biol. 45, 339–377. doi: 10.1016/0022-5193(74)90119-2

47. Perez-Lopez, U., Robredo, A., Lacuesta, M., Sgherri, C., Mena-Petite, A., Navari-Izzo, F., et al. (2010). Lipoic acid and redox status in barley plants subjected to salinity and elevated CO2. Physiol. Plant. 139, 256–268. doi: 10.1111/j.1399-3054.2010.01361.x

48. Pilalis, E., Chatziioannou, A., Thomasset, B., and Kolisis, F. (2011). An in silico compartmentalized metabolic model of Brassica napus enables the systemic study of regulatory aspects of plant central metabolism. Biotechnol. Bioeng. 108, 1673–1682. doi: 10.1002/bit.23107

49. Poolman, M. G., Kundu, S., Shaw, R., and Fell, D. A. (2013). Responses to light intensity in a genome-scale model of rice metabolism. Plant Physiol. 162, 1060–1072. doi: 10.1104/pp.113.216762

50. Poolman, M. G., Miguet, L., Sweetlove, L. J., and Fell, D. A. (2009). A genome-scale metabolic model of Arabidopsis and some of its properties. Plant Physiol. 151, 1570–1581. doi: 10.1104/pp.109.141267

51. Richter, A., Peter, E., Pors, Y., Lorenzen, S., Grimm, B., and Czarnecki, O. (2010). Rapid dark repression of 5-aminolevulinic acid synthesis in green barley leaves. Plant Cell Physiol. 51, 670–681. doi: 10.1093/pcp/pcq047

52. Rolletschek, H., Koch, K., Wobus, U., and Borisjuk, L. (2005). Positional cues for the starch/lipid balance in maize kernels and resource partitioning to the embryo. Plant J. 42, 69–83. doi: 10.1111/j.1365-313X.2005.02352.x

53. Rolletschek, H., Melkus, G., Grafahrend-Belau, E., Fuchs, J., Heinzel, N., Schreiber, F., et al. (2011). Combined noninvasive imaging and modeling approaches reveal metabolic compartmentation in the barley endosperm. Plant Cell 23, 3041–3054. doi: 10.1105/tpc.111.087015

54. Saha, R., Suthers, P. F., and Maranas, C. D. (2011). Zea mays iRS1563: a comprehensive genome-scale metabolic reconstruction of maize metabolism. PLoS ONE 6:e21784. doi: 10.1371/journal.pone.0021784

55. Satish Kumar, V., Dasika, M. S., and Maranas, C. D. (2007). Optimization based automated curation of metabolic reconstructions. BMC Bioinform. 8:212. doi: 10.1186/1471-2105-8-212

56. Saulnier, L., Marot, C., Chanilaud, E., and Thibault, J.-F. (1995). Cell wall polysaccharide interactions in maize bran. Carbohydr. Polym. 26, 279–287. doi: 10.1016/0144-8617(95)00020-8

57. Schnable, J. C., and Freeling, M. (2011). Genes identified by visible mutant phenotypes show increased bias toward one of two subgenomes of maize. PLoS ONE 6:e17855. doi: 10.1371/journal.pone.0017855

58. Schnable, J. C., Springer, N. M., and Freeling, M. (2011). Differentiation of the maize subgenomes by genome dominance and both ancient and ongoing gene loss. Proc. Natl. Acad. Sci. U.S.A. 108, 4069–4074. doi: 10.1073/pnas.1101368108

59. Schnable, P. S., Ware, D., Fulton, R. S., Stein, J. C., Wei, F., Pasternak, S., et al. (2009). The B73 maize genome: complexity, diversity, and dynamics. Science 326, 1112–1115. doi: 10.1126/science.1178534

60. Schwender, J., and Hay, J. O. (2012). Predictive modeling of biomass component tradeoffs in Brassica napus developing oilseeds based on in silico manipulation of storage metabolism. Plant Physiol. 160, 1218–1236. doi: 10.1104/pp.112.203927

61. Seaver, S. M., Gerdes, S., Frelin, O., Lerma-Ortiz, C., Bradbury, L. M., Zallot, R., et al. (2014). High-throughput comparison, functional annotation, and metabolic modeling of plant genomes using the PlantSEED resource. Proc. Natl. Acad. Sci. U.S.A. 111, 9645–9650. doi: 10.1073/pnas.1401329111

62. Shannon, J. C., Pien, F. M., and Liu, K. C. (1996). Nucleotides and nucleotide sugars in developing maize endosperms (synthesis of ADP-glucose in brittle-1). Plant Physiol. 110, 835–843.

63. Shi, J., Wang, H., Hazebroek, J., Ertl, D. S., and Harp, T. (2005). The maize low-phytic acid 3 encodes a myo-inositol kinase that plays a role in phytic acid biosynthesis in developing seeds. Plant J. 42, 708–719. doi: 10.1111/j.1365-313X.2005.02412.x

64. Simkin, A. J., Guirimand, G., Papon, N., Courdavault, V., Thabet, I., Ginis, O., et al. (2011). Peroxisomal localisation of the final steps of the mevalonic acid pathway in planta. Planta 234, 903–914. doi: 10.1007/s00425-011-1444-6

65. Skirycz, A., and Inze, D. (2010). More from less: plant growth under limited water. Curr. Opin. Biotechnol. 21, 197–203. doi: 10.1016/j.copbio.2010.03.002

66. Skogerson, K., Harrigan, G. G., Reynolds, T. L., Halls, S. C., Ruebelt, M., Iandolino, A., et al. (2010). Impact of genetics and environment on the metabolite composition of maize grain. J. Agric. Food Chem. 58, 3600–3610. doi: 10.1021/jf903705y

67. Sonnhammer, E. L., and Koonin, E. V. (2002). Orthology, paralogy and proposed classification for paralog subtypes. Trends Genet. 18, 619–620. doi: 10.1016/S0168-9525(02)02793-2

68. Spielbauer, G., Li, L., Romisch-Margl, L., Do, P. T., Fouquet, R., Fernie, A. R., et al. (2013). Chloroplast-localized 6-phosphogluconate dehydrogenase is critical for maize endosperm starch accumulation. J. Exp. Bot. 64, 2231–2242. doi: 10.1093/jxb/ert082

69. Stitt, M., Sulpice, R., and Keurentjes, J. (2010). Metabolic networks: how to identify key components in the regulation of metabolism and growth. Plant Physiol. 152, 428–444. doi: 10.1104/pp.109.150821

70. Sun, Q., Zybailov, B., Majeran, W., Friso, G., Olinares, P. D., and van Wijk, K. J. (2009). PPDB, the Plant Proteomics Database at Cornell. Nucleic Acids Res. 37, D969–D974. doi: 10.1093/nar/gkn654

71. Szal, B., Jolivet, Y., Hasenfratz-Sauder, M.-P., Dizengremel, P., and Rychter, A. M. (2003). Oxygen concentration regulates alternative oxidase expression in barley roots during hypoxia and post–hypoxia. Physiol. Plant. 119, 494–502. doi: 10.1046/j.1399-3054.2003.00161.x

72. Tanz, S. K., Castleden, I., Hooper, C. M., Vacher, M., Small, I., and Millar, H. A. (2013). SUBA3: a database for integrating experimentation and prediction to define the SUBcellular location of proteins in Arabidopsis. Nucleic Acids Res. 41, D1185–D1191. doi: 10.1093/nar/gks1151

73. Teas, H. J. (1954). B vitamins in starchy and sugary maize endosperms. Plant Physiol. 29, 190–194. doi: 10.1104/pp.29.2.190

74. Töpfer, N., Caldana, C., Grimbs, S., Willmitzer, L., Fernie, A. R., and Nikoloski, Z. (2013). Integration of genome-scale modeling and transcript profiling reveals metabolic pathways underlying light and temperature acclimation in Arabidopsis. Plant Cell 25, 1197–1211. doi: 10.1105/tpc.112.108852

75. Tumaney, A. W., Ohlrogge, J. B., and Pollard, M. (2004). Acetyl coenzyme A concentrations in plant tissues. J. Plant Physiol. 161, 485–488. doi: 10.1078/0176-1617-01258

76. Vilella, A. J., Severin, J., Ureta-Vidal, A., Heng, L., Durbin, R., and Birney, E. (2009). EnsemblCompara GeneTrees: complete, duplication-aware phylogenetic trees in vertebrates. Genome Res. 19, 327–335. doi: 10.1101/gr.073585.107

77. Wallace, N. H., and Kriz, A. L. (1991). Nucleotide sequence of a cDNA clone corresponding to the maize globulin-2 gene. Plant Physiol. 95, 973–975. doi: 10.1104/pp.95.3.973

78. Wang, Y., Eddy, J. A., and Price, N. D. (2012). Reconstruction of genome-scale metabolic models for 126 human tissues using mCADRE. BMC Syst. Biol. 6:153. doi: 10.1186/1752-0509-6-153

79. Weber, E. J. (1979). The lipids of corn germ and endosperm. J. Am. Oil Chem. Soc. 56, 637–641. doi: 10.1007/BF02679340

80. Weber, E. J. (1987). Carotenoids and tocols of corn grain determined by HPLC. J. Am. Oil Chem. Soc. 64, 1129–1134. doi: 10.1007/BF02612988

81. Wu, Y., Wang, W., and Messing, J. (2012). Balancing of sulfur storage in maize seed. BMC Plant Biol. 12:77. doi: 10.1186/1471-2229-12-77

82. Zhang, P., Dreher, K., Karthikeyan, A., Chi, A., Pujar, A., Caspi, R., et al. (2010). Creation of a genome-wide metabolic pathway database for Populus trichocarpa using a new approach for reconstruction and curation of metabolic pathways for plants. Plant Physiol. 153, 1479–1491. doi: 10.1104/pp.110.157396

83. Zur, H., Ruppin, E., and Shlomi, T. (2010). iMAT: an integrative metabolic analysis tool. Bioinformatics 26, 3140–3142. doi: 10.1093/bioinformatics/btq602

Figure 6 and some supplemental files are not available in this version of the article. To view this additional information, please use the citation on the first page of this chapter.

DT2008: A Promising New Genetic Resource for Improved Drought Tolerance in Soybean When Solely Dependent on Symbiotic N_2 Fixation

SAAD SULIEMAN, CHIEN VAN HA, MARYAM NASR ESFAHANI, YASUKO WATANABE, RIE NISHIYAMA, CHUNG THI BAO PHAM, DONG VAN NGUYEN, AND LAM-SON PHAN TRAN

9.1 INTRODUCTION

Soybean (*Glycine max* (L.) Merr.) has been classified among the most important commercial oilseed crops worldwide [1]. It can substantially provide oils, micronutrients, minerals, and vegetable proteins suitable for livestock feed and human consumption. In addition, soybean has supplied materials for industrial uses, such as biodiesel, plastics, lubricants, and hydraulic fluids. Currently, world production of soybean is greater than any other oilseed crop. Globally, it accounts for approximately 68% of global crop legume production and 57% of world oilseed production [2].

DT2008: A Promising New Genetic Resource for Improved Drought Tolerance in Soybean When Solely Dependent on Symbiotic N_2 Fixation. © Sulieman S, Ha CV, Esfahani MN, Watanabe Y, Nishiyama R, Pham CTB, Nguyen DV, and Tran L-SP. BioMed Research International **2015** (2015). http://dx.doi.org/10.1155/2015/687213. Licensed under a Creative Commons Attribution 3.0 Unported License, http://creativecommons.org/licenses/by/3.0/.

Collectively, soybean production occupies around 6% of the world's available land [3].

As a leguminous plant, soybean has a superior potential capability to fix atmospheric N_2 in association with highly specialized soil bacteria. Under most conditions, soybean meets 58–68% of its nitrogen (N) demand through symbiotic association, but it can fulfill up to 100% with the aid of this vital process [4–6]. Moreover, a large portion of the fixed N can be readily accessible for the subsequent crops in the rotation systems or the natural ecosystems. Therefore, the soybean-rhizobia relationship represents a vital option to sustain agricultural development due to its superior N_2 fixation, enabling us to reduce the dependence on N fertilizers and thus avoiding the overexploitation of natural resources. Optimizing this association can upgrade soybean production and enhance soil fertility, whilst reducing the production costs and environmental impacts associated with N-chemical fertilizers [5]. Nevertheless, nodulating soybean plants growth and production are highly sensitive to adverse environmental conditions, particularly water scarcity in soils [7, 8].

In Vietnam, soybean occupies an important front position in the structure of agricultural crops throughout the country [9]. Recently, Vietnam's soybean production continues to fall well below the demand for food, feed, and vegetable oil industry. According to the 2012's statistical data, Vietnam imported 1.29 million metric tonnes of soybeans which represents a 26% increase over the previous year [10]. Due to high prices in the global market, soybean importation value had reached $776 million in 2012 (41% increase over the past year). Currently, the Vietnamese Government's Master Plan for Oilseeds has further development priorities for the sector with an objective of 350000 ha of soybean-cultivated land and a production of 700000 metric tonnes by 2020 (http://www.thecropsite.com/reports/?id=3701&country=VN). However, drought has a tremendous effect on soybean growth and development, thus negatively affecting the projected expansion of crop production [11]. In recent years, drought has occurred more and more commonly as a result of global warming and climate change [12]. Therefore, selective breeding for high drought-tolerant soybean cultivars and investigating the mechanisms to improve the drought tolerance of soybean have become top priority for many scientific researchers. On this basis, the soybean breeders at the Agriculture Genetics Institute (AGI) of Vietnam have initiated a long-term soybean breeding program to construct and release various drought-tolerant soybean cultivars through conventional breeding and radiation-induced mutagenesis. One of the newly developed prospective cultivars, the DT2008, revealed enhanced drought tolerance capability and yield stability (~2–4 metric tonnes per ha) under various field growing conditions [13]. Thus, we have started a joint project to fully characterize this cultivar under drought and various N regimes. Under nonnodulation conditions, we have recently documented that DT2008 has higher drought tolerance ability against the soybean reference cultivar Williams 82 (W82) [14].

In this report, we have extended our previous approach by comparing the drought-tolerant cultivar DT2008 and W82 based on their potential symbiotic association under drought and rehydration treatments. Results of this study demonstrated that DT2008 has a better drought tolerance and higher recovery level than W82.

9.2 MATERIALS AND METHODS

9.2.1 BIOLOGICAL MATERIALS

The soybean variety DT2008 was basically created by multiple hybridizations of local cultivars and subsequent irradiation with gamma rays Co^{60} – 18 Gy + F4 (DT2001/IS10) [15]. It has a wide adaptability to various harsh conditions and is suitably cultivated in 3 crops per year with a growth duration ranging from 110 to 120 days and relatively higher yield potentiality (2.5–4.0 tonnes/ha). In addition to its superior drought and thermotolerance, DT2008 has comparatively higher level of resistance against three kinds of diseases, namely, rust, downy mildew, and bacterial pustule [15]. Thus, in this study, we have intended to examine this promising drought-tolerant variety versus the widely used soybean reference cultivar W82 that was used to produce the reference genome sequence of soybean [16]. Accordingly, these materials would provide an efficient platform for omic analyses to identify new single nucleotide polymorphisms (SNPs) and promising candidate genes for genetic engineering. For testing the potential symbiotic capability under drought and rehydration conditions, both

cultivars were inoculated with the microsymbiont Bradyrhizobium japonicum strain USDA110. Owing to its superior symbiotic N_2 fixation activity
and full determination of its genome sequence, *B. japonicum* USDA110
has been widely used for the purpose of physiology, molecular genetics,
and ecological studies [17].

9.2.2 GENERAL PLANT AND BACTERIAL GROWTH CONDITIONS

Seeds of DT2008 and W82 were separately germinated in 6-litre pots containing autoclaved vermiculite as rooting substrate in a controlled greenhouse conditions (continuous 30°C temperature, 60% relative humidity,
12/12 h photoperiod, and $150\,\mu mol\,m^{-2}\,s^{-1}$ photon flux density). Seeds
were inoculated with *B. japonicum* USDA110 grown in yeast mannitol
broth (YMB) (mannitol $2\,g\,L^{-1}$; yeast extract $0.4\,g\,L^{-1}$; K_2HPO_4 $0.5\,g\,L^{-1}$;
$MgSO_4\cdot 7H_2O$ 0.2 $g\,L^{-1}$; NaCl $0.1\,g\,L^{-1}$; pH 6.8) for 48 h at 28°C. Cultures were diluted with water and added at a rate of ~$10^8\,cells\,mL^{-1}$ after
the seeds were sown in the vermiculite. Plants were watered to field capacity three times a week with full-strength Herridge's nutrient solution
[18] until the stress treatments were imposed. The basal nutrient solution
contained $0.25\,mM$ $CaCl_2$; $0.25\,mM$ KCl; $0.5\,mM$ $MgSO_4\cdot 7H_2O$; $0.13\,mM$
KH_2PO_4; $0.13\,mM$ K_2HPO_4; $23.5\,\mu M$ Fe (III)-EDTA; $71.5 \times 10^{-2}\,mg\,L^{-1}$
H_3BO_3; $45.3 \times 10^{-2}\,mg\,L^{-1}$ $MnCl_2\cdot 4H_2O$; $2.8 \times 10^{-2}\,mg\,L^{-1}$ $ZnCl_2$; $1.3 \times$
$10^{-2}\,mg\,L^{-1}$ $CuCl_2\cdot 2H_2O$; and $0.6 \times 10^{-2}\,mg\,L^{-1}$ $NaMoO_4\cdot 2H_2O$.

9.2.3 DROUGHT AND RECOVERY TREATMENTS

Drought was imposed on 21-day-old plants by withholding water. The
plants were randomly separated into two main sets (control and drought)
containing four biological replicates each. Control (well-watered) plants
were watered every day, whereas drought was imposed by withholding
water for either 4 or 7 days (4 D or 7 D). Recuperation was carried out by
rewatering the stressed plants for 3 days (7 D + 3 W). Both water-stressed
(WS) and well-watered (WW) plants were harvested at set time points:

4, 7, and 10 D after the onset of drought-rehydration treatments. At each time point, soil volumetric moisture contents (VMC) were monitored using a HydroSense soil moisture probe (Campbell Scientific, Inc.). Measurements of various growth and nodulation parameters were performed at the end of the stress and rehydration periods. At each harvest, plants were fractioned into shoots, roots, and nodules. Shoot and root tissues were dried at 65°C for a minimum of 48 h and weighed for dry matter (DM) determination.

9.2.4 STATISTICAL ANALYSIS

Means and standard errors (SEs) were used to plot figures and evaluate treatment responses. All statistical analyses were performed using statistical tools imbedded in Microsoft Excel 2010. The significance of difference between means was determined by Student's t-test where the values of $P < 0.05$ were considered significant and those of $P < 0.01$ and $P < 0.001$ were highly significant.

9.3 RESULTS AND DISCUSSION

Drought is a recurring problem limiting nodulation and N_2 fixation in crop production particularly in tropical and semiarid tropical areas [19, 20]. Under the present scenarios of climate change, drought is more likely to occur, leading to ultimate growth and productivity reductions for most important economic crops, including soybean. Although access to irrigation can be used partially to alleviate drought impact, the usage of soybean drought-tolerant cultivars remains the most practically promising strategy to adopt. To cope with water deficit, drought-tolerant cultivars have developed a number of strategies that are genetically encoded [1, 9]. Thus, it is important to elucidate these striking adaptive mechanisms developed in such tolerant cultivars in order to improve the agronomic performance of soybean and other plant species by genetic engineering [21, 22]. Indeed, many physiological and biochemical responses to drought are shared amongst various tested plant species [23].

Our adopted strategy in the improvement of drought-tolerant cultivars is based on establishing an integrated approach involving conventional breeding and radiation-induced mutagenesis program and subsequently analyzing the internal adaptive mechanisms, which underlie plant responses to drought, through molecular biology techniques. A long-term, multidisciplinary research program was started to produce high-yielding adaptive cultivars for limited water conditions which exploited the drought-tolerant traits of DT2008. This biological resource was basically produced by multiple hybridizations of local cultivars and irradiation exposure [14]. A question was then raised whether DT2008 is able to maintain N_2 fixation at high level during drought, thus contributing to its improved productivity. To provide an answer to this question, in this study, we initially carried out a comprehensive comparative analysis between DT2008 and the reference cultivar W82, and whole genome was sequenced [16], thereby enabling us to identify potentially important mutations or SNPs responsible for enhanced N_2 fixation under drought in future molecular studies.

9.3.1 PLANT GROWTH AND BIOMASS PRODUCTION ARE LESS NEGATIVELY AFFECTED IN THE DROUGHT-TOLERANT CULTIVAR DT2008

The alteration in biomass allocation is a principle strategy for coping with progressive soil-drying conditions [14]. Several groups have reported that DM partitioning is very important in the determination of soybean productivity [19, 24]. Total biomass has been used as a selection criterion for assessing drought tolerance in soybean. Understanding assimilation and allocation processes affected by water deficit is a fundamental prerequisite step in identifying and improving soybean tolerance to drought [11].

In this study, soybean genotypes tested showed differential responses for growth traits examined. For example, the DT2008 plants exhibited more stable shoot growth in terms of shoot length (Figure 1), shoot fresh weight (FW), and dry weight (DW) (Table 1), when compared with W82, suggesting that DT2008 possesses a better shoot growth rate than W82 under the examined water deficit regimes (4 D, 7 D). In contrary, W82 exhibited a higher degree of susceptibility upon subjection to the equivalent

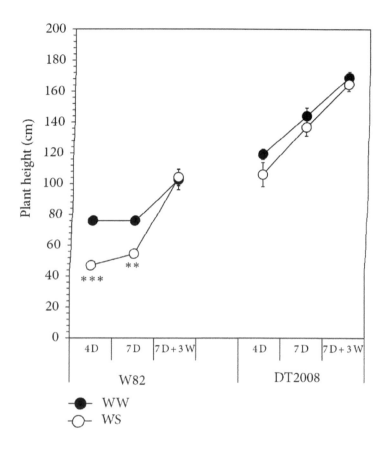

Figure 1: Comparison of plant height of the nodulated W82 and DT2008 plants after growth for 21 days in vermiculite soil and exposure to drought stress and recovery treatments. Well-watered (WW) plants were irrigated every day, whereas drought stress (WS) was imposed by withholding water for either four (4 D) or seven (7 D) days. Recovery treatment was experienced by withholding water for seven days followed by a subsequent rewatering for three days (7 D + 3 W). Error bars represent standard errors (plants/genotype). Asterisks indicate significant differences as determined by Student's t-test (**P < 0.01; ***P < 0.001).

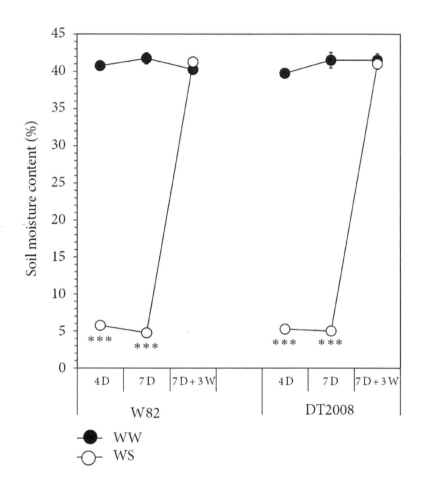

FIGURE 2: Monitoring of volumetric soil moisture contents during the drought stress and recovery treatments. Nodulated W82 and DT2008 plants were grown for 21 days in vermiculite soil and exposed to water deficit and rewatering treatments. Well-watered (WW) plants were irrigated every day, whereas drought stress (WS) was imposed by withholding water for either four (4 D) or seven (7 D) days. Recovery treatment was experienced by withholding water for seven days followed by a subsequent rewatering for three days (7 D + 3 W). Error bars represent standard errors. Asterisks indicate significant differences as determined by Student's t-test (***P < 0.001).

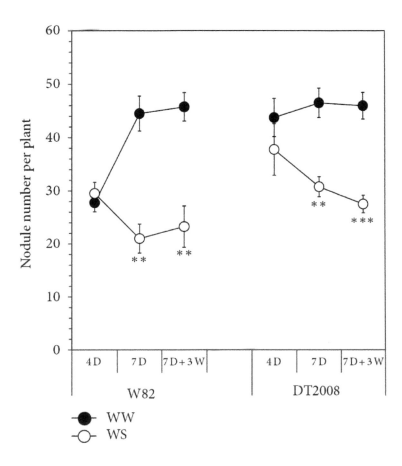

FIGURE 3: Comparison of nodule number per plant of the nodulated W82 and DT2008 plants after growth for 21 days in vermiculite soil and exposure to drought stress and recovery treatments. Well-watered (WW) plants were irrigated every day, whereas drought stress (WS) was imposed by withholding water for either four (4 D) or seven (7 D) days. Recovery treatment was experienced by withholding water for seven days followed by a subsequent rewatering for three days (7 D + 3 W). Error bars represent standard errors (n = 4 plants/genotype). Asterisks indicate significant differences as determined by Student's t-test (**P < 0.01; ***P < 0.001).

drought treatments as indicated by significant decreases in the shoot length (Figure 1), shoot FW, and DW (Table 1).

TABLE 1: Comparison of biomass production of the nodulated W82 and DT2008 plants after growth for 21 days in vermiculite soil and exposure to drought stress and recovery treatments. Well-watered (WW) plants were irrigated every day, whereas drought stress (WS) was imposed by withholding water for either four (4 D) or seven (7 D) days. Recovery treatment was experienced by withholding water for seven days followed by a subsequent rewatering for three days (7 D + 3 W). Data presented are the means ± SE of four replicates. Asterisks indicate significant differences as determined by Student's t-test (*P < 0.05; **P < 0.01; ***P < 0.001).

	Treatment	W82		DT2008	
		WW	WS	WW	WS
Fresh weight (g)					
Shoot	4 D	5.46 ± 0.42	3.72 ± 0.18**	6.92 ± 0.48	6.06 ± 0.53
	7 D	6.64 ± 0.54	3.89 ± 0.18**	9.94 ± 0.50	6.51 ± 0.39**
	7 D + 3 W	10.57 ± 0.61	7.10 ± 0.87*	11.03 ± 0.58	7.31 ± 0.30**
Root	4 D	2.32 ± 0.28	0.91 ± 0.09**	2.58 ± 0.19	1.78 ± 0.09*
	7 D	3.68 ± 0.13	1.64 ± 0.15***	4.11 ± 0.30	2.44 ± 0.33**
	7 D + 3 W	3.89 ± 0.22	2.31 ± 0.26**	5.62 ± 0.33	4.23 ± 0.10**
Nodules	4 D	0.44 ± 0.04	0.31 ± 0.03*	0.61 ± 0.02	0.36 ± 0.04**
	7 D	0.57 ± 0.05	0.27 ± 0.03**	0.71 ± 0.07	0.31 ± 0.03**
	7 D + 3 W	0.68 ± 0.06	0.35 ± 0.06**	0.55 ± 0.04	0.37 ± 0.02**
Dry weight (g)					
Shoot	4 D	0.74 ± 0.04	0.60 ± 0.03*	1.16 ± 0.06	1.11 ± 0.09
	7 D	1.36 ± 0.10	0.77 ± 0.04**	1.99 ± 0.10	1.30 ± 0.08**
	7 D + 3 W	2.11 ± 0.12	1.41 ± 0.18*	2.21 ± 0.12	1.46 ± 0.06**
Root	4 D	0.15 ± 0.02	0.06 ± 0.01**	0.18 ± 0.02	0.12 ± 0.01*
	7 D	0.17 ± 0.01	0.07 ± 0.01***	0.21 ± 0.02	0.12 ± 0.02**
	7 D + 3 W	0.23 ± 0.01	0.14 ± 0.02**	0.34 ± 0.02	0.25 ± 0.01**

Drought tolerance mechanisms in leguminous plants are closely related to the root traits of the cultivated genotypes [23, 25]. In comparison with W82, DT2008 was more tolerant to water deficit as judged by its higher root fresh and DM biomass accumulations (Table 1). Understandably, maintenance of DT2008 root growth under progressive decline in soil water content would enhance drought tolerance due to an increased capacity of water uptake. Importantly, differences between DT2008 and W82 were observed under conditions that ensured similar amounts of soil water as indicated by VMC measurements (Figure 2). These results further support that DT2008 is more strongly tolerant to drought than W82 as indicated recently under nonnodulation conditions [14]. Collectively, our results suggest that the enhanced root systems of DT2008 may significantly contribute to its improved drought tolerance in comparison with W82.

9.3.2 DROUGHT-INDUCED CHANGES IN NODULATION PATTERNS

Most leguminous plants, including soybean, have particular features in response to water deficit, such as reduced rates of nodulation and nitrogenase activity [26]. The acute sensitivity of nodulation to water deficit has been considered a major limiting factor towards improving soybean productivity. Despite several attempts and considerable research effort during the last decades, the molecular mechanism(s) underlining this sensitivity remains largely unidentified [27]. In this work, the effect of drought on the N_2 fixation was evaluated based on nodule growth and development, specifically, the number of nodules per plant (Figure 3) and total nodule FW (Table 1), which frequently correlate well with shoot DM, providing an acceptable basis of N_2-fixing efficiency [28].

In comparison with W82, DT2008 established relatively higher number of nodules per plant (Figure 3) and accumulated more nodule FW (Table 1) under sufficient water supply which might contribute to the higher growth rate of DT2008 versus that of W82 (Table 1). Upon exposure to water deficit (4 D and 7 D), nodulation pattern, in terms of nodule number, was found to be different between DT2008 and W82. The total nodule number per plant significantly reduced in both cultivars after 7 D of water stress; however, the nodule number in DT2008 was still significantly higher than that in W82 at 7 D ($P < 0.05$ as measured by Student's t-test) (Figure 3). In case of nodule FW, although the 4 D water stress regime resulted in a significant reduction of nodule FW in both W82 and DT2008, the nodule FW in the drought-tolerant DT2008 was still slightly higher than that of the sensitive W82 cultivar (Table 1). These results indicate that water stress has a certain varied effect on nodulation patterns between the two genotypes, which might contribute to the differential drought-tolerant levels of W82 and DT2008 in addition to the different root growth rate (Table 1). Moreover, one would also expect that the DT2008 would have additionally certain internal adaptation mechanisms that might enhance the symbiotic efficiency under stressful conditions.

9.3.3 DT2008 CULTIVAR HAS A HIGHER REACTIVATION CAPABILITY THAN THE MODEL W82 CULTIVAR UNDER RECOVERY CONDITIONS

In the field, plants often encounter unexpected cycles of progressive soil dryness. Under such conditions, plant survival and productivity rely very much on the internal acclimatization mechanisms, which reduce or even prevent cellular damage during the stress period, as well as on the potential capacity of the stressed plant to recover and maintain normal metabolic functioning [29]. Thus, plant recovery following rewatering is an essential trait for plant survival and reflects the balance between reconstruction of damaged structures and adequate metabolism restoration [23]. Obviously, much effort has been directed towards the response of N_2 fixation under drought conditions, while few investigations, if any, have considered the genotypic difference in nodulation and plant growth and development, particularly after recovery from progressive soil drying. Such investigation would be particularly useful for the analysis of the early changes occurring during the reactivation of normal nodule metabolic processes.

In the present report, the differential responses in the recovery from drought treatments were evidenced when DT2008 and W82 plants were subjected to 7 D of water withholding, followed by subsequent rewatering for 3 D. With the exception of nodule number per plant (Figure 3), plant rewatering was able to reduce or maintain the negative impact of drought on all nodulation and growth traits examined in the DT2008 genotype. Although in DT2008 the total nodule number per plant was not recovered in response to rewatering (Figure 3), the specific fixation per unit nodule mass would still have a chance to increase and compensate the observed reduction in nodulation number. Alternatively, the recovery time of 3 days (7 D + 3 W) was not sufficient for DT2008 to assume an effective recovery of nodule growth and development. Indeed, many small nodules were observed in DT2008 after 3 days of recovery. However, these nodules were too small in size to be considered in the data analysis. It should still be noticed that the total nodule number, as well as the total nodule FW, of DT2008 was still slightly higher than that of W82 following the recovery treatment (Figure 3 and Table 1). As for comparison of plant growth and development under recovery conditions, we found that rewatering drastically affected all of the parameters examined in W82 and DT2008. Although DT2008 and W82 were shown to be similarly affected at similar significant level upon being exposed to the recovery treatments, the DT2008 genotype remarkably exhibited a better performance when compared with W82 at the same equivalent treatment (Table 1).

9.4 CONCLUSIONS

This study aimed to characterize the newly developed soybean cultivar DT2008 when fully grown under symbiotic N_2 fixation conditions with principle objective to determine the divergent drought responses versus the reference cultivar W82 when both are subjected to drought and recovery treatments. Contrasting tolerant and sensitive symbiotic responses were identified for each genotype in association with the microsymbiont *B. japonicum* strain USDA110. The results reported here indicated that DT2008 has a superior nodule development under water deficit and recovery in comparison with W82, highlighting that it might be a heritable trait. In addition to difference in root growth rate (this work and [14]), difference in nodule development rate might contribute to differential drought-tolerant levels of DT2008 and W82 under symbiotic conditions. Thus, DT2008 and W82 genotypes can offer a genetic resource for comparative genomics, ultimately enabling soybean scientists to identify novel SNPs and genes underlining N_2 fixation under drought for development of soybean cultivars with improved drought tolerance. Strategies involving various omic approaches, such as transcriptomics, proteomics, and metabolomics, will be highly promising in this platform for determination of genes, mutations, and SNPs responsible for enhanced drought tolerance of

DT2008 for genetic engineering. In fact, a combination of the conventional breeding, marker-assisted breeding, and genetic engineering strategies will be necessary in soybean improvement under increasing water limitation in the near future. As such, the generation of novel improved soybean cultivars bearing drought-tolerant trait(s) is highly expected to cope with the current and future expected water limitations.

REFERENCES

1. L. P. Manavalan, S. K. Guttikonda, L.-S. P. Tran, and H. T. Nguyen, "Physiological and molecular approaches to improve drought resistance in soybean," Plant and Cell Physiology, vol. 50, no. 7, pp. 1260–1276, 2009.
2. D. F. Herridge, M. B. Peoples, and R. M. Boddey, "Global inputs of biological nitrogen fixation in agricultural systems," Plant and Soil, vol. 311, no. 1-2, pp. 1–18, 2008.
3. P. D. Goldsmith, "Economics of soybean production, marketing and utilization," in Soybeans Chemistry, Production, Processing, and Utilization, L. P. Johnson, P. A. White, and R. Galloway, Eds., pp. 117–150, AOCS Press, Urbana, Ill, USA, 2008.
4. F. Salvagiotti, K. G. Cassman, J. E. Specht, D. T. Walters, A. Weiss, and A. Dobermann, "Nitrogen uptake, fixation and response to fertilizer N in soybeans: a review," Field Crops Research, vol. 108, no. 1, pp. 1–13, 2008.
5. M. B. Peoples, J. Brockwell, D. F. Herridge et al., "The contributions of nitrogen-fixing crop legumes to the productivity of agricultural systems," Symbiosis, vol. 48, no. 1–3, pp. 1–17, 2009.
6. E. S. Jensen, M. B. Peoples, R. M. Boddey et al., "Legumes for mitigation of climate change and the provision of feedstock for biofuels and biorefineries. A review," Agronomy for Sustainable Development, vol. 32, no. 2, pp. 329–364, 2012.
7. R. Serraj, T. R. Sinclair, and L. C. Purcell, "Symbiotic N2 fixation response to drought," Journal of Experimental Botany, vol. 50, no. 331, pp. 143–155, 1999. ·
8. C. A. King and L. C. Purcell, "Inhibition of N2 fixation in soybean is associated with elevated ureides and amino acids," Plant Physiology, vol. 137, no. 4, pp. 1389–1396, 2005.
9. N. P. Thao, N. B. A. Thu, X. L. T. Hoang, C. V. Ha, and L.-S. P. Tran, "Differential expression analysis of a subset of drought-responsive GmNAC genes in two soybean cultivars differing in drought tolerance," International Journal of Molecular Sciences, vol. 14, no. 12, pp. 23828–23841, 2013.
10. USDA GAIN, "Vietnam—oilseeds and products annual 2013," Global Agricultural Information Network (GAIN) Report VM3018, USDA GAIN, Hanoi, Vietnam, 2013.
11. N. B. A. Thu, Q. T. Nguyen, X. L. T. Hoang, N. P. Thao, and L.-S. P. Tran, "Evaluation of drought tolerance of the vietnamese soybean cultivars provides potential

resources for soybean production and genetic engineering," BioMed Research International, vol. 2014, Article ID 809736, 9 pages, 2014.

12. A. Dai, "Increasing drought under global warming in observations and models," Nature Climate Change, vol. 3, no. 1, pp. 52–58, 2013.

13. M. Q. Vinh, P. T. B. Chung, N. V. Manh, and L. T. A. Hong, "Results of research, creation, drought-tolerant soybean variety, DT2008," Journal of Vietnamese Agricultural Science and Technology, vol. 6, pp. 46–50, 2010.

14. C. V. Ha, D. T. Le, R. Nishiyama et al., "Characterization of the newly developed soybean cultivar DT2008 in relation to the model variety W82 reveals a new genetic resource for comparative and functional genomics for improved drought tolerance," BioMed Research International, vol. 2013, Article ID 759657, 8 pages, 2013.

15. M. Q. Vinh, D. K. Thinh, D. T. Bang, D. H. At, L. H. Ham, and Q. Y. Shu, "Current status and research directions of induced mutation application to seed crops improvement in Vietnam," in Induced Plant Mutations in the Genomics Era, pp. 341–345, Food and Agriculture Organization of the United Nations, Rome, Italy, 2009.

16. J. Schmutz, S. B. Cannon, J. Schlueter et al., "Genome sequence of the palaeopolyploid soybean," Nature, vol. 463, no. 7278, pp. 178–183, 2010.

17. T. Kaneko, Y. Nakamura, S. Sato et al., "Complete genomic sequence of nitrogen-fixing symbiotic bacterium Bradyrhizobium japonicum USDA110," DNA Research, vol. 9, no. 6, pp. 189–197, 2002.

18. A. C. Delves, A. Mathews, D. A. Day, A. S. Carter, B. J. Carroll, and P. M. Gresshoff, "Regulation of the soybean-Rhizobium nodule symbiosis by shoot and root factors," Plant Physiology, vol. 82, no. 2, pp. 588–590, 1986.

19. T. R. Sinclair, L. C. Purcell, C. A. King, C. H. Sneller, P. Chen, and V. Vadez, "Drought tolerance and yield increase of soybean resulting from improved symbiotic N2 fixation," Field Crops Research, vol. 101, no. 1, pp. 68–71, 2007.

20. E. Gil-Quintana, E. Larrainzar, C. Arrese-Igor, and E. M. González, "Is N-feedback involved in the inhibition of nitrogen fixation in drought-stressed Medicago truncatula?" Journal of Experimental Botany, vol. 64, no. 1, pp. 281–292, 2013.

21. N. P. Thao and L.-S. P. Tran, "Potentials toward genetic engineering of drought-tolerant soybean," Critical Reviews in Biotechnology, vol. 32, no. 4, pp. 349–362, 2012.

22. C. V. Ha, D. T. Le, R. Nishiyama et al., "The auxin response factor transcription factor family in soybean: genome-wide identification and expression analyses during development and water stress," DNA Research, vol. 20, no. 5, pp. 511–524, 2013.

23. Y. Kang, Y. Han, I. Torres-Jerez et al., "System responses to long-term drought and re-watering of two contrasting alfalfa varieties," Plant Journal, vol. 68, no. 5, pp. 871–889, 2011.

24. J. G. Streeter, "Effects of drought on nitrogen fixation in soybean root nodules," Plant, Cell and Environment, vol. 26, no. 8, pp. 1199–1204, 2003.

25. A. Sarker, W. Erskine, and M. Singh, "Variation in shoot and root characteristics and their association with drought tolerance in lentil landraces," Genetic Resources and Crop Evolution, vol. 52, no. 1, pp. 89–97, 2005.

26. A. J. Gordon, F. R. Minchin, L. Skøt, and C. L. James, "Stress-induced declines in soybean N2 fixation are related to nodule sucrose synthase activity," Plant Physiology, vol. 114, no. 3, pp. 937–946, 1997. ·

27. E. Larrainzar, J. A. Molenaar, S. Wienkoop et al., "Drought stress provokes the down-regulation of methionine and ethylene biosynthesis pathways in Medicago truncatula roots and nodules," Plant, Cell and Environment, vol. 37, no. 9, pp. 2051–2063, 2014.

28. P. Somasegaran and H. J. Hoben, "Quantifying the growth of Rhizobia," in Handbook for Rhizobia: Methods in Legume-Rhizobium Technology, P. Somasegaran and H. J. Hoben, Eds., pp. 47–57, Springer, New York, NY, USA, 1994.

29. A. Blum, "Crop responses to drought and the interpretation of adaptation," Plant Growth Regulation, vol. 20, no. 2, pp. 135–148, 1996.

CHAPTER 10

Current Knowledge in Lentil Genomics and Its Application for Crop Improvement

SHIV KUMAR, KARTHIKA RAJENDRAN, JITENDRA KUMAR, ALADDIN HAMWIEH, AND MICHAEL BAUM

10.1 INTRODUCTION

Lentil (*Lens culinaris* ssp. *culinaris* Medikus) is a diploid (2n=2X=14) self-pollinating crop with a genome size of approximately 4 Gbp (Arumuganathan and Earle, 1991). It provides affordable source of dietary proteins (22–35%), minerals, fiber, and carbohydrates to poor people and plays a vital role in alleviating malnutrition and micronutrient deficiencies in developing countries. As it exhibits low glycemic index, it is highly recommended by physicians for the people suffering from diabetes, obesity, and cardiovascular diseases (Srivastava and Vasishtha, 2012). In fact, vegetable protein is gaining preference over the animal protein for consumption by the health conscious people in the present day. This could be one of

Current Knowledge in Lentil Genomics and its Application for Crop Improvement © Kumar S, Rajendran K, Kumar J, Hamwieh A, and Baum M. Frontiers in Plant Science **6**,78 (2015). http://dx.doi.org/10.3389/fpls.2015.00078. Licensed under a Creative Commons Attribution 4.0 International License, http://creativecommons.org/licenses/by/4.0/.

the reasons for increased per capita consumption (Vandenberg, 2009) and fivefold increase in global lentil production (from 0.85 to 4.43 Mt) during the last five decades, through a 155% increase in sown area and the doubling of average yields from 528 to 1068 kg ha^{-1} (FAOSTAT, 2014). Lentil cultivation often provides rotational benefits to cereal-based cropping systems through biological nitrogen fixation, carbon sequestration, and through effective control of weeds, diseases, and insect pests. It generates livelihood for the small-scale farmers practicing agriculture in the dryland agricultural ecosystems of South Asia, Sub-Saharan Africa, West Asia, and North Africa (Kumar et al., 2013). However, the lentil yields remain low in many developing countries as it is often cultivated as a rainfed crop under difficult edaphic conditions and subjected to terminal drought, heat stress, low soil fertility, and various diseases including ascochyta blight (*Ascochyta lentis*), fusarium wilt (*Fusarium oxysporum* f.sp. lentis), anthracnose (*Colletotrichum truncatum*), stemphylium blight (*Stemphylium botryosum*), rust (*Uromyces viciae-fabae*), collar rot (*Sclerotiun rolfsii*), root rot (*Rhizoctonia solani*), and white mold (*Sclerotinia sclerotiorum*) (Kumar et al., 2013; Sharpe et al., 2013). So far, the classical plant breeding approach of selection-recombination-selection has been successful in mainstreaming some of the easy-to-manage monogenic traits in lentil. However, this approach is less precise and time consuming when dealing with traits of breeders' interest which are often quantitative in nature and highly influenced by environment and genotype–environment (GE) interaction (Kumar and Ali, 2006). In order to identify, fix, and select superior recombinants more precisely and efficiently, there is a need to integrate biotechnological approaches such as marker assisted selection (MAS) and genetic engineering in lentil breeding program to mainstream new genetic variability in the cultivated gene pool.

The current lentil breeding programs are limited in their ability to implement MAS due to a lack of genomic resources. In comparison to major legume crops such as soybean, common bean, pigeon pea, and chickpea, the pace of development of genomic resources is slow in lentil (Kumar et al., 2014). Large genome size, narrow genetic base, lack of candidate genes, low density linkage map, and the difficulty in identifying beneficial alleles are the main limiting factors in genomics enabled improvement in lentil. Molecular tools have occasionally been used by lentil breeders and geneticists to understand the genetic basis of a few traits related to biotic (ascochyta blight, anthracnose, rust, fusarium wilt, stemphylium blight) and abiotic (drought, frost, cold, boron, salinity) stresses (Kumar et al., 2014). Recent developments in the next generation sequencing (NGS) technologies have facilitated the development of array-based high-throughput (HTP) genotyping platforms with SNP markers. Bett et al. (2014) have carried out large amounts of next-generation sequencing on lentil cultivar, CDC Redberry. An initial draft of 23x coverage produced scaffolds covering over half the genome (2.7 Gb of the expected 4.3 Gb) and recent additional 125x coverage is currently being assembled. Gene sequences for several traits of interest were identified using the initial 23x draft assembly and derived SNP markers are now available for MAS in the lentil breeding program (Bett et al., 2014). Besides, the close phylogenetic relationships with the model legumes such as *Medicago truncatula* and *Lotus japonicus* have provided ample opportunities for comparative genome mapping and identified putative orthologous gene sequence resources in these genomes (Weller et al., 2012; Kaur et al., 2014). These genomic tools and technologies have opened up new avenues for practicing genomics assisted selection in lentil. There is also a tremendous scope to develop lentil cultivars through reverse genetic approaches. In this context, this review has been made to evaluate the research progress achieved in lentil genomics along with the discussion on future prospective for genetic enhancement.

10.2 DEVELOPMENT OF GENOMICS RESOURCES

10.2.1 MOLECULAR MARKERS

The first genetic map of lentil was constructed using morphological and isozyme markers in early 1980's (Zamir and Ladizinsky, 1984; Tadmor et al., 1987). After the discovery of molecular markers starting from the restriction fragment length polymorphism (RFLP), significant progress has been made in molecular marker development and genotyping platforms in lentils. It began with the hybridization based DNA markers such as RFLP (Havey and Muehlbauer, 1989) and moved toward the use of PCR based markers such as random amplified polymorphic DNA (RAPD), amplified fragment length polymorphism (AFLP) and simple sequence repeats (SSR) markers for genotyping. The first comprehensive linkage map with 177 RAPD, AFLP, RFLP, and morphological markers was developed using interspecific recombinant inbred lines (RIL) population of a single cross of *L. culinaris* × *L. orientalis* (Eujayl et al., 1998a). Among the various PCR based markers, SSR markers have made significant contribution to the recent development of lentil genome maps. The first genomic library was constructed from a cultivated accession, ILL5588 using the restriction enzyme Sau3AI (*Staphylococcus aureus* 3A) and screened with (GT)10, (GA)10, (GC)10, (GAA)8, (TA)10, and (TAA) probes (Hamwieh et al., 2005). Using this library initially a set of 30 highly polymorphic SSR markers were developed. Since this study was aimed at isolating SSRs that are abundant and well distributed in the genome, a non-enriched library was used for screening purposes. Hamwieh et al. (2009) further developed an additional set of 14 SSR markers and used them for genetic diversity analysis of the lentil core set. A set of 122 functional SSR markers have recently been developed using a genomic library enriched for GA/CT motifs for utilization in the lentil breeding program (Verma et al., 2014).

Recently, the PCR-based markers are being rapidly replaced by the DNA chip based markers, particularly with SNPs. SNPs are abundant in nature and common even across legume genomes (Chagne et al., 2007). There are various technologies for evaluation of SNP loci and many of these are amenable to automation for allele calling and data collection. The availability of extensive sequence database has made a new beginning to exploit them as a HTP marker system for genome mapping studies. Recent efforts in re-sequencing alleles to discover SNPs in lentil have facilitated automated high-throughput genotyping platforms (HTP). As a result, SNPs have emerged as potential markers for NGS approaches. About 44,879 SNP markers have been identified in lentil using Illumina Genome Analyzer (Sharpe et al., 2013). Temel et al. (2014) have identified another set of 50,960 SNPs and constructed a SNP based linkage map in lentil. The recent discovery of high-density SNP markers has facilitated the establishment of ultra HTP genotyping technologies such as Illumina GoldenGate (GG), which can accommodate more than 1000 SNPs in GG platforms (Sharpe et al., 2013; Kaur et al., 2014). Since SNP discovery and genotyping require expensive and sophisticated platforms, the development and exploitation of SNP markers is still limited in lentil. There are techniques available to detect SNPs such as allele-specific PCR, single base extension and array hybridization methods. These are cost effective and through the use of allele-specific PCR (KASPar) markers, we can include small to moderate amount of SNPs for any specific application (Fedoruk et al., 2013; Sharpe et al., 2013).

10.2.2 TRANSCRIPTOME ASSEMBLIES

As the characterization of lentil whole genome is still in progress, transcriptome assemblies provide excellent opportunities to identify expressed sequenced tag (EST) derived SSR and SNP markers and intron-targeted primers (ITP). In the early days, the classical dideoxynucleotide chain termination method of Sanger has been used to sequence cDNA libraries and generate ESTs across various crops. ESTs are short DNA sequences of 150–400 bp from a cDNA clone that correspond to a particular mRNA. Development of HTP functional genomics approaches like serial analysis of gene expression (SAGE) has led to the generation of more ESTs. The first EST library was made from a mixture of eight cultivars with varying seed phenotypes (Vijayan et al., 2009). The second cDNA library was prepared from the leaflets of a Canadian cultivar 'Eston' inoculated with *Colletotrichum truncatum* (Kumar et al., 2014). The cDNA clones corresponding to the ESTs of interest can be used as RFLP or CAPS based markers (Varshney et al., 2005). The EST sequence data also serve the purpose of identifying SSRs and/or SNPs. Before the ESTs, development of SSR and SNP markers was expensive and required high resource laboratories, but presently any user can download them from the database and use some special bioinformatic programs like MISA for SSR detection (Thiel et al., 2003; Varshney et al., 2005) and Snipper for SNP discovery (Kota et al., 2003; Varshney et al., 2005). As on January 2015, there are about 10,341 ESTs available for lentil (NCBI, 2015).

Kaur et al. (2011) carried out transcriptome sequencing of lentil based on the second-generation technology which permits large-scale unigene assembly and SSR marker discovery. They used tissue-specific cDNA samples from six genotypes (Northfield, ILL2024, Indianhead, Digger, ILL6788, and ILL7537) using Roche 454 GS-FLX Titanium technology, and generated c. 1.38×106 ESTs. De novo assembly generated 15,354 contigs and 68,715 singletons. Out of huge ESTs produced, 3,470 SNP and EST-SSRs have been identified. Development of genomic resources has become cost effective with the advent of NGS of ESTs. Validation of a subset of 192 EST-SSR markers across a panel of 12 cultivated genotypes showed 47.5% polymorphism from a set of 2,393 EST-SSR markers developed in lentil (Kaur et al., 2011). In recent times, transcriptome cDNA library sequencing using Illumina GA/GAIIx system has provided a potential alternative. Sharpe et al. (2013) developed 3'-cDNA reads from nine *L. culinaris* and two *L. ervoides* accessions using 454 pyrosequencing technology, identified SNPs, selected the sub-set of SNP for the development of a 1536 SNP Illumina GG array and used the array to construct a SNP based genetic map of *L. culinaris* mapping population. Similarly, Verma et al. (2013) used the short reads obtained from Illumina GAII and developed de novo transcriptome assemblies of lentil, developed SSR markers and utilized them in diversity analysis. Temel et al. (2014) used two lentil cultivars, Precoz and WA8649041 and their RILs using Illumina CASAVA pipelines, detected SNP markers, and generated a SNP based linkage map. As a result of transcriptome sequencing, massive data have been obtained in the form of about 847,824 high quality sequence reads and the transcriptome assemblies with 84,074 unigenes (Sharpe et al., 2013; Verma et al., 2013).

10.2.3 BI-PARENTAL MAPPING POPULATIONS

Efforts have been made at International Center for Agricultural Research in the Dry Areas (ICARDA) and national programs to develop mapping populations for key traits in lentil (Table 1). RIL populations have been developed from the crosses made between contrasting parents for the traits of interest through single seed descent method. Indian Institute of Pulses Research (IIPR) has recently developed RIL population from a cross between ILL6002 and ILL7663 in order to identify and map early growth vigor genes in lentil. Identification of markers linked to the gene(s)/QTL governing these traits will help in development of genotype having high biomass at early stage. For tagging and mapping of genes of earliness, another mapping population has been developed from a cross between Precoz (Medium early) and L4603 (early) at IIPR, Kanpur, India. Another mapping population segregating for earliness with a cross made between ILL5588 (late flowering) and ILL6005 (early flowering) is available in University of Tasmania, Hobart, TAS, Australia (Weller et al., 2012). It has the loci ELF3 (EARLY FLOWERING 3) which involved in circadian clock function and contribute to reduce the photoperiod response in cultivars to be grown under short season environmental conditions. CSK Himachal Pradesh Agricultural University, Palampur, India has developed RIL populations involving both intra and intersubspecific crosses that differ for rust reaction, drought tolerance, flowering time, plant vigor, shattering tolerance, seed size, and seed weight. Two mapping populations one each with the University of Saskatchewan, Saskatoon, SK, Canada (ILL4605 × ILL5888) and PAU (L-9-12 × FLIP-2004-7L) have been used for molecular mapping (Saha et al., 2010b; Mekonnen et al., 2014). With the rapid generation advancement technology (Mobini et al., 2014) which allows 4–5 generations per year in lentil will boost the development of much needed genetic resources for genomics enabled improvement.

TABLE 1: Mapping populations developed for various traits in lentil at International Center for Agricultural Research in the Dry Areas (ICARDA).

Trait	Cross	Population size
Drought	ILL 7946 × ILL 7979	174
Cold	ILL 4605 × ILL 10657	153
Earliness	ILL 7115 × ILL 8009	150
Rust	ILL 5888 × ILL 6002	152
Fusarium wilt	ILL 213 × ILL 5883	150
	Precoz × Idleb 2	
Zn content	ILL 5722 × ILL 9888	177
	ILL 9888 × ILL 5480	149
Fe content	ILL 9932 × ILL 9951	193

10.2.4 GENETIC LINKAGE MAPS

In the past, both inter- and intra-specific mapping populations were used for the construction of linkage maps in lentil. The first genetic mapping (linkage analysis) was began by Zamir and Ladizinsky (1984) and the first map comprising DNA based markers was produced by Havey and Muehlbauer (1989). Subsequent maps were published by several workers. With the development of PCR based markers, the number of available markers across the *Lens* genome increased dramatically (Kumar et al., 2011, 2014). The first extensive map comprised of RAPD, AFLP, RFLP, and morphological markers was constructed using a RIL population from a cross between a cultivated *L. culinaris* ssp. *culinaris* cultivar and a *L. culinaris* ssp. *orientalis* accession (Eujayl et al., 1998a). As lentil has low level of polymorphism in the cultivated gene pool the inter-varietal linkage maps were developed through the use of diverge parents from the wild and cultivated species. However, such molecular maps derived from these populations often result low recombination rate and smaller map size. Intra-specific mapping populations have more practical utility in QTL identification and to tag desirable genes of interest than the previous kind of mapping population. Rubeena et al. (2003) published the first intraspecific lentil map comprising 114 RAPD, inter simple sequence repeat (ISSR) and resistance gene analog (RGA) markers. Rubeena et al. (2006) reported F_2 map comprising 72 markers (38 RAPD, 30 AFLP, 3 ISSR, and one morphological) spanning 412.5 cM. The first *Lens* map to include SSR markers was that of Duran et al. (2004). Hamwieh et al. (2005) added 39 SSR and 50 AFLP markers to the map constructed by Eujayl et al. (1998a) to produce a comprehensive *Lens* map comprising 283 genetic markers covering 715 cM. Subsequently, the first lentil map that contained 18 SSR and 79 cross genera ITAP gene-based markers was constructed using a F_5 RIL population developed from a cross between ILL5722 and ILL5588 (Phan et al., 2007). The map comprised seven linkage groups (LGs) that varied from 80.2 to 274.6 cM in length and spanned a total of 928.4 cM. Gupta et al. (2012a) used 196 markers including new 15 *M. truncatula* EST-SSR/SSR in a population of 94 RILs produced from a cross between ILL5588 and ILL5722 and generated 11 LGs covering 1156.4 cM. An intersubspecific F_2 *Lens* linkage map consisting of 199 PCR-based markers (28 SSRs, 9

ISSRs and 162 RAPDs) mapped on to 11 LGs covering a distance of 3847 cM has been constructed (Gupta et al., 2012b). Recently, population specific linkage maps are developed by Perez de la Vega et al. (2011) and Andeden et al. (2013). A list of comprehensive linkage maps in lentil is provided in Table 2.

10.2.5 COMPARATIVE GENOME MAPPING

Comparative genome mapping has demonstrated different levels of genome conservation among crop species during the course of evolution (Choi et al., 2004; Zhu et al., 2005). The lentil genome has shown different degrees of synteny with other legume crops (Weeden et al., 1992; Simon and Muehlbauer, 1997; Phan et al., 2007; Choudhary et al., 2009). Development of PCR-based markers has improved transferability of genetic information among species through comparative genomics and has facilitated the establishment of phylogenetic relationship in plants species. Since the availability of SSR markers in lentil is limited, other legumes offer great scope of marker transferability for genome-wide coverage. Pandian et al. (2000) observed 5% transferability of chickpea-specific STMS primers in lentil while Reddy et al. (2010) observed successful amplification of 62% *Trifolium* markers followed by *Medicago* (36%) and *Pisum* (25%). Datta et al. (2011) reported transferability of 19 STMS markers in lentil from common bean, chickpea, pigeon pea, and soybean. The lack of lentil-specific SSR markers propelled the mining and transfer of EST-SSR sequences from the model genome *M. truncatula* to enrich an existing intraspecific lentil genetic map (Gupta et al., 2012a). They published 21 clear and reproducible SSR markers showing polymorphism between parents, Northfield and Digger. EST-based ITAP markers have recently been developed from related crops and applied to lentil. ESTs were compared for phylogenetic distant from *M. truncatula*, *Lupinus albus*, and *G. max* to produce 500 ITAP markers that could be applied to lentil (Phan et al., 2007). Also, 126 *M. truncatula* cross-species markers were used to generate comparative genetic maps of lentil and white lupin and macrosyntenic relationships between lentil and field pea was observed. The techniques of comparative genomics provided significant opportunities for genetic diversity studies in lentil. The conserved primers (CPs) based on *M. trun-catula* EST sequences flanking one or more introns were used to sequence amplicons in 175 wild and 133 domesticated lentil accessions (Alo et al., 2011). The analysis of the sequences confirmed that *L. nigricans* and *L. ervoides* are well-defined between the species at the DNA sequence level. The availability of draft genome sequences of *M. truncatula*, *L. japonicus*, and *Glycine max* have increased the possibilities of deriving more genomic resources by exploring new molecular markers through bioinformatics platforms which are capable of transfer across the species, belong to the Galegoid clade. Weller et al. (2012) identified two major loci controlling differences in photoperiod response between wild and domesticated pea *HR* (High response to photoperiod) and *ELF3* and identified orthologous gene loci of *ELF3* in lentil. Recently, Kaur et al. (2014) made a comparison of the flanking markers SNP_20002998 and SNP_20000246 in lentil for boron tolerance with the *Arabidopsis thaliana* and *M. trucatuala* genome sequences and identified candidate genes associated with boron tolerance.

10.2.6 FUNCTIONAL GENOMICS

Genomic maps are useful to identify gene(s)/QTL responsible for controlling the variation for the underlying trait of interest. Gene cloning approach helps to characterize and reveal the function of the gene/QTL being identified. The knowledge of genes cloned in legumes can facilitate the development of functional markers for MAS. Many functionally known resistance gene analogs (RGA) have been cloned in lentil (Yaish et al., 2004). Likewise the numerous genes coding transcription factors (TFs) are identified in *Arabidopsis* in a large scale. As the distribution of TF genes does not significantly differ between legume and non-legume species, TF genes have been identified in legumes on the basis of sequence homology with *Arabidopsis* genes. Using functional genomics approaches, genes expressing differentially in contrasting genotypes can also be identified. Differential gene transcript profiles were assessed among resistant (ILL7537) and susceptible (ILL6002) lentil genotypes at 6, 24, 48, 72, and 96 h after inoculation with *Ascochyta lentis* (AL4 isolate; Ford et al., 2007). The non-redundant differentially expressed genes for each accession and time points were hierarchically clustered using Euclidean metrics. In total, 25 differentially expressed sequences were up-regulated and 56 down-regulated in ILL7537 whereas 26 were up-regulated and 44 down-regulated in ILL6002. Several candidate defense genes were characterized from lentil including a b-1, 3-glucanase, a pathogenesis-related protein from the Bet v I family, a pea disease resistance response protein 230 (DRR230-a), a disease resistance response protein (DRRG49-C), a PR4 type gene and a gene encoding an antimicrobial SNAKIN2 protein, all of which have been fully sequenced. Several TFs were also recovered at 6 h after inoculation and future aim is to further biologically characterize these and earlier responses to gain a comprehensive understanding of the key pathogen recognition and defense pathways to *A. lentis* in lentil. Also, the full-length gene sequences will be used in transgenic studies to further characterize their functions. Microarrays play important role in identifying gene networks underlying the expression of important plant traits. A DNA pulse chip made up of 565 ESTs from a chickpea cDNA library enriched for reaction to *A. rabiei,* 156 ESTs from a *Lathyrus* cDNA library enriched for reaction to *A. pinodes* and 41 lentil ESTs and RGAs from the GenBank database (Coram and Pang, 2005) was employed to study expression profiles for ascochyta blight resistant (ILL7537) and susceptible (ILL6002) cultivars (Mustafa et al., 2006).

10.3 APPLICATION OF GENOMIC RESOURCES FOR LENTIL IMPROVEMENT

10.3.1 GENETIC FINGER PRINTING

Genetic diversity analysis has been studied among a set of cultivated and wild lentils using various molecular marker system and genetic materials. Earlier studies have used RFLP, AFLP, and RAPD markers to assess genetic diversity and phylogenetic analyses within and among Lens species (Havey and Muehlbauer, 1989; Aboelwafa et al., 1995; Sharma et al., 1995, 1996; Ahmad and McNeil, 1996; Ford et al., 1997) and gene mapping (Eujayl et al., 1998b; Tullu et al., 2003; Duran et al., 2004; Kahraman et al., 2004; Hamwieh et al., 2005). As a part of the CGIAR's Generation

Challenge Program (GCP), ICARDA has identified a composite collection of lentil germplasm and characterized them by using SSR markers. ICARDA holds the largest global collection of lentil with >11,000 accessions. From this collection, a global composite collection of 960 accessions (Table 3) representing landraces, wild relatives, elite breeding lines, and cultivars was established (Furman, 2006). The results indicated two major clusters separating south Asia (Nepal, India, Pakistan, and Afghanistan) from the Middle East and western countries (Figure 1). The major output of this study was a reference set which represents around 15% (135 accessions) of the global composite collection representing all the geographical regions. This set has been phenotyped for different biotic and abiotic stresses, and emerged as a useful genetic resource to start with (Kumar et al., 2014). Recently, a set of SSR markers was used to study the genetic diversity of lentil mini core set. The mini core collection comprised 109 accessions from 15 countries representing 57 cultigens (including 18 breeding lines) from 8 countries to 52 wild accessions (*L. culinaris* ssp. *orientalis*, *L. culinaris* ssp. *tomentosus* and *L. culinaris* ssp. *odemensis*) from 11 countries. The total alleles detected across the SSR loci were 182, with a mean of 13 alleles per locus. Wild accessions were rich in allelic variation (151 alleles) compared to cultigens (114 alleles). The genetic diversity index for the SSR loci in the wild accessions ranged from 0.16 (SSR28 in *L. culinaris* ssp. *odemensis*) to 0.93 (SSR66 in *L. culinaris* ssp. *orientalis*) with a mean of 0.66, while in the cultigens, genetic diversity varied between 0.03 (SSR28) and 0.87 (SSR207) with a mean of 0.65. Cluster analysis indicated two major clusters (Figure 2), mainly one with the cultigens and the other with wild accessions (Hamwieh et al., 2009). The recent techniques of comparative genomics also provided significant opportunities for genetic diversity studies in lentil. The CPs based on *M. truncatula* EST sequences flanking one or more introns were used to sequence amplicons in 175 wild and 133 domesticated accessions. This analysis of the sequences confirmed that *L. nigricans* and *L. ervoides* are well-defined species at the DNA sequence level. *L. culinaris* ssp. *orientalis* is the progenitor of domesticated lentil, *L. culinaris* ssp. *culinaris*, but a more specific area of origin can be suggested in southern Turkey. The study detected the divergence, following domestication, of the domesticated gene pool into overlapping large seeded (megasperma) and small-seeded (microsperma) groups and observed that lentil domestication led to a loss of genetic diversity of approximately 40% (Alo et al., 2011).

10.3.2 HYBRID TESTING

Making crosses between diverse parents is difficult in practice in lentil because of very small flowers leading to increase the chances of selfing. In addition to this, differentiating F1 plants from selfed ones also becomes difficult due to low phenotypic diversity between the parents. Hence molecular markers have been found very useful to detect the hybridity of F1 plants in lentil. Solanki et al. (2010) used molecular markers in lentil and detected only 21% plants as true hybrids. These results suggest that molecular markers can reduce the time and money required to grow a population from selfed or admixed plants and increase the efficiency of plant breeders in selection of recombinant plants.

10.3.3 MARKER ASSISTED SELECTION

Molecular markers linked to desirable gene(s)/QTL have been reported for marker-assisted selection in lentil (Table 4). Morphological markers viz., cotyledon (Yc), anthocyanin in stem (Gs), pod indehiscence (Pi), seed coat pattern (Scp), flower color (W), radiation frost tolerance locus (Rf), early flowering (Sn), and ground color of the seed (Gc) were mapped as qualitative markers because they exhibited monogenic dominant mode of inheritance (Eujayl et al., 1998a; Duran et al., 2004; Hamwieh et al., 2005; Tullu et al., 2008). Further analysis for the association between DNA markers and Fusarium wilt resistance (Fw) gene was confirmed (Eujayl et al., 1998b; Hamwieh et al., 2005). However, only SSR59-2B was closely linked with Fw at 19.7 cM (Hamwieh et al., 2005). Anthracnose disease resistance (Lct-2) was mapped by Tullu et al. (2003). To date, quantitatively inherited traits have been mapped by Duran et al. (2004) who detected five QTL each for the height of the first ramification and flowering time, three for plant height, seven for pod dehiscence, and one each for shoot number and seed diameter. Five and four QTL were identified for winter survival and winter injury, using a RIL population of 106 lines derived from WA8649090 × Precoz (Kahraman et al., 2004). In this study, experiments were conducted at multiple locations and only one of five QTL was expressed in all environments. Mapping of Ascochyta blight resistance using an F2 population derived from ILL7537 × ILL6002 identified three QTL accounting for 47% (QTL-1 and QTL-2) and 10% (QTL-3) of disease variation. Recently, QTL conferring resistance to Stemphylium blight and rust diseases using RIL populations were identified in lentil (Saha et al., 2010a,b). Though the use of F2 populations in identification of QTL has been done widely in lentil, their use in marker-trait analysis has led to identification of only major QTL. Thus, several minor QTL were overlooked in such populations and identification of environmental responsive QTL was difficult. Because quantitative traits are influenced by both genetic and environmental effects, RILs or near isogenic lines (NILs) are more suitable populations to accurately dissect their components. For ascochyta blight, three QTL each were detected for resistance at seedling and pod/maturity stages (Gupta et al., 2012a). Together these accounted for 34 and 61% of the total estimated phenotypic variation and demonstrated that resistance at different growth stages is potentially conditioned by different genomic regions. Kaur et al. (2014) identified QTL for boron tolerance in Cassab × ILL2024 mapping population. Both simple interval mapping (SIM) and composite interval mapping (CIM) confirmed the presence of QTL in LG4.2 between SNP_20002998 and SNP_20000246. The flanking markers identified may be useful for MAS and pyramiding of potentially different resistance genes into elite backgrounds that are resistant throughout the cropping season. While using QTL pyramiding approach Taran et al. (2003) identified lines with combined resistance to ascochyta blight resistance (AbR1 and ral1) and Anthracnose (OPO61250) in CDC Robin and 964a-46 RIL population for developing cultivars resistance to both ascochyta blight and anthracnose in lentil.

10.3.4 GENE-TRAIT ASSOCIATION ANALYSIS USING NATURAL DIVERSE POPULATION

Bi-parental mapping approach causes more chances for segregation distortion through favoring of one parental allele over another. Also, the molecular markers which can be polymorphic within the interspecific populations might not be polymorphic at the species level as genetic background affects their utility in MAS process. Association mapping is an alternative approach that can address these shortcomings of bi-parental linkage mapping. While using historical recombination in natural populations, landraces, breeding material and varieties, association mapping does marker-trait association and identifies QTL with high resolution. There are two different types of association mapping which can be done on any crop species: genome-wide association studies (GWAS) and candidate gene association mapping. However, to date there are very few reported studies about association mapping in lentil. It is mainly due to the lack of genomic resources available for lentil. After identification of 1536-SNP Illumina GG array (Lc1536) by Sharpe et al. (2013), the Lc1536 array was used in GWAS. The linkage disequilibrium (LD) in lentil may occur similar to that in barley, soybean, and *M. truncatula* (Branca et al., 2011). Fedoruk et al. (2013) used association mapping in lentil to identify QTL for seed size and seed shape. As the properly designed association panels have a greater frequency of alleles encompassing the genetic variation of a crop, it can greatly facilitate to save time and cost while performing MAS in lentil.

10.3.5 GENETIC TRANSFORMATIONS

Transgenic approach uses functional genes which are not available within the crossable gene pool. Thus cloned genes are important genomic resources for making genetic manipulation through transformation. Commonly, the particle bombardment and the *Agrobacterium tumefaciens* infection methods have been used to introduce genes with novel functions. With the explosion of sequence information available in the databases, transformation systems have also become useful tools to study gene function via RNA interference 'knockout,' T-DNA insertion or transforming a genotype lacking a particular gene. Thus a robust, reproducible and efficient transformation system combined with a protocol to regenerate complete fertile plants from transformed cells is essential to fully study plant gene functions.

Following the initial report of shoot regeneration (Bajaj and Dhanju, 1979) from apical meristems, it has been achieved routinely with different explants such as apical meristems (Bajaj and Dhanju, 1979), stem nodes (Polanco et al., 1988; Singh and Raghuvanshi, 1989; Ahmad et al., 1997), cotyledonary node (Warkentin and McHughen, 1992), epicotyls (Williams and McHughen, 1986), decapitated embryo, embryo axis and immature seeds (Polanco and Ruiz, 2001), and cotyledonary petioles (Khawar and Özcan, 2002). The induction of functional roots on in vitro-developed shoots has been the major challenge in lentil micro propagation. The difficulty to induce roots is thought to be associated with the use of cytokinin to obtain multiple shoots from the initial explants (Mohamed et al., 1992). Among the several studies conducted on root induction from shoots, Fratini and Ruiz (2003) reported 95% rooting efficiency from nodal segments cultured in an inverted orientation in media with 5 μM indole acetic acid (IAA) and 1 μM kinetin (KN). Sarker et al. (2003) reported 30% rooting efficiency on MS medium supplemented with 25 mg/l indole butyric acid (IBA).

To date, transformation of lentil has been reported through *A. tumefaciens*-mediated gene transfer (Lurquin et al., 1998) and biolistic transformation including electroporation (Chowrira et al., 1996) and particle bombardment (Gulati et al., 2002; Mahmoudian et al., 2002). Warkentin and McHughen (1992) reported the susceptibility of lentil to *A. tumefaciens* and later evaluated a number of explant types including shoot apices, epicotyl, root, cotyledons, and cotyledonary nodes. All explants showed transient b-glucuronidase (GUS) expression at the wound sites except cotyledonary nodes, which were subsequently transformed by Sarker et al. (2003). Oktem et al. (1999) reported the first transient and stable chimeric transgene expression on cotyledonary lentil nodes using particle bombardment. Gulati et al. (2002) reported regeneration of the first fertile transgenic lentil plants on MS medium with 4.4 μM benzyladenine (BA), 5.2 μM gibberellic acid (GA3), and chlorsulfuron (5 nM for 28 days and 2.5 nM for the rest of the culture period), followed by micrografting and transplantation in soil. The first successful work was reported by Barton et al. (1997), using pCGP1258 plasmid construct on four lentil genotypes. Khatib et al. (2007) have developed herbicide-resistant lentil through *A. tumefaciens* mediated transformation. This was achieved with the same plasmid construct pCGP1258, harboring the bar gene conferring resistance to the herbicide glufosinate ammonium that was transformed using *A. tumefaciens* strain AgL0. Three lentil lines, ILL5582, ILL5883, and ILL5588, were used and a high selection pressure of 20 mg/l of glufosinate was applied to the explants for 18 weeks. Surviving shoots were subsequently grafted onto non-transgenic rootstock and plantlets were transferred to soil and acclimatized. The presence of the transgene was confirmed by PCR and the gene function was confirmed via herbicide application. Recently, Akcay et al. (2009) reported the production of transgenic lentil plants via Agrobacterium-mediated transformation and the stable transmission of the nptII and gusA genes in the subsequent generations. However, these studies were mostly confined to establish transformation techniques rather than the introduction of genes into improved varieties. Khatib et al. (2011) reported for the first time the introduction of the DREB1A gene into lentil for enhancing drought and salinity tolerance. The PCR results confirmed the insertion and stable inheritance of the gene of interest and bar marker gene in the plant genome. The Southern blot analysis revealed integration of a single copy of the transgene. The DREB1A gene driven by rd29A promoter transcribed in the transgenic plants by inducing salt stress in form of sodium chloride solution. The results showed that mRNA was accumulated and thus the DREB1A gene was expressed in the transgenic plants.

10.4 FUTURE PERSPECTIVES

Application of MAS is still limited in lentil. The NGS technology has opened up new opportunity for the fast development of sequence based markers. Access to HTP genotyping and sequencing technologies is expected to speed up the genetic gain across the target environments in lentil. These developments ultimately will increase the utilization of genomic resources in genetic improvement of lentil and will lead fast track development of improved cultivars. Further, increasing number of re-sequencing database in coming days will allow identification of more SNPs and consequently, HTP cost-effective genotyping assays using only informative SNPs would become available for the development of high density linkages for MAS. Recent collaborations among the labs in Canada, Australia, Czech Republic, Spain, USA, ICARDA, and Kenya will facilitate further assembly and annotation of the draft genome, as well as add to the growing database of genetic diversity in the global lentil germplasm. This will include use of long reads based on PacBio sequencing to assemble smaller scaffolds into larger assemblies. Key mapping populations would be genotyped using GBS technology to anchor scaffolds into chromosomal pseudo-molecules and selected lentil genotypes need to be re-sequenced to reveal the genomic diversity in lentil germplasm and provide a road map for future breeding activities. These advances also simultaneously encourage the lentil breeders to develop specialized mapping population such as nested association mapping (NAM) and multi-parents advanced generation inter-cross (MAGIC) populations to generate the genome-wide allelic and haplotype data. Likewise, non-transgenic techniques such as target-induced local lesion in genomes (TILLING) and RNA interference (RNAi) also have demonstrated potential scope for lentil improvement. TILLING has significantly contributed to the understanding of function of pea subtilase (SBT1.1) and tendril-less (tl) genes which control the seed size and tendril formation (D'Erfurth et al., 2012). At ICARDA, mutagenic lentil populations have been recently developed using the mutagen, ethyl methane sulfonate (EMS) in order to identify any point and knock-out mutations for tendril formation and other traits such as pod shattering, herbicide tolerance and Orobanche tolerance. Likewise the other non-transgenic approaches including RNAi technology and virus-induced gene silencing (VIGS) will help understand the molecular mechanisms of biological nitrogen fixation in lentil. The coming years would provide more opportunities to integrate GAB tools in the conventional breeding program. At the same time, more concerted efforts are required to develop other genomic resources such as BAC libraries and other transcriptome assemblies.

10.5 CONCLUSION

Identifying the desired variability for target traits, utilizing the variability in breeding programs, and selecting and advancing the targeted recombinants are the major steps in a breeding program. Conventional breeding approaches are helpful to utilize the available genetic variability in the cultivated germplasm, resulting in the development of several red and yellow cotyledon varieties of lentil with tolerance/resistance to cold, ascochyta blight, rust, and wilt. In the last decade, several linkage maps have been developed and QTL/genes identified for the traits of interest in lentil. This has opened up the scope for mainstreaming genomics enabled improvement in lentil breeding programs. It will get further boost once the draft genome sequence and resequencing of the reference set of lentil is completed.

REFERENCES

1. Aboelwafa, A., Murai, K., and Shimada, T. (1995). Intra-specific and inter-specific variations in Lens revealed by RAPD markers. Theor. Appl. Genet. 90, 335–340.
2. Ahmad, M., Fautrier, A.G., McNeil, D.L., Hill, G.D., and Burritt, D.J. (1997). In vitro propagation of Lens species and their F1 interspecific hybrids. Plant Cell Tiss. Org. Cult. 47, 169–176. doi: 10.1007/bf02318954
3. Ahmad, M., and McNeil, D.L. (1996). Comparison of crossability, RAPD, SDS-PAGE and morphological markers for revealing genetic relationships within and among Lens species. Theor. Appl. Genet. 93, 788–793. doi: 10.1007/bf00224077
4. Akcay, U.C., Mahmoudian, M., Kamci, H., Yucel, M., and Oktem, H.A. (2009). Agrobacterium tumefaciens-mediated genetic transformation of a recalcitrant grain legume, lentil (*Lens culinaris* Medik). Plant Cell Rep. 28, 407–417. doi: 10.1007/s00299-008-0652-4
5. Aldemir, S.B., Sever, T., Ates, D., Yagmur, B., Kaya, H.B., Temel, H.Y.,et al. (2014). "QTL mapping of genes controlling Fe uptake in lentil (*Lens culinaris* L.) seed using recombinant inbred lines," in Proceedings of the Plant and Animal Genome Conference XXII P3360, San Diego, CA.

6.   Alo, F., Furman, B.J., Akhunov, E., Dvorak, J., and Gepts, P. (2011). Leveraging genomic resources of model species for the assessment of diversity and phylogeny in wild and domesticated lentil. J.Hered. 102, 315–329. doi: 10.1093/jhered/esr015

7.   Andeden, E.E., Derya, M., Baloch, F.S., Kilian, B., and Ozkan, H. (2013). "Development of SSR markers in lentil," in Proceedings of Plant and Animal Genome Conference XXI P0351, San Diego, CA.

8.   Arumuganathan, K., and Earle, E.D. (1991). Nuclear DNA content of some important plant species. Plant Mol. Biol. Rep. 9, 208–218. doi: 10.1007/BF02672069

9.   Bajaj, Y.P.S., and Dhanju, M.S. (1979). Regeneration of plants from apical meristem tips of some legumes. Curr. Sci. 48, 906–907.

10. Barton, J., Klyne, A., Tennakon, D., Francis, C., and Hamblin, J. (1997). "Development of a system for gene transfer to lentils," in Proceedings of International Food Legume Research Conference III, Adelaide, SA.

11. Bett, K., Ramsay, L., Sharpe, A., Cook, D., Penmetsa, R.V., Verma, N.,et al. (2014). "Lentil genome sequencing: establishing a comprehensive platform for molecular breeding," in Proceedings of International Food Legumes Research Conference (IFLRC-VI) and ICCLG-VII (Saskatoon, SK: Crop Development Center), 19.

12. Branca, A., Paape, T.D., Zhou, P., Briskine, R., Farmer, A.D., Mudge, J.,et al. (2011). Whole genome nucleotide diversity, recombination and linkage disequilibrium in the model legume *Medicago truncatula*. Proc. Nat. Acad. Sci. U.S.A. 108, E864–E870. doi: 10.1073/pnas.1104032108

13. Chagne, D., Carlisle, C.M., Blond, C., Volz, R.K., Whitworth, C.J., Oraguzie, N.C.,et al. (2007). Mapping a candidate gene (MdMYB10) for red flesh and foliage colour in apple. BMC Genomics 8:212. doi: 10.1186/1471-2164-8-212

14. Choi, H.K., Mun, J.H., Kim, D.J., Zhu, H.Y., Baek, J.M., Mudge, J.,et al. (2004). Estimating genome conservation between crop and model legume species. Proc. Natl. Acad. Sci. U.S.A. 101, 15289–15294. doi: 10.1073/pnas.0402251101

15. Choudhary, S., Sethy, N.K., Shokeen, B., and Bhatia, S. (2009). Development of chickpea EST-SSR markers and analysis of allelic variation across related species. Theor. Appl. Genet. 118, 591–608. doi: 10.1007/s00122-008-0923-z

16. Chowrira, G.M., Akella, V., Fuerst, P.E., and Lurquin, P.F. (1996). Transgenic grain legumes obtained by in planta electroporation-mediated gene transfer. Mol. Biotechnol. 5, 85–96. doi: 10.1007/bf02789058

17. Coram, T.E., and Pang, E.C.K. (2005). Expression profiling of chickpea genes differentially regulated during a resistance response to Ascochyta rabiei. Plant Biotechnol. J. 4, 647–666. doi: 10.1111/j.1467-7652.2006.00208.x

18. Datta, S., Tiwari, S., Kaashyap, M., Gupta, P.P., Choudhury, P.R., Kumari, J.,et al. (2011). Genetic similarity analysis in Lentil using cross-genera legume sequence tagged Microsatellite site markers. Crop Sci. 51, 2412–2422. doi: 10.2135/cropsci2010.12.0743

19. de la Puente, R., Garcia, P., Polanco, C., and Perez de la Vega, M. (2013). An improved intersubspecific genetic map in Lens including functional markers. Span. J. Agric. Res. 11, 132–136. doi: 10.5424/sjar/2013111-3283

20. D'Erfurth, I., Le Signor, C., Aubert, G., Sanchez, M., Vernoud, V., Darchy, B.,et al. (2012). A role for an endosperm-localized subtilase in the control of seed size in legumes. New Phytol. 196, 738–751. doi: 10.1111/j.1469-8137.2012.04296

21. Duran, Y., Fratini, R., Garcia, P., and Pérez de la Vega, M. (2004). An intersub-specific genetic map of Lens. Theor. Appl. Genet. 108, 1265–1273. doi: 10.1007/s00122-003-1542-3

22. Eujayl, I., Baum, M., Powell, W., Erskine, W., and Pehu, E. (1998a). A genetic link-age map of lentil (Lens sp.) based on RAPD and AFLP markers using recombinant inbred lines. Theor. Appl. Genet. 97, 83–89. doi: 10.1007/s001220050869

23. Eujayl, I., Erskine, W., Bayaa, B., Baum, M., and Pehu, E. (1998b). Fusarium vascu-lar wilt in lentil: inheritance and identification of DNA markers for resistance. Plant Breed. 117, 497–499. doi: 10.1111/j.1439-0523.1998.tb01982.x

24. FAOSTAT. (2014). Available at: http://faostat3.fao.org/browse/Q/QC/E [accessed on January 12, 2015].

25. Fedoruk, M.J., Vandenberg, A., and Bett, K.E. (2013). Quantitative trait loci analysis of seed quality characteristics in lentil using single nucleotide polymorphism mark-ers. Plant Genome 6, 37–39. doi: 10.3835/plantgenome2013.05.0012

26. Ford, R., Pang, E.C.K., and Taylor, P.W.J. (1997). Diversity analysis and species identification in Lens using PCR generated markers. Euphytica 96, 247–255. doi: 10.1023/a:1003097600701

27. Ford, R., Pang, E.C.K., and Taylor, P.W.J. (1999). Genetics of resistance to ascochy-ta blight (Ascochyta lentis) of lentil and the identification of closely linked RAPD markers. Theor. Appl. Genet. 98, 93–98. doi: 10.1007/s001220051044

28. Ford, R., Viret, L., Pang, E.C.K., Taylor, P.W.J., Materne, M., and Mustafa, B. (2007). "Defense responses in lentil to Ascochyta lentis," in Proceedings of the Australian Plant Pathology Conference (Adelaide, SA: Australian Plant Pathology Society), 64.

29. Fratini, R., Duran, Y., Garcia, P., and Pérez de la Vega, M. (2007). Identification of quantitative trait loci (QTL) for plant structure, growth habit and yield in lentil. Span. J. Agric. Res. 5, 348–356. doi: 10.5424/sjar/2007053-255

30. Fratini, R., and Ruiz, M.L. (2003). A rooting procedure for lentil (Lens culinaris Medik.) and other hypogeous legumes (pea, chickpea and lathyrus) based on explant polarity. Plant Cell Rep. 21, 726–732. doi: 10.1007/s00299-003-0603-z

31. Fratini, R., Ruiz, M.L., and Pérez de la Vega, M. (2004). Intra-specific and inter-sub-specific crossing in lentil (Lens culinaris Medik.). Can. J. Plant Sci. 84, 981–986. doi: 10.4141/p03-201

32. Furman, B.J. (2006). Methodology to establish a composite collection: case study in lentil. Plant Genetic Resour. 4, 2–12. doi: 10.1079/PGR200599

33. Gulati, A., Schryer, P., and McHughen, A. (2002). Production of fertile transgenic lentil (Lens culinaris Medik) plants using particle bombardment. In Vitro Cell. Dev. Biol. Plant 38, 316–324. doi: 10.1079/ivp2002303

34. Gupta, D., Taylor, P.W.J., Inder, P., Phan, H.T.T., Ellwood, S.R., Mathur, P.N.,et al. (2012a). Integration of EST-SSR markers of Medicago truncatula into intraspecific linkage map of lentil and identification of QTL conferring resistance to ascochyta blight at seedling and pod stages. Mol. Breed. 30, 429–439. doi: 10.1007/s11032-011-9634-2

35. Gupta, M., Verma, B., Kumar, N., Chahota, R.K., Rathour, R., Sharma, S.K.,et al. (2012b). Construction of intersubspecific molecular genetic map of lentil based on ISSR, RAPD and SSR markers. J.Genet. 91, 279–287. doi: 10.1007/s12041-012-0180-4

36. Hamwieh, A., Udupa, S., Choumane, W., Sarker, A., Dreyer, F., Jung, C.,et al. (2005). A genetic linkage map of Lens sp. based on microsatellite and AFLP markers and the localization of fusarium vascular wilt resistance. Theor. Appl. Genet. 110, 669–677. doi: 10.1007/s00122-004-1892-5

37. Hamwieh, A., Udupa, S.M., Sarker, A., Jung, C., and Baum, M. (2009). Development of new microsatellite markers and their application in the analysis of genetic diversity in lentils. Breed. Sci. 59, 77–86. doi: 10.1270/jsbbs.59.77

38. Havey, M.J., and Muehlbauer, F.J. (1989). Linkages between restriction fragment length, isozyme and morphological markers in lentil. Theor. Appl. Genet. 77, 395–401. doi: 10.1007/bf00305835

39. Kahraman, A., Demirel, U., Ozden, M., and Muehlbauer, F.J. (2010). Mapping of QTLs for leaf area and the association with winter hardiness in fall-sown lentil. Afr. J. Biotechnol. 9, 8515–8519.

40. Kahraman, A., Kusmenoglu, I., Aydin, N., Aydogan, A., Erskine, W., and Muehlbauer, F.J. (2004). QTL mapping of winter hardiness genes in lentil. Crop Sci. 44, 13–22. doi: 10.2135/cropsci2004.0013

41. Kaur, S., Cogan, N.I., Stephens, A., Noy, D., Butsch, M., Forster, J.,et al. (2014). EST-SNP discovery and dense genetic mapping in lentil (*Lens culinaris* Medik.) enable candidate gene selection for boron tolerance. Theor. Appl. Genet. 127, 703–713. doi: 10.1007/s00122-013-2252-0

42. Kaur, S., Cogan, N.O.I., Pembleton, L.W., Shinozuka, M., Savin, K.W., Materne, M.,et al. (2011). Transcriptome sequencing of lentil based on second-generation technology permits large-scale unigene assembly and SSR marker discovery. BMC Genomics 12:265. doi: 10.1186/1471-2164-12-265

43. Khatib, F., Koudsieh, S., Ghazal, B., Barton, J., Tsujimoto, H., and Baum, M. (2007). Developing herbicide resistant lentil (*Lens culinaris* Medikus subsp. culinaris) through Agrobacterium-mediated transformation. Arab. J. Plant Protect. 25, 185–192.

44. Khatib, F., Makris, A., Yamaguchi-Shinozaki, K., Kumar, S., Sarker, A., Erskine, W.,et al. (2011). Expression of the DREB1A gene in lentil (*Lens culinaris* Medik. subsp culinaris) transformed with the Agrobacterium system. Crop Pasture Sci. 62, 488–495. doi: 10.1071/cp10351

45. Khawar, K.M., and Özcan, S. (2002). Effect of indole-3-butyric acid on in vitro root development in lentil (*Lens culinaris* Medik.). Turk. J. Bot. 26, 109–111.

46. Kota, R., Rudd, S., Facius, A., Kolesov, G., Thiel, T., Zhang, H.,et al. (2003). Snipping polymorphisms from large EST collections in barley (Hordeum vulgare L.). Mol. Genet. Genomics 270, 24–33. doi: 10.1007/s00438-003-0891-6

47. Kumar, J., Pratap, A., Solanki, R.K., Gupta, D.S., Goyal, A., Chaturvedi, S.K.,et al. (2011). Advances in genomics resources for improving food legume crops. J.Agril. Sci. 150, 289–318. doi: 10.1017/S0021859611000554

48. Kumar, S., and Ali, M. (2006). GE interaction and its breeding implications in pulses. Botanica 56, 31–36.

49. Kumar, S.K., Barpete, S., Kumar, J., Gupta, P., and Sarker, A. (2013). Global lentil production: constraints and strategies. SATSA Mukhapatra – Annu. Tech. Issue 17, 1–13.

50. Kumar, S., Hamweih, A., Manickavelu, A., Kumar, J., Sharma, T.R., and Baum, M. (2014). "Advances in lentil genomics," in Legumes in Omics Era, eds S. Gupta, N. Nadarajan, and D.S. Gupta (New York: Springer Science+Business Media), 111–130.

51. Lurquin, P.F., Cai, Z., Stiff, C.M., and Fuerst, E.P. (1998). Half-embryo cocultivation technique for estimating the susceptibility of pea (Pisum sativum L.) and lentil (*Lens culinaris* Medik.) cultivars to Agrobacterium tumefaciens. Mol. Biotechnol. 9, 175–179. doi: 10.1007/BF02760819

52. Mahmoudian, M., Yucel, M., and Oktem, H.A. (2002). Transformation of lentil (*Lens culinaris* M.) cotyledonary nodes by vacuum infiltration of Agrobacterium tumefaciens. Plant Mol. Biol. Rep. 20, 251–257. doi: 10.1007/bf02782460

53. Mekonnen, F., Mekbib, F., Kumar, S., Ahmed, S., Chahota, R.K., Sharma, T.R.,et al. (2014). Identification of molecular markers associated with rust resistance genes in lentil (*Lens culinaris* sub sp. culinaris). Can. J. Plant Prot. 2, 27–36.

54. Mobini, S.H., Lulsdorf, T., Warkentin, D., and Vanderberg, A. (2014). Plant growth regulators improve in vitro flowering and rapid generation advancement in lentil and faba bean. In Vitro Cell. Dev. Biol. Plant doi: 10.1007/s11627-014-9647-8

55. Mohamed, M.F., Read, P.E., and Coyne, D.P. (1992). Plant regeneration from in vitro culture of embryonic axis explants in common and tepary beans. J.Am. Soc. Hort. Sci. 117, 332–336.

56. Mustafa, B.M., Coram, T.E., Pang, E.C.K., Taylor, P.W.J., and Ford, R. (2006). "Unraveling *Ascochyta lentis* resistance in lentil,"in Proceedings of the Ascochyta 2006 Conference, 2nd–5th July, France.

57. NCBI. (2015). Available at: http://www.ncbi.nlm.nih.gov/nucest/?term=lentil [accessed on January 12, 2015].

58. Oktem, H., Mahmoudian, M., and Yucel, M. (1999). GUS gene delivery and expression in lentil cotyledonary nodes using particle bombardment. LENS Newslett. 26, 3–6.

59. Pandian, A., Ford, R., and Taylor, P.W. (2000). Transferability of sequence tagged microsatellite site (STMS) primers across four major pulses. Plant Mol. Biol. Rep. 18, 395–395. doi: 10.1007/BF02825069

60. Perez de la Vega, M., Fratini, R.M., and Muehlbauer, F.J. (2011). "Lentil," in Genetics, Genomic and Breeding of Cool Season Grain Legumes (Genetics, Genomics and Breeding in Crop Plants), eds M. Perez de la Vega, A. M. Torres, J. I. Cubero, and C. Kole (Boca Raton, FL: Science Pubs), 98–150.

61. Phan, H.T., Ellwood, S.R., Hane, J.K., Ford, R., Materne, M., and Oliver, R.P. (2007). Extensive macrosynteny between *Medicago truncatula* and *Lens culinaris* ssp. culinaris. Theor. Appl. Genet. 114, 549–558. doi: 10.1007/s00122-006-0455-3

62. Polanco, M.C., Peláez, M.I., and Ruiz, M.L. (1988). Factors affecting callus and shoot formation from in vitro cultures of *Lens culinaris* Medik. Plant Cell Tiss. Org. Cult. 15, 175–182. doi: 10.1007/BF00035759

63. Polanco, M.C., and Ruiz, M.L. (2001). Factors that affect plant regeneration from in vitro culture of immature seeds in four lentil cultivars. Plant Cell Tiss. Org. Cult. 66, 133–139. doi: 10.1023/A:1010652818812

64. Reddy, M.R.K., Rathour, R., Kumar, N., Katoch, P., and Sharma, T.R. (2010). Cross-genera legume SSR markers for analysis of genetic diversity in Lens species. Plant Breed. 129, 514–518. doi: 10.1111/j.1439-0523.2009.01723.x

65. Rubeena, Ford, R., and Taylor, P.W.J. (2003). Construction of an intraspecific linkage map of lentil (*Lens culinaris* ssp. culinaris). Theor. Appl. Genet. 107, 910–916. doi: 10.1007/s00122-003-1326-9

66. Rubeena, A., Taylor, P.W.J., Ades, P.K., and Ford, R. (2006). QTL mapping of resistance in lentil (*Lens culinaris* ssp. culinaris) to ascochyta blight (*Ascochyta lentis*). Plant Breed. 125, 506–512. doi: 10.1111/j.1439-0523.2006.01259.x

67. Saha, G.C., Sarker, A., Chen, W., Vandemark, G.J., and Muehlbauer, F.J. (2010a). Inheritance and linkage map positions of genes conferring resistance to stemphylium blight in lentil. Crop Sci. 50, 1831–1839. doi: 10.2135/cropsci2009.12.0709

68. Saha, G.C., Sarker, A., Chen, W.D., Vandemark, G.J., and Muehlbauer, F.J. (2010b). Identification of markers associated with genes for rust resistance in *Lens culinaris* Medik. Euphytica 175, 261–265. doi: 10.1007/s10681-010-0187-y

69. Saha, G.C., Sarker, A., Chen, W., Vandemark, G.J., and Muehlbauer, F.J. (2013). Inheritance and linkage map positions of genes conferring agromorphological traits in *Lens culinaris* Medik. Int. J. Agron. 2013, 9. doi: 10.1155/2013/618926

70. Sarker, R., Mustafa, B.M., Biswas, A., Mahbub, S., Nahar, M., Hashem, R.,et al. (2003). In vitro regeneration in lentil (*Lens culinaris* Medik.). Plant Tissue Cult. 13, 155–163.

71. Sever, T., Ates, D., Aldemir, S.B., Yagmur, B., Kaya, H.B., Temel, H.Y.,et al. (2014). "Identification QTLs controlling genes to se uptake in lentil seeds," in Proceedings of the Plant and Animal Genome XXII Conference, San Diego, CA.

72. Sharma, S.K., Dawson, I.K., and Waugh, R. (1995). Relationships among cultivated and wild lentils revealed by rapd analysis. Theor. Appl. Genet. 91, 647–654. doi: 10.1007/BF00223292

73. Sharma, S.K., Knox, M.R., and Ellis, T.H.N. (1996). AFLP analysis of the diversity and phylogeny of Lens and its comparison with RAPD analysis. Theor. Appl. Genet. 93, 751–758. doi: 10.1007/bf00224072

74. Sharpe, A.G., Ramsay, L., Sanderson, L.A., Fedoruk, M.J., Clarke, W.E., Rong, L.,et al. (2013). Ancient orphan crop joins modern era: gene-based SNP discovery and mapping in lentil. BMC Genomics 14:192. doi: 10.1186/1471-2164-14-192

75. Simon, C.J., and Muehlbauer, F.J. (1997). Construction of a chickpea linkage map and its comparison with maps of pea and lentil. J.Hered. 88, 115–119. doi: 10.1093/oxfordjournals.jhered.a023068

76. Singh, R., and Raghuvanshi, S. (1989). Plantlet regeneration from nodal segment and shoot tip derived explants of lentil. LENS Newslett. 16, 33–35.

77. Solanki, R., Singh, S., and Kumar, J. (2010). Molecular marker assisted testing of hybridity of F1 plants in lentil. Food Legumes 23, 21–24.

78. Srivastava, R.P., and Vasishtha, H. (2012). Saponins and lectins of Indian chickpeas (Cicer arietinum) and lentils (Lens culinaris). Indian J. Agric. Biochem. 25, 44–47.

79. Tadmor, Y., Zamir, D., and Ladizinsky, G. (1987). Genetic mapping of an ancient translocation in the genus Lens. Theor. Appl. Genet. 73, 883–892. doi: 10.1007/bf00289394

80. Tahir, M., and Muehlbauer, F.J. (1994). Gene-mapping in lentil with recombinant inbred lines. J.Hered. 85, 306–310.

81. Tanyolac, B., Ozatay, S., Kahraman, A., and Muehlbauer, F. (2010). Linkage mapping of lentil (*Lens culinaris* L.) genome using recombinant inbred lines revealed by AFLP, ISSR, RAPD and some morphologic markers. J. Agric. Biotechnol. Sustain. Dev. 2, 1–6.

82. Taran, B., Buchwaldt, L., Tullu, A., Banniza, S., Warkentin, T.D., and Vandenberg, A. (2003). Using molecular markers to pyramid genes for resistance to ascochyta blight and anthracnose in lentil (*Lens culinaris* Medik.). Euphytica. 134, 223–230. doi: 10.1023/B:EUPH.0000003913.39616.fd

83. Temel, H.Y., Gol, D., Kahriman, A., and Tanyolac, M.B. (2014). Single nucleotide polymorphism discovery through Illumina- based transcriptome sequencing and mapping in lentil. Turk. J. Agric. For. 38, 1–19. doi: 10.3906/tar-1409-70

84. Thiel, T., Michalek, W., Varshney, R., and Graner, A. (2003). Exploiting EST databases for the development and characterization of gene-derived SSR-markers in barley (Hordeum vulgare L.). Theor. Appl. Genet. 106, 411–422.

85. Tullu, A., Buchwaldt, L., Warkentin, T., Taran, B., and Vandenberg, A. (2003). Genetics of resistance to anthracnose and identification of AFLP and RAPD markers linked to the resistance gene in PI 320937 germplasm of lentil (*Lens culinaris* Medikus). Theor. Appl. Genet. 106, 428–434.

86. Tullu, A., Tar'an, B., Breitkreutz, C., Buchwaidt, L., Banniza, S., Warkentin, T.D.,et al. (2006). A quantitative-trait locus for resistance to ascochyta blight *Ascochyta lentis* maps close to a gene for resistance to anthracnose *Colletotrichum truncatum* in lentil. Can. J. Plant Pathol. 28, 588–595. doi: 10.1080/07060660609507337

87. Tullu, A., Tar'an, B., Warkentin, T., and Vandenberg, A. (2008). Construction of an Intraspecific linkage map and QTL analysis for earliness and plant height in lentil. Crop Sci. 48, 2254–2264. doi: 10.2135/cropsci2007.11.0628

88. Vandenberg, A. (2009). "Lentil expansion in Canada," in Milestones in Legume Research, eds M. Ali and S. Kumar (Kanpur: Indian Institute of Pulses Research), 58–72.

89. Varshney, R.K., Graner, A., and Sorrells, M.E. (2005). Genic microsatellite markers in plants: features and applications. Trends Biotechnol. 23, 48–55. doi: 10.1016/j.tibtech.2004.11.005

90. Verma, P., Shah, N., and Bhatia, S. (2013). Development of an expressed gene catalogue and molecular markers from the de novo assembly of short sequence reads of the lentil (*Lens culinaris* Medik.) transcriptome. Plant Biotechnol. J. 11, 894–905. doi: 10.1111/pbi.12082

91. Verma, P., Sharma, T.R., Srivastava, P.S., Abdin, M.Z., and Bhatia, S. (2014). Exploring genetic variability within lentil (*Lens culinaris* Medik.) and across related legumes using a newly developed set of microsatellite markers. Mol. Biol. Rep. 41, 5607–5625. doi: 10.1007/s11033-014-3431-z

92. Vijayan, P., Vandenberg, A., and Bett, K.E. (2009). A Mixed Genotype Lentil EST Library Representing the Normalized Transcriptome of Different Seed Development Stages. Available at: http://www.ncbi.nlm.nih.gov/nucest/?term=lens%20culinaris

93. Warkentin, T.D., and McHughen, A. (1992). Agrobacterium tumefaciens-mediated beta-glucuronidase (GUS) gene expression in lentil (*Lens culinaris* Medik.) tissues. Plant Cell Rep. 11, 274–278.

94. Weeden, N.F., Muehlbauer, F.J., and Ladizinsky, G. (1992). Extensive conservation of linkage relationships between pea and lentil genetic maps. J.Hered. 83, 123–129.

95. Weller, J.L., Liew, L.C., Hecht, V.F.G., Rajandran, V., Laurie, R.E., Ridge, S.,et al. (2012). A conserved molecular basis for photoperiod adaptation in two temperate legumes. Proc. Natl. Acad. Sci. U.S.A. 109, 21158–21163. doi: 10.1073/pnas.1207943110

96. Williams, D.J., and McHughen, A. (1986). Plant regeneration of the legume *Lens culinaris* Medik. (lentil) in vitro. Plant Cell Tiss. Org. Cult. 7, 149–153. doi: 10.1007/BF00043039

97. Yaish, M.W., Saenz de Miera, L.E., and Perez de La Vega, M. (2004). Isolation of a family of resistance gene analogue sequences of the nucleotide binding site (NBS) type from Lens species. Genome 47, 650–659. doi: 10.1139/g04-027

98. Zamir, D., and Ladizinsky, G. (1984). Genetics of allozyme variants and linkage groups in lentil. Euphytica 33, 329–336. doi: 10.1007/bf00021129

99. Zhu, H., Choi, H.K., Cook, D.R., and Shoemaker, R.C. (2005). Bridging model and crop legumes through comparative genomics. Plant Physiol. 137, 1189–1196. doi: 10.1104/pp.104.058891

Figures 1 and 2 and Tables 2–4 are not available in this version of the article. To view this additional information, please use the citation on the first page of this chapter.

Author Notes

CHAPTER 1

Conflict of Interest Statement

The author declares that the research was conducted in the absence of any commercial or financial relationships that could be construed as a potential conflict of interest.

CHAPTER 2

Acknowledgments

We are very grateful for Jim Barlow for assistance with setting up the differential thermostats, and Thurston Heaton, Dave Grantham and Bill Bailey for setting up the experimental plots. Jenny Rowntree, Ruth Lopez, Avanti Wadugodapitiya, YolandSavriama and Tilly Eldridge provided invaluable help in planting.

CHAPTER 3

Acknowledgments

Melodie McGeoch and two anonymous reviewers provided helpful comments on the manuscript. This work arose from a workshop sponsored by CSIRO's Cutting Edge Symposium series and is linked to research being undertaken under the Science Industry Endowment Fund. SLC is supported by Australian Research Council Grant DP140102815.

CHAPTER 4

Conflict of Interest
The authors declare that the research was conducted in the absence of any commercial or financial relationships that could be construed as a potential conflict of interest.

Acknowledgments
Sonali Sengupta thanks the Fast-Track Young Scientist Award Program of the Department of Science and Technology and the Department of Biotechnology, Government of India, for support. Arun Lahiri Majumder is a Raja Ramanna Fellow of the Department of Atomic Energy, Government of India. We cordially thank Dr. Harald Keller, Senior Scientist, INRA, France, for his kind permission to reproduce the lignin biosythetic pathway figure from his publication, appropriately cited. We further thank the reviewers for their valuable comments which helped us to improve the manuscript.

CHAPTER 5:

Acknowledgments
This work is supported by the Hong Kong RGC Collaborative Research Fund (CUHK3/CRF/11G), the Hong Kong RGC General Research Fund (468610), and funding from the Lo Kwee-Seong Biomedical Research Fund and Lee Hysan Foundation. Jee Yan Chu copy-edited this manuscript.

Conflict of Interest
The authors declare no conflict of interest.

CHAPTER 6

Acknowledgments
The authors are grateful to Andrzej Walichnowski for help with paper editing, Joanne Schiavoni for formatting, and Michael Shillinglaw for figure preparation. This chapter was written within the scope of the Genome

Canada TUFGEN project, and support from all funding partners is gratefully acknowledged.

CHAPTER 7

Competing Interests
The authors declare that they have no competing interests.

Acknowledgments
This whole work was initiated and largely supported by SOFIPROTEOL under the FASO (Fonds d'Action Stratégique des Oléoprotéagineux) project "PEAPOL", to answer the increasing demand from French breeders for a massive development of markers in pea, allowing a breakthrough in MAS. The validation step in a KASP™ genotyping assay was partly funded by the European Community's Seventh Framework Programme (FP7/ 2007–2013) under the grant agreement n°FP7-613551, LEGATO project. We thank the GenOuest (genouest.org) cluster team and especially Olivier Collin who allowed us to perform all the bioinformatics analysis, and Delphine Naquin who performed preliminary bioinformatics tests. We also thank the CNRGV (French Plant Genomic Resource Center) and especially Arnaud Bellec, who allowed us to perform preliminary molecular biology tests. We greatly acknowledge Jerôme Gouzy, Jean-Pierre Martinant, Grégoire Aubert and Erwan Corre for useful discussions on methodological developments that have made this project a success. We greatly acknowledge Jeroen Wilmer for critical review of the manuscript. We thank Leigh Gebbie, LKG SCIENTIFIC EDITING & TRANSLATION for her assistance in correcting the English version of the manuscript.

Authors' Contributions
GB conceived and coordinated the study, carried out all the genetic and statistical analyses, and wrote the manuscript. SAC carried out all the bioinformatics analysis and co-wrote the manuscript. MF carried out the construction of the 64,263 marker genetic map. PP supervised the discoSnp analysis. EL carried out all the molecular biology experiments. OB coordinated the sequencing experiments. CL participated in the construction of the 914 markers genetic map. MLPN participated in the production and selected the genotyping material. NRI co-coordinated the study. AB

co-coordinated the study and co-wrote the manuscript. All authors read and approved the final manuscript.

CHAPTER 8

Conflict of Interest
The authors declare that the research was conducted in the absence of any commercial or financial relationships that could be construed as a potential conflict of interest.

Acknowledgments
This work was supported by National Science Foundation Grant Number IOS-1025398, by an endowment from the C.V. Griffin Sr. Foundation, and by the Office of Science, Office of Biological and Environmental Research, of the US Department of Energy under Contract Number DE-ACO2-06CH11357, as part of the DOE Systems Biology Knowledgebase.

CHAPTER 9

Conflict of Interest
The authors declare that there is no conflict of interests regarding the publication of this paper.

Acknowledgments
The authors are grateful to P. M. Gresshoff (University of Queensland, Australia) for providing theBradyrhizobium strain (USDA110). Saad Sulieman was supported by a postdoc fellowship from the Japan Society for the Promotion of Science (JSPS), and Chien Van Ha is supported by a Ph.D. fellowship from "International Program Associate" of RIKEN, Japan. This work was also supported in part by a grant (Project code 03/2012/ HĐ-ĐTĐL) from the Vietnam Ministry of Science and Technology to the Research Group of Dong Van Nguyen.

CHAPTER 10

Conflict of Interest

The authors declare that the research was conducted in the absence of any commercial or financial relationships that could be construed as a potential conflict of interest.

Acknowledgments
The authors acknowledge CGIAR Research Program on Grain Legumes for providing financial and research support.

Index

Milton Keynes UK
Ingram Content Group UK Ltd.
UKHW022059141024
449569UK00031B/1698